TNF Pathophysiology. Molecular and Cellular Mechanisms

Current Directions in Autoimmunity

Vol. 11

Series Editor

A.N. Theofilopoulos La Jolla, Calif.

TNF Pathophysiology

Molecular and Cellular Mechanisms

Volume Editors

G. Kollias Vari
P.P. Sfikakis Athens

15 figures, 2 in color, and 5 tables, 2010

KARGER

Basel · Freiburg · Paris · London · New York · Bangalore · Bangkok · Shanghai · Singapore · Tokyo · Sydney

George Kollias, PhD
Institute of Immunology
Biomedical Sciences Research Center
'Alexander Fleming'
Vari, Greece

Petros P. Sfikakis, MD, PhD
First Department of Propedeutic and
Internal Medicine
Laikon Hospital
Athens University Medical School
Athens, Greece

Library of Congress Cataloging-in-Publication Data

TNF pathophysiology : molecular and cellular mechanisms / volume editors, G. Kollias, P.P. Sfikakis.
p. ; cm. – (Current directions in autoimmunity, ISSN 1422–2132; v. 11)
Includes bibliographical references and index.
ISBN 978–3–8055–9383–0 (hard cover: alk. paper)
1. Tumor necrosis factor – Pathophysiology. I. Kollias, G. (George) II. Sfikakis, P.P. (Petros P.)
III. Title: Tumor necrosis factor pathophysiology.
IV. Series: Current directions in autoimmunity, v. 11. 1422–2132;
[DNLM: 1. Tumor Necrosis Factors – immunology. 2. Tumor Necrosis Factors – physiology.
3. Transcription, Genetic – immunology.
W1 CU788DR v.11 2010 / QW 630 T6267 2010]
QR185.8.T84T543 2010
616.07'9–dc22

2009051929

Bibliographic Indices. This publication is listed in bibliographic services, including Current Contents® and PubMed/MEDLINE.

www.karger.com
Printed in Switzerland on acid-free and non-aging paper (ISO 9706) by Reinhardt Druck, Basel
ISSN 1422–2132
ISBN 978–3–8055–9383–0
e-ISBN 978–3–8055–9384–7

Contents

Preface

TNF is a pleiotropic cytokine central to the development and homeostasis of the immune system and a regulator of cell activation, differentiation and death. TNF is involved in a multitude of biological processes, such as acute and chronic inflammation, autoimmunity, infection and tumor responses. In the last decades, there has been an enormous scientific and clinical interest in understanding TNF's function in physiology and disease, and a vast amount of data has accumulated at the biochemical, molecular and cellular level, establishing TNF as a prototype for in-depth understanding of a cytokine's physiological and pathogenic functions. Perturbations of TNF and its signals in transgenic models have provided a wealth of information about its function at the organism level as well as created unique animal models for chronic inflammatory disorders. Collectively, this knowledge primed the successful development of anti-TNF therapies for several human diseases and opened new avenues for safer and more effective drug discovery.

The chapters in this volume cover recent developments in TNF regulation and function from a basic molecular and cellular level to whole organism perspectives and their clinical implications. Thorough understanding of the mechanisms by which this key molecular player is produced and functions to regulate cell biology, immunity and disease should set novel paradigms on how genes contribute to biological system development and physiology.

We wish to thank the series editor as well as all the contributing authors for giving us the opportunity to assemble this excellent review volume on TNF biology.

George Kollias, Vari
Petros P. Sfikakis, Athens

Kollias G, Sfikakis PP (eds): TNF Pathophysiology. Molecular and Cellular Mechanisms.
Curr Dir Autoimmun. Basel, Karger, 2010, vol 11, pp 1–26

Cellular Mechanisms of TNF Function in Models of Inflammation and Autoimmunity

Maria Apostolaki · Maria Armaka · Panayiotis Victoratos · George Kollias

Institute of Immunology, Biomedical Sciences Research Center 'Alexander Fleming', Vari, Greece

Abstract

The TNF/TNF receptor (TNFR) system has a prominent role in the pathogenesis of chronic inflammatory and autoimmune disorders. Extensive research in animal models with deregulated TNF expression has documented that TNF may initiate or sustain inflammatory pathology, while at the same time may exert immunomodulatory or disease-suppressive activities. The TNF/TNFR system encompassing both the soluble and the transmembrane form of TNF with differential biological activities, as well as the differential usage of its receptors, mediating distinct functions, appears to confer complexity but also specificity in the action of TNF. The inherent complexity in TNF-mediated pathophysiology highlights the requirement to address the role of TNF taking into account both proinflammatory tissue-damaging and immunomodulatory functions in a cellular and receptor-specific manner. In this review, we discuss our current understanding of the involvement of TNF in chronic inflammation and autoimmunity, focusing on TNF-mediated cellular pathways leading to the pathogenesis or progression of joint and intestinal inflammatory pathology. Knowledge of the mechanisms by which TNF either initiates or contributes to disease pathology is fundamentally required for the design of safe and effective anti-TNF/TNFR therapies for human inflammatory and autoimmune disorders.

TNF-α is a pleiotropic, proinflammatory mediator whose function is implicated in a wide range of inflammatory, infectious, autoimmune and malignant conditions. TNF is produced in response to infection to confer immunity to the host. While the effect of the TNF in infection is beneficial, tight regulation of TNF production is required to protect the host from the detrimental activities of TNF. Deregulated TNF overexpression can give rise to chronic inflammatory and autoimmune disorders, such as chronic inflammatory arthritis, inflammatory bowel disease (IBD) and multiple sclerosis (MS). Currently, TNF-blocking agents are widely used and have shown encouraging results for the treatment of rheumatoid arthritis (RA) and other inflammatory disorders including Crohn's disease (CD), ankylosing spondylitis and psoriasis [1–3]. By contrast, anti-TNF therapy in patients with MS led to the adverse outcome worsening disease symptoms [4, 5]. Thus, blocking the TNF activity is not always beneficial. Moreover, anti-TNF therapy has led to side effects including opportunistic infections, demyelination, systemic lupus erythematosus symptoms and increased risk for lymphoma [6–9].

TNF is produced in response to bacterial, inflammatory and other stimuli primarily by cells of the immune system, such as macrophages and T and B lymphocytes, but also by additional cell types, including endothelial cells, mast cells and neuronal tissues [10]. Although TNF is initially synthesized as a transmembrane molecule [11], upon cleavage by the metalloprotease TNF-α-converting enzyme (TACE or ADAM17), the secreted monomers that are generated form biologically active homotrimers [12]. Both the soluble and transmembrane forms of TNF are biologically active in their trimeric forms [12]. TNF exerts its biological functions following interaction with its cognate membrane receptors (TNFR), p55TNFR (TNFR1) and p75TNFR (TNFR2) [13], which can additionally be released from the cell surface by proteolysis to produce soluble forms suggested to neutralize the action of TNF [14, 15]. Although most cell types appear to express TNFR1, TNFR2 is preferentially expressed in hematopoietic cells and is more efficiently activated by transmembrane as opposed to soluble TNF [16]. Both opposing and overlapping effects are mediated following activation of TNFR1 or TNFR2 by TNF. Signaling through TNFR1 leads to the activation of the transcription factor nuclear factor-κB (NF-κB) and mitogen-activated protein kinase pathways, and has been prominently associated with proinflammatory, cytotoxic and apoptotic responses [17, 18]. In turn, TNFR2 lacks an intracellular death domain, which is present in TNFR1, and appears to mediate signals promoting cellular activation, proliferation and migration [19, 20].

The generation of animal models with engineered defects in TNF or TNFR expression has been pivotal in our current understanding or TNF/TNFR function. Early studies in TNF-deficient mice revealed the physiological role of TNF in secondary lymphoid organ microarchitecture and function, and in the host defense response [21]. These properties have been attributed to TNFR1 [22–24]. With relevance to disease pathogenesis, studies in mice with perturbed TNF expression have advanced our understanding of the detrimental activities of TNF leading to inflammatory pathology, but have additionally revealed a critical immunomodulatory function for TNF in inhibiting autoimmunity. Deregulated TNF overproduction in transgenic mice is sufficient to initiate multi-organ or tissue-specific inflammation, leading to the spontaneous pathology resembling RA [25, 26], IBD [26] or MS [27, 28]. The ensuing pathology that develops appears to be determined by the locality, cellular context, bioactivity, and chronicity of TNF production. In addition to the above models in which pathology develops as a result of TNF overexpression, mice expressing non-sheddable TNFR1 exhibit increased host defense responses but develop spontaneous liver pathology and enhanced susceptibility to inflammation and autoimmunity, indicating that TNFR1 receptor shedding may regulate TNF activity in vivo by defining thresholds of TNF function [29]. In humans, mutations affecting the shedding of TNFR1 have been associated with the development of TNFR-associated periodic syndrome, characterized by episodes of fever and localized inflammation [30]. It appears therefore that detrimental TNF activities may also arise when mechanisms that aim to control the opposing beneficial and hazardous functions of TNF are eliminated. Still, the paradigm of sustained TNF activity resulting in organ-specific autoimmune or inflammatory pathology is not always followed. In models of systemic autoimmunity or autoimmune diabetes, TNF appears to either promote or inhibit autoimmune pathology depending on factors such as developmental stage, background genetic susceptibility and timing of TNF expression [31–34].

Mechanistically, the contrasting proinflammatory and disease-suppressing activities of TNF may be partly attributed to the diverse functions of its receptors, as well as the differential bioactivities of its soluble and transmembrane forms. Thus transgenic TNF overexpression in the central nervous system results in spontaneous inflammatory demyelinating disease [27, 28], whereas TNF appears to promote the initiation phase in the antigen-induced experimental encephalomyelitis (EAE) model [35], in line with a harmful proinflammatory TNF activity. Most interestingly however, TNF-deficient mice immunized with myelin oligodendrocyte glycoprotein display prolonged self-reactivity to myelin, resulting in exacerbated EAE at a time point when remission would normally take place [36]. In the context of TNFR deficiency, in TNFR1-deficient mice although the initial phase of EAE is suppressed as in TNF-deficient mice, regression to myelin autoreactivity is preserved [36]. By contrast, in double-deficient TNFR1, $-2^{-/-}$ mice, disease is exacerbated in a manner similar to $TNF^{-/-}$ mice, resulting in chronic EAE and late autoimmune reactivity [36]. These data indicate an important role for TNFR1 in mediating the detrimental effects of TNF in the initial stages of the disease, whereas TNFR2 appears sufficient in mediating TNF suppression of autoimmune reactivity. In the same disease setting using the EAE model and mice generated to express an uncleavable mutant TNF protein, we have shown that transmembrane TNF can actively suppress both the inflammatory and autoimmune phase of disease [37]. Furthermore, although transmembrane TNF fails to support splenic structure and function [37, 38], it is capable of supporting host defense responses against *Listeria monocytogenes* [37]. Notably, transmembrane TNF is not adequate to support development of arthritis in the TNF-dependent tristetraprolin (TTP)-deficient model [37]. Therefore, transmembrane TNF may preserve some of the beneficial activities of TNF while lacking detrimental functions.

In view of the above therapeutic approaches, aiming to block TNFR1 or soluble TNF may be preferable to the complete blockade of TNF in the treatment of chronic inflammation and autoimmunity. On this basis, in the following paragraphs we discuss current knowledge derived from animal models on the cellular and receptor-specific functions of TNF in arthritis and IBD.

Cellular Mechanisms of TNF Function in Models of Rheumatoid Arthritis

RA is traditionally described as a pattern of arthropathy involving a prolonged synovitis of multiple diarthodial joints. The synovitis leads to pain, soft tissue swelling and stiffness resulting in loss of joint function. RA is characterized by the presence of inflammatory infiltrate in the synovium, the thin membrane predominated mostly by resident fibroblasts and scarcely detected macrophages located adjacent to and in direct contact with the intra-articular cavity of the joint. During the course of disease, the synovial membrane gradually increases in thickness, transforming itself into an aggressive cellular mass called pannus that invades and destroys articular structures [39]. The disease affects about 1% of the population worldwide, thus creating a substantial personal, social and economical burden [40].

Research in the last decades has focused on unraveling the pathogenesis of RA either by applying molecular techniques and genetic analysis on human tissue or by generating animal models of RA for identifying and analyzing the pathogenic pathways that lead to all aspects of disease: inflammation, cartilage breakdown and bone erosion. Considerable genetic knowledge suggests that the genetic susceptibility and linkage for RA is rather complicated, as several genes and loci are acknowledged to be linked with disease in population studies [41–43]. Interestingly, due to the genetic association of HLA-DR genes with RA [44], a significant role for T cell-dependent mechanisms had also been proposed. However, the lack of abundant detection of T cell-derived products in RA synovium and synovial fluid [45] and the low clinical efficacy of a nondepleting anti-CD4 antibody (keliximab) [46] challenged the notion of a primarily pathogenic role for T cells in RA. Early theories on etiopathogenesis also focused on analyzing clinical findings, such as deregulated autoantibodies and immune complexes, and these observations led to the conclusion that RA is an autoimmune disease. The detection of rheumatoid factor [47], although not a very specific finding for RA, as well as high titers of other autoantibodies signified the role of B cells in RA pathology, and this is further emphasized by clinical improvement in patients receiving rituximab, an anti-CD20 antibody [48]. In the 1990s, Firestein and Zvaifler [49] and Firestein [50] reviewing the current RA literature, disputed the acquired immunity-based orchestration of the inflammatory response in RA and suggested that, most probably, innate signals govern and perpetuate the disease manifestations. This was in line with the detection of innate cytokine networks in joint tissue cultures from RA patients [51]. The experimental paradigm on the role of TNF in arthritic pathology became apparent by the generation of human TNF transgenic mice (hTNF-Tg or Tg197); hTNF-Tg mice express high levels of human TNF transgene due to genetic modification of the 3′ prime UTR of human TNF gene. The mice develop inflammatory polyarthritis with all characteristics of RA, and disease is abrogated by the anti-human TNF regime [25]. The clinical efficacy of anti-TNF therapy in patients with RA indeed showed significant results in dampening disease activity [1], emphasizing the importance of TNF in human disease.

Based on the possible role of 3′UTR in the translational repression of TNF mRNA, a targeted mutant lacking endogenous 3′UTR ARE elements of murine TNF mRNA ($TNF^{\Delta ARE}$ mice) develops arthritis and Crohn's-like IBD, further confirming TNF-mediated mechanisms in orchestrating the arthritogenic response in mice [26]. In both animal models, the arthritogenic potential of TNF is mediated through TNFR1 [25, 26]. The role of TNFR2 could only be demonstrated for $TNF^{\Delta ARE}$ mice, since human TNF does not signal through murine TNFR2 [52]. Thus, genetic deficiency of TNFR2 in $TNF^{\Delta ARE}$ results in a more aggressive form of arthritis, implying that TNFR2 may act to counterbalance the pathogenic TNF signals [26]. The negative control of TNFR2-mediated signaling in modifying disease phenotype could perhaps be associated with the immunoregulatory function of this receptor. Therefore, the $TNF^{\Delta ARE}$ and hTNF-Tg mice appear to be most informative animal models as to the TNF-mediated mechanisms operating in arthritis.

Importantly, the beneficial effect of TNF neutralization and the key role of this cytokine in arthritic disease have been further demonstrated in animal models other than the hTNF-Tg mice. The widely used collagen-induced arthritis model (CIA), generated by heterologous CII immunization in animals, may be treated effectively with anti-TNF antibody or other TNF inhibitors administered prior to disease onset [53, 54]. In addition, anti-TNF monoclonal antibody treatment was used successfully after disease onset in CIA and resulted in reduced inflammation [55]. Experiments with TNF-deficient animals showed that TNF is crucial but not dominant in the CIA model, as pathology develops with delayed onset and milder symptoms [56]. Similarly, TNFR1 deficiency delays the onset but does not ameliorate the clinical signs of disease in this model [57]. The SKG strain carries a natural point mutation affecting the gene encoding an SH2 domain of ZAP-70, a key signal transduction molecule in T cells, and spontaneously develops arthropathy [58]. TNF deficiency retarded the onset and substantially reduced disease incidence and severity in this model [59]. As other proinflammatory cytokines are implicated in arthritis, such as interleukin (IL)-1 which stands in a leading position at the cytokine cascade of RA, it is worth mentioning that arthropathy developing in IL-1 receptor antagonist-deficient mice due to uncontrolled IL-1 signaling [60] could be rescued in a TNF null background [61]. Even though TTP-deficient animals do not carry any mutations in cytokine-related genes, they develop a systemic inflammatory syndrome with severe polyarticular arthritis and autoimmunity, as well as medullary and extramedullary myeloid hyperplasia [62]. TTP is a zinc finger protein involved in ARE-containing mRNA degradation; in TTP deficiency, ARE-containing TNF mRNA cannot be degraded. Apparently, the phenotype of TTP-deficient mice is TNF/TNFR1 dependent [62, 63]. A recently developed animal model provided evidence that defective apoptosis in macrophages via inducible ablation of DNase II leads to the development of a severe inflammatory polyarthropathy [64]. Pathology is abrogated by anti-TNF administration, implying that the synovial tissue is extremely susceptible to aberrant innate TNF signaling events

[64]. It is therefore evident that TNF/TNFR1 signaling interferes actively with the arthritogenic process at multiple levels regulating immune reactivity and cellular fate, independently of the animal model employed.

Mesenchymal Cell-Specific Role of TNFR1 in the Pathogenesis of TNF-Driven Inflammatory Arthritis

Despite the debates on the autoimmune or autoinflammatory nature of RA, and the plethora of described mechanisms leading to pathogenic cytokine disbalances in RA, experimental evidence indicates that chronic innate immune activation of the synovial fibroblast (SF) could be a dominant pathogenic event in RA. SFs are resident mesenchymal cells of the synovial membrane and their origin is still debatable; they represent a heterogeneous population of cells in terms of tissue localization, physiology (intimal and subintimal) and derivation (nonepithelial, mesenchymal cells) and display differential activation and differentiation properties [65]. The primary physiological role of the SF is to provide a nourishing environment for the cartilage and to lubricate the articular surfaces through production of hyalorunan, lubricin, and collagens. They lack expression of MHC class II antigens, CD68 and do not present any phagocytic activity. Notably, intimal SFs express CD55, ICAM-1, and increased VCAM-1 levels, as compared to subintimal SFs and other types of fibroblasts [66], enabling them to interact with other cell types, such as mononuclear lymphocytes, T and B cells, and to modulate leukocyte trafficking.

Several lines of evidence have indicated the autonomous arthritogenic function of SFs both in vitro and in vivo. Cultured RA-SFs can proliferate in an anchorage-independent manner, escape contact inhibition growth arrest, and express a variety of transcription factors and matrix metalloproteinases (MMPs) [67], while they exhibit deregulated expression of Wnt-related molecules potentially indicating that they may have reacquired the primordial phenotype, accounting for their hyperproliferation and aggressive invasiveness, properties usually detected in tumors [68]. Notwithstanding the notion that cytokines like TNF can trigger SF activation and proliferation [69, 70], it seems that SFs can maintain their activation status without the need for continuous stimulation from the proinflammatory microenvironment. The most convincing evidence on the autonomous nature of the RA-SFs has been provided by Muller-Ladner et al. [71] in an elegant study showing RA-SFs cotransplanted with human cartilage into immunodeficient mice to grow invasively into adjacent cartilage even in the absence of other cells of the human immune system. The arthritogenicity of murine SFs derived from hTNF-Tg mice was also exhibited when intraarticularly injected in immunodeficient mice [72]. The innate activation of SFs may be explained either by continuous stimulation from paracrine proinflammatory mediators (e.g. TNF), or by chronic innate signals through pattern recognition receptors on their surface [73]. Proinflammatory cytokine and chemokine production, as well as upregulated expression of adhesion molecules by the activated SF, may in turn promote the recruitment and retention of immune cells in the synovium [74].

More recent concepts suggest the cytoskeletal control machinery as a target of TNF in partly regulating the inflammatory and apoptotic phenomena [75]. In agreement with this concept, TNF-induced activation of NF-κB and cytokine secretion in human cultured RA-SFs is dependent on the activation of RhoA, a GTPase which promotes actin polymerization (F-actin formation), through p65/RelA NF-κB subunit binding to newly formed F-actin, suggesting a central role of this GTPase in the arthritic inflammatory response [76, 77]. Interestingly, stress fiber formation in SFs from the hTNF-Tg mice is significantly more intense [78], and hTNF-Tg SFs show increased proliferative, migratory and adherence capacity compared to WT cells [79]. This phenotype of the hTNF-Tg SF is not reversed by short-term anti-TNF treatment hinting an imprinted phenotype of the murine cells derived from an overexpressing human TNF environment [79], as suggested for RA-SFs [67]. Additionally, it was recently shown that a number of deregulated genes, known to be involved in actin filament and cytoskeleton organization, such as *gsn*, *aqp1*, *cdc42hom*, *eef1a1*, *tuba1*, *rab14*, *lsp1*, *lst1*, *mylc2b*, *pitpnm* and *pstpip1*, were strongly deregulated in the hTNF-Tg animal model [78]. Importantly, the genetic ablation of gelsolin, encoded by *gsn*, a gene found downregulated in hTNF-Tg SFs, resulted in exacerbation of the arthritic disease in hTNF-Tg mice [78], validating the functional significance of the actin cytoskeleton rearrangements in the pathophysiology of the disease. More recently, cadherin-11, a junction molecule ubiquitous to many tissues, was shown to function as a major mediator of synovial architecture by organizing SFs via formation of cell-to-cell adherent junctions [80] and remodeling of the actin cytoskeleton [81]. The stromal cell signature of cadherin-11 in arthritis was confirmed when the passive K/BxN model was applied to cadherin-11-deficient recipients, resulting in suppression of autoimmune arthritis pathology [82]. Interestingly, TNF has been reported to drive cell-cell adhesion molecule cadherin-11 expression in the rheumatic synovium [83] and to promote the invasive behavior of RA-SFs [84]. In view of the above, these data provide further evidence to support the concept that TNF may promote structural changes in the synovial lining leading to the activation and pathogenic function of the SF through direct or indirect modulation of actin cytoskeleton dynamics.

TNF-modeled arthritis offers an adequate system to decipher the cellular requirements for the induction of TNF-mediated arthritic pathology and to explore the potential of cell-specific therapeutic approaches. Remarkably, we have previously shown that, in both the hTNF-Tg and $TNF^{\Delta ARE}$ models, inflammatory arthritis develops in the absence of the adaptive immune response (RAG1 deficiency), implying that either innate immunity or other immune or nonimmune mechanisms could be responsible for initiation and perpetuation of disease [26, 85]. In these models, we have used reciprocal bone marrow transplantation experiments in TNF-overexpressing mice ($TNF^{\Delta ARE}$ or hTNF-Tg mice) and TNFR-deficient mice to decipher the cellular sources and targets of pathogenic TNF. With this approach, we have shown that in

$TNF^{\Delta ARE}$ mice the pathogenic TNF source is located in the radiosensitive bone marrow compartment, whereas in the hTNF-Tg mice, pathogenic human TNF derives from radioresistant-stromal cells [Armaka, unpubl. obs.]. While the $TNF^{\Delta ARE}$ and hTNF-Tg mice do not share the same arthritogenic TNF pool, the bone marrow engraftment experiments indicated that the cellular target of pathogenic TNF is a shared hallmark; arthritic pathology develops exclusively by TNFR1-mediated signaling in radioresistant stromal cells in both models [86]. Furthermore, early activation of the SF, evidenced by the misbalanced production of MMPs and their inhibitors TIMPs (tissue inhibitor of MMPs) prior to the appearance of inflammatory infiltrate in joint area, indicated the early proinflammatory triggering of the SF in TNF-driven arthritis. In this context, the in vivo validation of the SF as the mesenchymal component sufficient to elicit TNF/TNFR1-mediated disease was confirmed by the selective mesenchymal expression of the TNFR1 allele in both TNF-overexpressing murine models using Cre/LoxP technology; the mice develop full-blown arthritis under the restricted SF expression of TNFR1 [86]. These data clearly establish the importance of the SF not only as primary target but also as coordinator of all aspects of the arthritic phenotype in mice. Accordingly, it would be extremely interesting to investigate whether TNFR1 signaling in the SF is an absolute requirement for the induction of disease and further analyze in a cell-specific manner the molecular pathways implicated and their contribution to the course of disease.

In light of the evidence discussed so far, SFs can initiate the pathogenic cascade through sensing of pathogenic triggers such as TNF, promote the disease through tissue destruction and recruitment of inflammatory cells, and thus amplify and sustain the immune response constituting a key cell type in disease pathogenesis and perpetuation (fig. 1).

Cellular Mechanisms of TNF Function in Models of Inflammatory Bowel Disease

IBD is a chronic inflammatory disorder of unknown etiology that affects the gastrointestinal tract. The prevailing concept regarding the etiopathogenesis of both subtypes of IBD, CD and ulcerative colitis (UC), is that disease pathogenesis involves dysregulated immune responses against antigens of the intestinal flora influenced by genetic and environmental factors [87]. Although this concept applies to both subtypes of IBD, these are characterized by distinct localization and histopathological features. In CD, inflammation is primarily manifested in the terminal ileum, but can affect any region of the gastrointestinal track, whereas in UC inflammation is restricted to the colon. In addition, the presence of transmural inflammation often associated with granulomas is characteristic of CD, whereas in UC inflammation is typically restricted to the superficial mucosal and submucosal layers. Despite these distinct features, both CD and UC are considered predominantly T cell-mediated processes. Recent genome-wide association studies have identified genetic variation in the

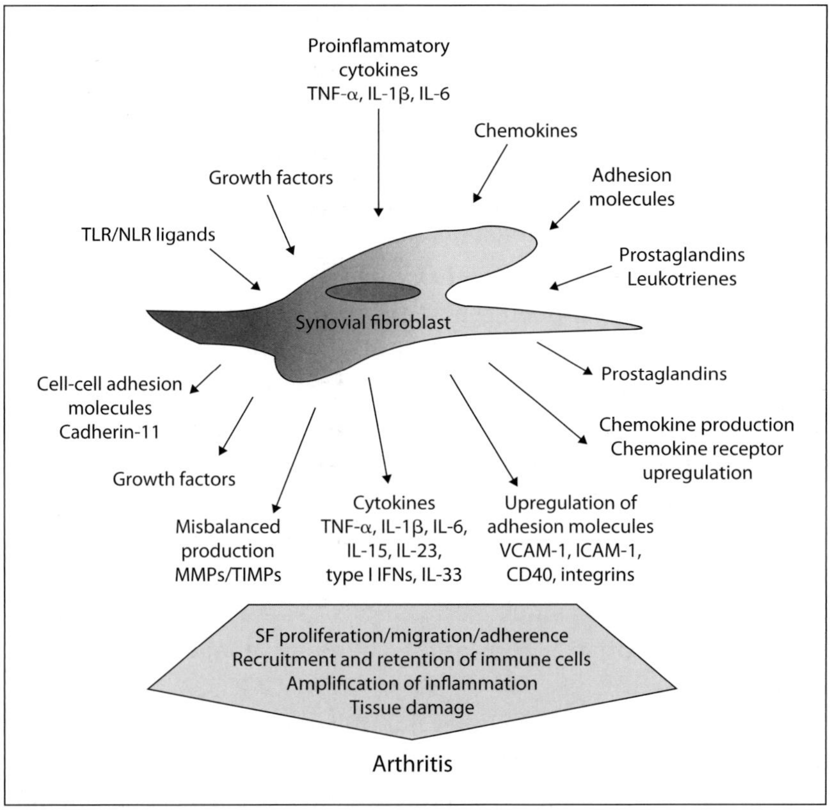

Fig. 1. SF activators, products and effector functions.

innate immune system gene NOD2, and the autophagy genes ATG16L1 and IRGM to be associated with CD, whereas genetic variation in the gene for the IL-23 receptor (IL-23R), or in the gene regions of the common IL-12/23 cytokine subunit p40 (IL-12/23 p40), the cytokine TNFSF15, and the NKX2–3 gene involved in mucosal tissue architecture were associated with both CD and UC [88–91]. These associations provide further support to the hypothesis that innate and adaptive immune responses to intestinal microbiota are involved in IBD pathogenesis and play an emerging role in the autophagy pathway in CD.

A wide collection of animal models generated over the years, either inducible or following genetic gene targeting resulting in spontaneous phenotypes, have proven essential in our current understanding of IBD pathogenesis [92]. TNF has a prominent role in many of these models. Importantly, antibodies against TNF have proven to be effective in the treatment of CD [2], but also more recently in the treatment of UC [93]. The dominant role of TNF as an initiating factor of intestinal inflammation, and CD in particular, was exemplified by the generation of the $TNF^{\Delta ARE}$ mice carrying a genetic deletion in the ARE elements of the TNF mRNA, resulting to chronic TNF overproduction and the spontaneous development of Crohn's-like IBD pathology and inflammatory arthritis [26]. Remarkably, intestinal pathology in these mice develops primarily in the terminal ileum and only occasionally in the proximal colon. Histological features include intestinal villous blunting and broadening, transmural inflammation, and the formation of granulomas, which in addition to the ileal localization highlight the unique resemblance of intestinal pathology in the $TNF^{\Delta ARE}$ model to human CD. Intestinal pathology develops in the $TNF^{\Delta ARE}$ mice as a result of spontaneous TNF overproduction from multiple sources including myeloid and lymphoid cells, but also stromal cells, such as fibroblasts [26]. Restricting TNF overproduction in myeloid cells or T lymphocytes is sufficient to drive intestinal inflammation, indicating that chronic TNF overproduction from either innate or adaptive immune effectors can support the development of pathology in this model [94]. TNFR1 appears dominant in mediating TNF pathogenic signals, as IBD pathology fails to develop in TNFR1-deficient $TNF^{\Delta ARE}$ mice. By contrast, $TNF^{\Delta ARE}$ mice genetically deficient for TNFR2 display attenuated but not neutralized inflammation, suggesting that TNFR2 contributes but is clearly less important than TNFR1 in TNF-driven intestinal pathology [26]. Therefore, TNF appears to act as an initiating factor that orchestrates the inflammatory response leading to intestinal inflammation in the ileum.

A key role for TNF has been established in several IBD models, in which regardless of the underlying pathogenic mechanisms, approaches such as TNF/TNFR1-genetic inactivation or the administration of anti-TNF antibodies reduce or ameliorate inflammation [95–99]. Importantly, TNF may be involved in various aspects of the disease process. Evidence on this can be provided in mice with intestinal epithelial cell-specific inhibition of NF-κB, which develop colon inflammation [99]. In this model, compromised epithelial integrity occurs due to increased TNF-mediated apoptosis in epithelial cells with impaired NF-κB signaling. Bacterial translocation in the mucosa as a result of the barrier defect induces proinflammatory TNF and IL-1β overexpression, promoting immune cell activation and recruitment. Pathology was ameliorated in TNFR1-deficient mice [99]. Thus, TNF appears to both induce epithelial apoptosis and amplify the subsequent inflammatory response. However, in contrast to the above, TNF may additionally mediate processes that oppose mucosal inflammation, as evidenced in dextran sodium sulfate-induced colitis, which is aggravated in TNF-deficient mice [100]. TNFR1 signaling in myeloid cells has been reported to contribute to suppression of pathology through the control of epithelial cell apoptosis in this model [101]. TNFR2, however, contributes to exacerbation of colitis through the innate response [101]. A similar contribution for TNFR2 through the adaptive system has been described, as more severe colitis was induced by the reconstitution of severe combined immunodeficient mice with TNFR2-overexpressing CD4+CD62L+ T cells [102]. Therefore, when trying to understand the mode of action of TNF in IBD pathogenesis we should aim to identify the cell- and receptor-specific mechanisms by which TNF contributes to both the initiation and/or progression of disease. These may include direct effects of TNF on cellular subsets critical for the pathogenesis of intestinal inflammation, but also TNF modulation of the cellular and molecular pathways relevant to the ensuing adaptive immune response.

Adaptive Immune Responses in TNF-Driven Intestinal Inflammation

An integral characteristic of IBD is the development of excessive effector T lymphocyte responses. Potential mechanisms that are thought to account for such responses include both cytokine overproduction resulting in exaggerated effector T cell responses and/or defective suppression of mucosal effector T cell responses by regulatory T lymphocytes [92]. Notably, intestinal inflammation in the $TNF^{\Delta ARE}$ mice does not develop in the genetic absence of the mature T and B lymphocytes in RAG1-deficient $TNF^{\Delta ARE}$ mice [26]. By contrast, inflammatory arthritis still develops in these mice [26], indicating the differential requirement of the adaptive immune response in TNF-driven mucosal inflammation as opposed to arthritis. In subsequent studies, we established a critical role for CD8+ T lymphocytes as the pathogenic effectors in this model, whereas CD4+ T lymphocytes appear to exert a protective effect, evidenced by exacerbated intestinal pathology in their genetic absence [94]. Recently, we have identified the prominent deregulation of intestinal intraepithelial lymphocyte (IEL) populations in the $TNF^{\Delta ARE}$ model [103]. The lymphocytes present in the epithelium of normal mice can be broadly categorized into two subsets, conventional CD4+ and CD8αβ+ T lymphocytes that have migrated to the intestinal epithelium following initial priming by cognate antigen in the periphery, and unconventional T lymphocytes bearing either TCRαβ or TCRγδ receptors characterized by the coexpression of CD8αα molecules [104]. This latter subset consists of long-term residents of the epithelium attributed with an important role in maintaining mucosal tissue homeostasis [105–107]. Interestingly, intestinal inflammation in the $TNF^{\Delta ARE}$ mice was associated with the early decline of CD8αα-expressing IELs preceding the inflammatory infiltration of the lamina propria [103]. During the advanced disease stage, almost the entire IEL compartment was found to consist of conventional lymphocytes [103]. The requirement for CD8+ T lymphocyte effector function for intestinal pathology highlights the significance of these findings in the $TNF^{\Delta ARE}$ model. At present, it is unclear whether the perturbation of intestinal T lymphocytes is a direct effect of TNF on these cells, or an effect on other cell types of the mucosal tissue that could ultimately result in this process. Interestingly, however, CD8αβ T lymphocyte recruitment in the epithelium, as well as inflammatory infiltration in the lamina propria of $TNF^{\Delta ARE}$ mice, do not require the function of the chemokine CCL25 or its receptor CCR9 [103]. Despite the prominent association of this particular chemokine axis with lymphocyte recruitment in the small intestine [108, 109], intestinal lymphocyte recruitment and inflammatory pathology develop unperturbed in the genetic absence of CCL25 or in CCR9 in $TNF^{\Delta ARE}$ mice. By contrast, genetic ablation of β_7-integrin in $TNF^{\Delta ARE}$ mice results in ameliorated pathology, establishing a critical requirement for β_7-integrin-mediated interactions in promoting lymphocyte recruitment and pathology in TNF-driven inflammation [103].

Given the dominant role of T lymphocytes in IBD pathogenesis, mucosal T lymphocyte cytokine responses have been a subject of intense study. Data from both murine models and human samples have supported that CD is mediated by T helper (Th) 1 responses [110]. Recently, however, a novel Th lymphocyte subset, termed Th17 cells, has been identified, which produces a range of cytokines including IL-17(A), IL-17F, IL-21 and IL-22, and has been associated with autoimmune disease [111]. In addition, the receptor for the cytokine IL-23, which shares a subunit with the cytokine IL-12 [112], and mediates IL-17 production, has been associated with CD susceptibility [91], indicating that the IL-23-IL-17 axis may be involved in CD pathogenesis. The high expression of both interferon (IFN)-γ and IL-17 in the mucosa of CD patients has been reported [113, 114]. Studies on the role of IL-23 and Th17 cells in IBD in different mouse models have yielded conflicting results, reporting either a requirement for this axis for colitis [115, 116] or exacerbated pathology in IL-17-deficient mice [117]. An alternative hypothesis is that both IFN-γ and IL-17 may be synergistically involved in IBD pathology. However, it appears that Th1 and Th17 cells cross-regulate each other, and IL-17A can directly inhibit Th1 cell development mediating a protective effect in T cell-mediated colitis [118]. Functional data on the role of IL-17 in models of small intestinal inflammation are currently lacking. In the $TNF^{\Delta ARE}$ mouse, the genetic deficiency of IFN-γ or IL-12/23p40 subunit resulted in attenuated pathology [94]. Intriguingly, in TNF-driven intestinal inflammation we have shown increased Th17 and decreased Th1 cytokine responses by lamina propria CD4+ T lymphocytes [103]. Additionally, we have identified CD8αβ intraepithelial and lamina propria lymphocytes as the principle IFN-γ-producers, suggesting that the IFN-γ-dependence of the $TNF^{\Delta ARE}$ model relies on CD8+ T lymphocyte IFN-γ production [103]. As the role of CD4+ T lymphocytes appears to be protective in $TNF^{\Delta ARE}$ intestinal pathology, it remains to be determined whether CD4+ T cell-mediated protection relies on their ability to produce IL-17. Therefore, TNF appears to promote Th17 at the expense of Th1 cells in the inflamed lamina propria. Whether or not this effect requires the function of IL-23 needs to be further addressed, although it will be important to discriminate additional IL-23-mediated effects, as IL-23 has been shown to promote innate immune activation and recruitment of neutrophils to inflamed tissues [119].

Mesenchymal Cell-Specific Role of TNFR1 in the Pathogenesis of TNF-Driven Crohn's-Like Inflammatory Bowel Disease

A wide range of experimental data generated from different IBD models already discussed has led to the hypothesis that TNF mediates its pathogenic signals in IBD primarily through innate immune activation associated with aberrant lamina propria T lymphocyte responses, enhanced lymphocyte recruitment through the upregulation of chemokine and adhesion molecules, and the induction of intestinal epithelial apoptosis and epithelial barrier dysfunction. These deleterious processes have indeed been attributed to TNF and are evidently implicated in the pathogenesis of IBD. However, the pleiotropic function of TNF hinders the identification of the early and perhaps sufficient cellular pathways that may be responsible for the initiation of the full spectrum of pathogenic cascades implicated in intestinal inflammation. We have recently identified using the $TNF^{\Delta ARE}$ model of Crohn's-like IBD and taking advantage of mice bearing mesenchymal cell-restricted expression of TNFR1 the intestinal subepithelial myofibroblast as a primary responder cell type sufficient for full pathogenic TNF/TNFR1 signaling in Crohn's-like IBD [86]. Intestinal myofibroblasts have a central role in maintaining mucosal tissue architecture and controlling inflammatory and repair processes [120]. Importantly, TNFR1-mediated signaling restricted in mesenchymal cells was shown to result in the development of Crohn's-like IBD pathology, with similar characteristics as described in the $TNF^{\Delta ARE}$ model, including ileal localization of inflammation, intestinal villous blunting and broadening, mucosal and submucosal infiltration of inflammatory cells, and transmural inflammation. Early activation of intestinal myofibroblasts by TNF was evidenced prior to the onset of intestinal inflammation, as shown by the deregulated expression of MMPs, MMP3 and MPP9, and their inhibitor, TIMP1 [86]. Notably, increased levels of MMPs, and MMP3 in particular have been reported in the mucosa of CD patients [121]. Although at present we cannot state whether additional cell types may be sufficient, or to what extend they can contribute as TNF targets to intestinal inflammation, the identification of the intestinal subepithelial myofibroblast as a cell type unique in the capability to initiate TNF-mediated pathways leading to intestinal pathology provides novel mechanistic insight into the role of TNF in IBD pathogenesis and alternative hypothesis for the mode of action of anti-TNF antibodies. The effectiveness of anti-TNF therapy, and in particular of the chimeric anti-TNF antibody infliximab, in CD patients has been associated with the induction of apoptosis of peripheral blood and lamina propria lymphocytes, by initiating reverse signaling through transmembrane TNF expressed on the surface of inflammatory cells [122]. On this basis, the mechanistic rationale of anti-TNF therapies was focused on the apoptosis-inducing potential of anti-TNF agents and distracted from the actual function of TNF as an initiating and perpetuating factor in inflammation. Importantly, a study addressing the modulation of intestinal myofibroblast function by anti-TNF therapy provided evidence for enhanced TIMP1 and, as a result, decreased MMP activity in CD myofibroblasts following treatment with infliximab [123]. Thus, combined data from mouse models and clinical samples support a dominant role for the modulation of the intestinal myofibroblast by TNF in CD pathogenesis.

Most interestingly, intestinal and peripheral arthritis pathology develops in the $TNF^{\Delta ARE}$ mice with combined features of ankylosing spondylitis, such as arthritis, bilateral inflammation of the sacroiliac joints, and enthesitis [86]. Emerging evidence in rheumatology suggest a possible connection of clinical or subclinical intestinal inflammation with arthritic manifestations in a group of diseases entitled as spondyloarthropathies (SpAs), which includes anklylosing spondylitis, reactive and psoriatic arthritis and undifferentiated spondyloarthritis [124]. Notably, a common feature of IBD and SpAs is the successful application of anti-TNF therapy [3]. Previous experimental evidence on hTNF-Tg mice confirmed a role for TNF in the development of bilateral sacroiliitis [125], a hallmark of SpA. In $TNF^{\Delta ARE}$ mice, arthritis-spondyloarthritis appears to be combined with intestinal pathology [86], as typically occurs in patients suffering from SpAs. Importantly, intact TNF/TNFR1 signaling in mesenchymal cells was proved to suffice for the induction of SpA-like pathology, including peripheral arthritis, sacroiliitis, and Crohn's-like IBD in the $TNF^{\Delta ARE}$ model [86]. These data establish an early and dominant role for mesenchymal cells as TNF responders in the pathogenesis of SpAs in a novel mechanistic perspective that may also explain the common occurrences of these pathologies in humans, as well as the remarkable response of a significant number of patients to anti-TNF therapies.

On the Role of TNF in Follicular Dendritic Cell Network Development, Antibody Responses and Autoimmune Arthritis

As antigen infiltrates through the cellular architecture of secondary lymphoid organ, it encounters different types of cells that are segregated into distinct anatomical regions, follicles consisting of B lymphocytes and a network of follicular dendritic cells (FDCs), and T cell areas consisting of T lymphocytes and a network of dendritic cells. Since establishment of acquired immune responses against self-antigens is a constant risk, the fine structure of secondary lymphoid organs places certain restrictions on the kind of cellular interactions that can take place. The dynamic microenvironment of B cell follicles and the germinal centers (GCs) are exceptional examples of a fine-tuning between mechanisms that favor a robust immune response and mechanisms that authenticate the antigen specificity of this response. Primary follicular structures and GCs provide a specialized microenvironment essential for capturing native forms of the antigen, the cautious selection and propagation of antigen-specific B cell clones and the elimination of nonspecific autoreactive clones that are accidentally generated by the rather stochastic process of somatic hypermutation. These important processes that occur within GCs produce a pool of memory B cells and a massive number of antibody-producing cells of increased affinity compared to their ancestor B cell clones. FDCs provide both organizing signals for the proper structure of the B cell follicles and regulatory signals to support the generation of GCs. The dependence of FDC development on TNF signaling illustrates another aspect of TNF function with implications on immunity and autoimmunity.

FDCs are radioresistant, nonhematopoietic lineage cells that are located exclusively within B cell follicles forming a dense network of dendritic processes. By the virtue of CXCL13, FDCs invite naïve CXCR5-bearing B cells to their vicinity resulting in the assembly of primary B cell follicles [126–129]. Once they are clustered, FDCs act as accessory cells that trap large amounts of immune complexes via complement and Fc receptors, providing niches for antigen engagement and selection of B cells with respect to the specificity and the affinity of B cell receptor [130, 131]. In contrast to widespread developmental effects of other homeostatic factors [132], the function of TNF and its receptor TNFR1 is considerably devoted on the development of mature FDCs. In both TNF- and TNFR1-deficient mice, analysis of spleen, lymph nodes and Peyer's patches has shown that lymphocyte segregation occurs normally with the absence of FDC networks to be the most striking and profound defect [21, 133–135]. Consequently, the typical clusters of B cell follicles fail to form and B cells are distributed in a ring-like structure around T cell area. In agreement with the lack of FDC networks and organized B cell follicles, the production of CXCL13 is greatly reduced, whereas the production of T cell area-derived CCL19 and CCL21 is not affected [136]. Moreover, the failure of FDCs to develop in the absence of TNF or TNFR1 causes profound defects in GC formation, antibody production and recall responses to T cell-dependent antigens [21, 133]. Through irradiation chimera and adoptive transfer experiments, it was established that the development of FDCs requires TNF production by lymphocytes, in particular B cells [135]. TNF is expressed on the plasma membrane and can be shed to a soluble form. Mice overexpressing membrane TNF appear fully capable of supporting TNFR1-dependent formation of FDCs and B cell follicles [137]. Further studies with mutant TNF mice revealed that the physiological levels of membrane TNF are inefficient in supporting primary B cell follicles and their associated FDC networks [37, 38]. Together, these findings showed that soluble TNF is essential for the generation of FDCs, but membrane TNF may contribute to this process. While studies with reciprocal reconstitution experiments in gene-targeted mice established that TNFR1 needs to operate in radioresistant stroma cells for the development of FDCs [138, 139], the possibility that the effect of TNFR1 signaling is mediated indirectly through a non-FDC cell that provides an essential trophic signal to FDCs had remained a strong argument. To address this issue, we employed a Cre-loxP genetic approach that restricts the expression of TNFR1 in FDCs and we showed that TNFR1 acts in a cell-autonomous fashion for the development of FDCs [140]. The expression of TNFR1 in FDCs was essential for the generation of FDC networks, the restoration of CXCL13 production and the subsequent correct organization of primary B cell follicles. Upon immunization, TNFR1-bearing FDCs were fully competent to upregulate adhesion molecules participating in interactions between FDCs and GC B cells, to generate strong GC responses and to restore antibody production. Taken together, these studies have unequivocally established that: (a) TNF-TNFR1 signaling acts directly on FDCs to induce their clustering within B cell follicles and to activate their ability to support GC responses, and (b) FDCs are critical cells organizing the B cell follicle and the mature GC responses.

The persistent activation of B lymphocytes that associates with enduring circulation of autoantibodies and the successful therapeutic intervention of B cell depletion in RA [141] emphasize the pivotal role of pathogenic B cell responses in several immune disorders [142, 143]. The longevity and the high affinity of plasma cells and memory B cells that emerge from GCs represent a great threat for the development of autoimmunity. In the course of many autoimmune disorders, the accumulation of lymphoid cells in the inflamed tissues establishes microenvironments that are rich in organogenic and inflammatory cytokines and chemokines driving the generation of lymphoid-like structures [144, 145]. These highly ordered structures have long been recognized to harbor ectopic GCs that contain FDCs, T cells and GC B cells with ongoing proliferation, affinity maturation and isotype switching. Importantly, the production of tissue-specific, disease-relevant autoantibodies by ectopic GCs [146–149] strongly suggests a pathogenic role of these GCs, at least in perpetuating autoimmune responses. Moreover, spontaneous GC formation has been described in secondary lymphoid organs of many autoimmune-prone mice, and these GCs appear at the time of autoantibody production and disease onset [150]. In healthy mice, autoreactive B cells are normally excluded from entering into B-cell follicles; this follicular exclusion, however, breaks down when T cell help is provided [151, 152]. In human patients, exclusion of autoreactive B cells has been shown to be defective and pathogenic B cells successfully participate in GC reactions and expand within the post-GC IgG memory and plasma cell compartments [153]. Blockade of CD40-CD40L interactions prevents GC formation and autoantibody production and ameliorates disease development in human patients with SLE [154] as well as in mouse models of lupus [150].

To examine the role of FDCs as a relevant cell type linking GC B cell autoreactivity with disease development, we took advantage of their dependence on TNFR1 signaling and we switched on/off their development by using the same Cre-loxP approach [140] in the K/BxN model of arthritis. K/BxN mice develop an aggressive form of arthritis that recapitulates autoantibody-mediated pathology of RA in humans [155, 156]. In the context of I-Ag7 molecule [155], transgenic T cell receptor CD4 T cells recognize glucose-6-phosphate isomerase (GPI) [157, 158], a ubiquitously expressed self-antigen that is also present on the surface of inflamed joints [159, 160]. Primed CD4 T cells collaborate with GPI-reactive B cells for the production of arthritogenic GPI antibodies with autoantibody-mediated inflammatory mechanisms to be the hallmark features of the effector phase of the disease [161–164]. We showed that the TNFR1-mediated lack of FDCs prevents the development of the disease because the formation of GPI-reactive GCs is compromised and the production of arthritogenic antibodies is drastically reduced [165]. We also found that the differentiation of arthritogenic CD4 T cells to a Tfh phenotype and their CXCR5-dependent migration into B cell follicles are critical steps for autoreactive GC formation, autoantibody production and disease development. The TNFR1-mediated integrity of FDCs is essential for the follicular relocation of arthritogenic Tfh cells, most probably by establishing the appropriate CXCL13 gradient. In addition to the FDC-dependent recruitment of Tfh cells, we showed that the deposition of immune complexes on FDCs provides niches for autoantigen engagement promoting Bcl-6 expression and aberrant positive selection of autoreactive GC B cells. Recently, the pathogenic function of Tfh cells was also documented in sanroque mice [166]. In these mutant mice, the excessive generation of Tfh cells causes the aberrant production of spontaneous GCs that are responsible for lupus-like pathology. Together, these studies [165, 166] demonstrate that the prevention of T cells from acquiring a follicular phenotype and entering follicles or helping GC B cells are important checkpoints to prohibit GC B cell autoimmunity. Of note, another aspect of our research is that the TNFR1-dependent disruption of FDCs by nongenetic tools efficiently suppresses the inductive mechanisms of autoantibody-mediated arthritis. Importantly, we showed that treatment with p75TNFR:Fc (etanercept) inhibits the maintenance of FDC networks resulting in decreased GC reactivity and autoantibody production with subsequent amelioration of autoantibody-mediated arthritis [165]. Considering the widespread clinical use of TNF antagonists for treatment of RA and other diseases, these findings substantiate the notion that the reduction of FDC function and thus abnormal B cell responses are among the beneficial effects of these treatments.

Conclusions

TNF appears to be a common pathogenic determinant in many inflammatory and autoimmune disorders. In a simplified view, pathology may develop as a consequence of the inability to regulate TNF expression, which in the context of an immunological response, promotes chronic innate activation and/or immune reactivity, amplifies inflammation, and eventually results in tissue damage. This view is significantly advanced in the current knowledge obtained from animal models, that TNF may orchestrate the inflammatory pathogenic cascade leading to the development of both joint and intestinal pathology by specifically targeting mesenchymal cell types, such as SFs and intestinal myofibroblasts. Fundamental differences in these pathologies exist, as exemplified in the requirement for the adaptive immune response in modeled TNF-mediated Crohn's-like IBD, but not in TNF-mediated arthritis. Mesenchymal cells responding to TNF become early activated, mediate tissue remodeling and in the case of IBD appear sufficient to initiate events leading to inflammatory mucosal innate and adaptive immune responses, required for intestinal pathology. These functions appear to critically require the function of TNFR1. Most interestingly, the physiological requirement for TNF in immune system structure and function may also become relevant with regard to disease pathogenesis. Thus, the expression of TNFR1 in FDCs is required for the generation of FDC networks and humoral antibody responses. FDC networks are in turn essential to support GC B cell development and the recruitment of arthritogenic Tfh cells, processes required for autoimmune-mediated arthritic pathology developing in the K/BxN model. Pharmacological inhibition of FDC maintenance with p75TNFR:Fc (etanercept) ameliorates disease development, supporting the importance of FDCs as a relevant target in autoimmunity, and indicating an additional potential mechanism by which anti-TNF therapy may contribute to the treatment of RA. Indeed, although anti-TNF therapy is widely used in arthritis and IBD patients, the mode of action of anti-TNF agents has remained a source of controversy. Furthermore, pharmacological inhibition of TNF has been associated with adverse effects. In this context defining the specific TNFRs and cellular subsets relevant to the ensuing pathology becomes essential for the design of more effective and safe therapeutic approaches. Still, further to the identification of direct, early, sufficient and/or required targets of TNF in disease pathogenesis, it remains critical to define the cellular and molecular interactions that consequently occur and may determine or influence progression of pathology. Animal models of TNF-driven pathologies may provide valuable information with regard to the above, which can be extrapolated in the clinic both through the identification of novel therapeutic targets and in terms of assessing the potential effectiveness of anti-TNF therapy in alleviating these processes.

Acknowledgements

Supported by the European Commission grants 028190 (TB REACT), F2-2008-223404 (Masterswitch Health) and by the MUGEN Network of Excellence LSHG-CT-2005-005203.

References

1 Elliott MJ, Maini RN, Feldmann M, Long-Fox A, Charles P, Katsikis P, Brennan FM, Walker J, Bijl H, Ghrayeb J, et al: Treatment of rheumatoid arthritis with chimeric monoclonal antibodies to tumor necrosis factor alpha. Arthritis Rheum 1993;36:1681–1690.

2 Targan SR, Hanauer SB, van Deventer SJ, Mayer L, Present DH, Braakman T, DeWoody KL, Schaible TF, Rutgeerts PJ: A short-term study of chimeric monoclonal antibody cA2 to tumor necrosis factor alpha for Crohn's disease. Crohn's Disease cA2 Study Group. N Engl J Med 1997;337:1029–1035.

3 Strand V, Kimberly R, Isaacs JD: Biologic therapies in rheumatology: lessons learned, future directions. Nat Rev Drug Discov 2007;6:75–92.

4 van Oosten BW, Barkhof F, Truyen L, Boringa JB, Bertelsmann FW, von Blomberg BM, Woody JN, Hartung HP, Polman CH: Increased MRI activity and immune activation in two multiple sclerosis patients treated with the monoclonal anti-tumor necrosis factor antibody cA2. Neurology 1996;47:1531–1534.

5 TNF neutralization in MS: results of a randomized, placebo-controlled multicenter study. The Lenercept Multiple Sclerosis Study Group and The University of British Columbia MS/MRI Analysis Group. Neurology 1999;53:457–465.

6 Slifman NR, Gershon SK, Lee JH, Edwards ET, Braun MM: *Listeria monocytogenes* infection as a complication of treatment with tumor necrosis factor alpha-neutralizing agents. Arthritis Rheum 2003;48:319–324.

7 Sicotte NL, Voskuhl RR: Onset of multiple sclerosis associated with anti-TNF therapy. Neurology 2001;57:1885–1888.

8 Shakoor N, Michalska M, Harris CA, Block JA: Drug-induced systemic lupus erythematosus associated with etanercept therapy. Lancet 2002;359:579–580.

9 Brown SL, Greene MH, Gershon SK, Edwards ET, Braun MM: Tumor necrosis factor antagonist therapy and lymphoma development: twenty-six cases reported to the Food and Drug Administration. Arthritis Rheum 2002;46:3151–3158.

10 Vassalli P: The pathophysiology of tumor necrosis factors. Annu Rev Immunol 1992;10:411–452.
11 Kriegler M, Perez C, DeFay K, Albert I, Lu SD: A novel form of TNF/cachectin is a cell surface cytotoxic transmembrane protein: ramifications for the complex physiology of TNF. Cell 1988;53:45–53.
12 Black RA, Rauch CT, Kozlosky CJ, Peschon JJ, Slack JL, Wolfson MF, Castner BJ, Stocking KL, Reddy P, Srinivasan S, Nelson N, Boiani N, Schooley KA, Gerhart M, Davis R, Fitzner JN, Johnson RS, Paxton RJ, March CJ, Cerretti DP: A metalloproteinase disintegrin that releases tumour-necrosis factor-alpha from cells. Nature 1997;385:729–733.
13 Vandenabeele P, Declercq W, Beyaert R, Fiers W: Two tumour necrosis factor receptors: structure and function. Trends Cell Biol 1995;5:392–399.
14 Pinckard JK, Sheehan KC, Arthur CD, Schreiber RD: Constitutive shedding of both p55 and p75 murine TNF receptors in vivo. J Immunol 1997; 158:3869–3873.
15 Aderka D: The potential biological and clinical significance of the soluble tumor necrosis factor receptors. Cytokine Growth Factor Rev 1996;7:231–240.
16 Grell M, Douni E, Wajant H, Lohden M, Clauss M, Maxeiner B, Georgopoulos S, Lesslauer W, Kollias G, Pfizenmaier K, Scheurich P: The transmembrane form of tumor necrosis factor is the prime activating ligand of the 80 kDa tumor necrosis factor receptor. Cell 1995;83:793–802.
17 Chen G, Goeddel DV: TNF-R1 signaling: a beautiful pathway. Science 2002;296:1634–1635.
18 Wajant H, Pfizenmaier K, Scheurich P: Tumor necrosis factor signaling. Cell Death Differ 2003;10: 45–65.
19 Grell M, Becke FM, Wajant H, Mannel DN, Scheurich P: TNF receptor type 2 mediates thymocyte proliferation independently of TNF receptor type 1. Eur J Immunol 1998;28:257–263.
20 Arnett HA, Mason J, Marino M, Suzuki K, Matsushima GK, Ting JP: TNF alpha promotes proliferation of oligodendrocyte progenitors and remyelination. Nat Neurosci 2001;4:1116–1122.
21 Pasparakis M, Alexopoulou L, Episkopou V, Kollias G: Immune and inflammatory responses in TNF alpha-deficient mice: a critical requirement for TNF alpha in the formation of primary B cell follicles, follicular dendritic cell networks and germinal centers, and in the maturation of the humoral immune response. J Exp Med 1996;184:1397–1411.
22 Rothe J, Lesslauer W, Lotscher H, Lang Y, Koebel P, Kontgen F, Althage A, Zinkernagel R, Steinmetz M, Bluethmann H: Mice lacking the tumour necrosis factor receptor 1 are resistant to TNF-mediated toxicity but highly susceptible to infection by *Listeria monocytogenes*. Nature 1993;364:798–802.
23 Pfeffer K, Matsuyama T, Kundig TM, Wakeham A, Kishihara K, Shahinian A, Wiegmann K, Ohashi PS, Kronke M, Mak TW: Mice deficient for the 55 kd tumor necrosis factor receptor are resistant to endotoxic shock, yet succumb to *L. monocytogenes* infection. Cell 1993;73:457–467.
24 Pasparakis M, Kousteni S, Peschon J, Kollias G: Tumor necrosis factor and the p55TNF receptor are required for optimal development of the marginal sinus and for migration of follicular dendritic cell precursors into splenic follicles. Cell Immunol 2000; 201:33–41.
25 Keffer J, Probert L, Cazlaris H, Georgopoulos S, Kaslaris E, Kioussis D, Kollias G: Transgenic mice expressing human tumour necrosis factor: a predictive genetic model of arthritis. EMBO J 1991;10: 4025–4031.
26 Kontoyiannis D, Pasparakis M, Pizarro TT, Cominelli F, Kollias G: Impaired on/off regulation of TNF biosynthesis in mice lacking TNF AU-rich elements: implications for joint and gut-associated immunopathologies. Immunity 1999;10:387–398.
27 Probert L, Akassoglou K, Pasparakis M, Kontogeorgos G, Kollias G: Spontaneous inflammatory demyelinating disease in transgenic mice showing central nervous system-specific expression of tumor necrosis factor alpha. Proc Natl Acad Sci USA 1995;92:11294–11298.
28 Akassoglou K, Probert L, Kontogeorgos G, Kollias G: Astrocyte-specific but not neuron-specific transmembrane TNF triggers inflammation and degeneration in the central nervous system of transgenic mice. J Immunol 1997;158:438–445.
29 Xanthoulea S, Pasparakis M, Kousteni S, Brakebusch C, Wallach D, Bauer J, Lassmann H, Kollias G: Tumor necrosis factor (TNF) receptor shedding controls thresholds of innate immune activation that balance opposing TNF functions in infectious and inflammatory diseases. J Exp Med 2004;200:367–376.
30 McDermott MF, Aksentijevich I, Galon J, et al: Germline mutations in the extracellular domains of the 55 kDa TNF receptor, TNFR1, define a family of dominantly inherited autoinflammatory syndromes. Cell 1999;97:133–144.
31 Grewal IS, Grewal KD, Wong FS, Picarella DE, Janeway CA Jr, Flavell RA: Local expression of transgene encoded TNF alpha in islets prevents autoimmune diabetes in nonobese diabetic (NOD) mice by preventing the development of auto-reactive islet-specific T cells. J Exp Med 1996;184:1963–1974.
32 Kontoyiannis D, Kollias G: Accelerated autoimmunity and lupus nephritis in NZB mice with an engineered heterozygous deficiency in tumor necrosis factor. Eur J Immunol 2000;30:2038–2047.

33 Cope A, Ettinger R, McDevitt H: The role of TNF alpha and related cytokines in the development and function of the autoreactive T-cell repertoire. Res Immunol 1997;148:307–312.

34 Green EA, Eynon EE, Flavell RA: Local expression of TNFalpha in neonatal NOD mice promotes diabetes by enhancing presentation of islet antigens. Immunity 1998;9:733–743.

35 Kassiotis G, Pasparakis M, Kollias G, Probert L: TNF accelerates the onset but does not alter the incidence and severity of myelin basic protein-induced experimental autoimmune encephalomyelitis. Eur J Immunol 1999;29:774–780.

36 Kassiotis G, Kollias G: Uncoupling the proinflammatory from the immunosuppressive properties of tumor necrosis factor (TNF) at the p55 TNF receptor level: implications for pathogenesis and therapy of autoimmune demyelination. J Exp Med 2001;193: 427–434.

37 Alexopoulou L, Kranidioti K, Xanthoulea S, Denis M, Kotanidou A, Douni E, Blackshear PJ, Kontoyiannis DL, Kollias G: Transmembrane TNF protects mutant mice against intracellular bacterial infections, chronic inflammation and autoimmunity. Eur J Immunol 2006;36:2768–2780.

38 Ruuls SR, Hoek RM, Ngo VN, McNeil T, Lucian LA, Janatpour MJ, Korner H, Scheerens H, Hessel EM, Cyster JG, McEvoy LM, Sedgwick JD: Membrane-bound TNF supports secondary lymphoid organ structure but is subservient to secreted TNF in driving autoimmune inflammation. Immunity 2001;15: 533–543.

39 Firestein GS: Evolving concepts of rheumatoid arthritis. Nature 2003;423:356–361.

40 Silman AJ, Pearson JE: Epidemiology and genetics of rheumatoid arthritis. Arthritis Res 2002;4 S265–S272.

41 Coenen MJ, Gregersen PK: Rheumatoid arthritis: a view of the current genetic landscape. Genes Immun 2009;10:101–111.

42 Gregersen PK, Amos CI, Lee AT, Lu Y, Remmers EF, Kastner DL, Seldin MF, Criswell LA, Plenge RM, Holers VM, Mikuls TR, Sokka T, Moreland LW, Bridges SL, Jr., Xie G, Begovich AB, Siminovitch KA: REL, encoding a member of the NF-kappaB family of transcription factors, is a newly defined risk locus for rheumatoid arthritis. Nat Genet 2009; 41:820–823.

43 Barton A, Thomson W, Ke X, Eyre S, Hinks A, Bowes J, Plant D, Gibbons LJ, Wilson AG, Bax DE, Morgan AW, Emery P, Steer S, Hocking L, Reid DM, Wordsworth P, Harrison P, Worthington J: Rheumatoid arthritis susceptibility loci at chromosomes 10p15, 12q13 and 22q13. Nat Genet 2008;40: 1156–1159.

44 Gregersen PK, Silver J, Winchester RJ: The shared epitope hypothesis. An approach to understanding the molecular genetics of susceptibility to rheumatoid arthritis. Arthritis Rheum 1987;30:1205–1213.

45 Firestein GS, Alvaro-Gracia JM, Maki R: Quantitative analysis of cytokine gene expression in rheumatoid arthritis. J Immunol 1990;144:3347–3353.

46 Choy EH, Panayi GS, Emery P, Madden S, Breedveld FC, Kraan MC, Kalden JR, Rascu A, Brown JC, Rapson N, Johnston JM: Repeat-cycle study of high-dose intravenous 4162W94 anti-CD4 humanized monoclonal antibody in rheumatoid arthritis. A randomized placebo-controlled trial. Rheumatology (Oxford) 2002;41:1142–1148.

47 Franklin EC, Holman HR, Muller-Eberhard HJ, Kunkel HG: An unusual protein component of high molecular weight in the serum of certain patients with rheumatoid arthritis. J Exp Med 1957;105:425–438.

48 Smolen JS, Aletaha D, Koeller M, Weisman MH, Emery P: New therapies for treatment of rheumatoid arthritis. Lancet 2007;370:1861–1874.

49 Firestein GS, Zvaifler NJ: How important are T cells in chronic rheumatoid synovitis? Arthritis Rheum 1990;33:768–773.

50 Firestein GS: Invasive fibroblast-like synoviocytes in rheumatoid arthritis. Passive responders or transformed aggressors? Arthritis Rheum 1996;39:1781–1790.

51 Brennan FM, Chantry D, Jackson A, Maini R, Feldmann M: Inhibitory effect of TNF alpha antibodies on synovial cell interleukin-1 production in rheumatoid arthritis. Lancet 1989;2:244–247.

52 Lewis M, Tartaglia LA, Lee A, Bennett GL, Rice GC, Wong GH, Chen EY, Goeddel DV: Cloning and expression of cDNAs for two distinct murine tumor necrosis factor receptors demonstrate one receptor is species specific. Proc Natl Acad Sci USA 1991;88: 2830–2834.

53 Thorbecke GJ, Shah R, Leu CH, Kuruvilla AP, Hardison AM, Palladino MA: Involvement of endogenous tumor necrosis factor alpha and transforming growth factor beta during induction of collagen type II arthritis in mice. Proc Natl Acad Sci USA 1992;89:7375–7379.

54 Piguet PF, Grau GE, Vesin C, Loetscher H, Gentz R, Lesslauer W: Evolution of collagen arthritis in mice is arrested by treatment with anti-tumour necrosis factor (TNF) antibody or a recombinant soluble TNF receptor. Immunology 1992;77:510–514.

55 Williams RO, Feldmann M, Maini RN: Anti-tumor necrosis factor ameliorates joint disease in murine collagen-induced arthritis. Proc Natl Acad Sci USA 1992;89:9784–9788.

56 Campbell IK, O'Donnell K, Lawlor KE, Wicks IP: Severe inflammatory arthritis and lymphadenopathy in the absence of TNF. J Clin Invest 2001;107:1519–1527.
57 Mori L, Iselin S, De Libero G, Lesslauer W: Attenuation of collagen-induced arthritis in 55-kDa TNF receptor type 1 (TNFR1)-IgG1-treated and TNFR1-deficient mice. J Immunol 1996;157:3178–3182.
58 Sakaguchi N, Takahashi T, Hata H, Nomura T, Tagami T, Yamazaki S, Sakihama T, Matsutani T, Negishi I, Nakatsuru S, Sakaguchi S: Altered thymic T-cell selection due to a mutation of the ZAP-70 gene causes autoimmune arthritis in mice. Nature 2003;426:454–460.
59 Hata H, Sakaguchi N, Yoshitomi H, Iwakura Y, Sekikawa K, Azuma Y, Kanai C, Moriizumi E, Nomura T, Nakamura T, Sakaguchi S: Distinct contribution of IL-6, TNF-alpha, IL-1, and IL-10 to T cell-mediated spontaneous autoimmune arthritis in mice. J Clin Invest 2004;114:582–588.
60 Horai R, Saijo S, Tanioka H, Nakae S, Sudo K, Okahara A, Ikuse T, Asano M, Iwakura Y: Development of chronic inflammatory arthropathy resembling rheumatoid arthritis in interleukin 1 receptor antagonist-deficient mice. J Exp Med 2000; 191:313–320.
61 Horai R, Nakajima A, Habiro K, Kotani M, Nakae S, Matsuki T, Nambu A, Saijo S, Kotaki H, Sudo K, Okahara A, Tanioka H, Ikuse T, Ishii N, Schwartzberg PL, Abe R, Iwakura Y: TNF-alpha is crucial for the development of autoimmune arthritis in IL-1 receptor antagonist-deficient mice. J Clin Invest 2004; 114:1603–1611.
62 Taylor GA, Carballo E, Lee DM, Lai WS, Thompson MJ, Patel DD, Schenkman DI, Gilkeson GS, Broxmeyer HE, Haynes BF, Blackshear PJ: A pathogenetic role for TNF alpha in the syndrome of cachexia, arthritis, and autoimmunity resulting from tristetraprolin (TTP) deficiency. Immunity 1996;4:445–454.
63 Carballo E, Blackshear PJ: Roles of tumor necrosis factor-alpha receptor subtypes in the pathogenesis of the tristetraprolin-deficiency syndrome. Blood 2001;98:2389–2395.
64 Kawane K, Ohtani M, Miwa K, Kizawa T, Kanbara Y, Yoshioka Y, Yoshikawa H, Nagata S: Chronic polyarthritis caused by mammalian DNA that escapes from degradation in macrophages. Nature 2006;443:998–1002.
65 Iwanaga T, Shikichi M, Kitamura H, Yanase H, Nozawa-Inoue K: Morphology and functional roles of synoviocytes in the joint. Arch Histol Cytol 2000;63:17–31.
66 Edwards JC: Fibroblast biology. Development and differentiation of synovial fibroblasts in arthritis. Arthritis Res 2000;2:344–347.
67 Karouzakis E, Gay RE, Gay S, Neidhart M: Epigenetic control in rheumatoid arthritis synovial fibroblasts. Nat Rev Rheumatol 2009;5:266–272.
68 Sen M: Wnt signalling in rheumatoid arthritis. Rheumatology (Oxf) 2005;44:708–713.
69 Butler DM, Piccoli DS, Hart PH, Hamilton JA: Stimulation of human synovial fibroblast DNA synthesis by recombinant human cytokines. J Rheumatol 1988;15:1463–1470.
70 Dayer JM, Krane SM, Russell RG, Robinson DR: Production of collagenase and prostaglandins by isolated adherent rheumatoid synovial cells. Proc Natl Acad Sci U S A 1976;73:945–949.
71 Muller-Ladner U, Kriegsmann J, Franklin BN, Matsumoto S, Geiler T, Gay RE, Gay S: Synovial fibroblasts of patients with rheumatoid arthritis attach to and invade normal human cartilage when engrafted into SCID mice. Am J Pathol 1996;149:1607–1615.
72 Aidinis V, Plows D, Haralambous S, Armaka M, Papadopoulos P, Kanaki MZ, Koczan D, Thiesen HJ, Kollias G: Functional analysis of an arthritogenic synovial fibroblast. Arthritis Res 2003;5:R140–R157.
73 Ospelt C, Gay S: The role of resident synovial cells in destructive arthritis. Best Pract Res Clin Rheumatol 2008;22:239–252.
74 Noss EH, Brenner MB: The role and therapeutic implications of fibroblast-like synoviocytes in inflammation and cartilage erosion in rheumatoid arthritis. Immunol Rev 2008;223:252–270.
75 Mathew SJ, Haubert D, Kronke M, Leptin M: Looking beyond death: a morphogenetic role for the TNF signalling pathway. J Cell Sci 2009;122:1939–1946.
76 Xu H, Liu P, Liang L, Danesh FR, Yang X, Ye Y, Zhan Z, Yu X, Peng H, Sun L: RhoA-mediated, tumor necrosis factor alpha-induced activation of NF-kappaB in rheumatoid synoviocytes: inhibitory effect of simvastatin. Arthritis Rheum 2006;54:3441–3451.
77 Are AF, Galkin VE, Pospelova TV, Pinaev GP: The p65/RelA subunit of NF-kappaB interacts with actin-containing structures. Exp Cell Res 2000;256:533–544.
78 Aidinis V, Carninci P, Armaka M, Witke W, Harokopos V, Pavelka N, Koczan D, Argyropoulos C, Thwin MM, Moller S, Kazunori W, Gopalakrishnakone P, Ricciardi-Castagnoli P, Thiesen HJ, Hayashizaki Y, Kollias G: Cytoskeletal rearrangements in synovial fibroblasts as a novel pathophysiological determinant of modeled rheumatoid arthritis. PLoS Genet 2005;1:e48.

79 Vasilopoulos Y, Gkretsi V, Armaka M, Aidinis V, Kollias G: Actin cytoskeleton dynamics linked to synovial fibroblast activation as a novel pathogenic principle in TNF-driven arthritis. Ann Rheum Dis 2007;66(suppl 3):iii23–iii28.
80 Valencia X, Higgins JM, Kiener HP, Lee DM, Podrebarac TA, Dascher CC, Watts GF, Mizoguchi E, Simmons B, Patel DD, Bhan AK, Brenner MB: Cadherin-11 provides specific cellular adhesion between fibroblast-like synoviocytes. J Exp Med 2004;200:1673–1679.
81 Kiener HP, Stipp CS, Allen PG, Higgins JM, Brenner MB: The cadherin-11 cytoplasmic juxtamembrane domain promotes alpha-catenin turnover at adherens junctions and intercellular motility. Mol Biol Cell 2006;17:2366–2376.
82 Lee DM, Kiener HP, Agarwal SK, Noss EH, Watts GF, Chisaka O, Takeichi M, Brenner MB: Cadherin-11 in synovial lining formation and pathology in arthritis. Science 2007;315:1006–1010.
83 Vandooren B, Cantaert T, ter Borg M, Noordenbos T, Kuhlman R, Gerlag D, Bongartz T, Reedquist K, Tak PP, Baeten D: Tumor necrosis factor alpha drives cadherin 11 expression in rheumatoid inflammation. Arthritis Rheum 2008;58:3051–3062.
84 Kiener HP, Niederreiter B, Lee DM, Jimenez-Boj E, Smolen JS, Brenner MB: Cadherin 11 promotes invasive behavior of fibroblast-like synoviocytes. Arthritis Rheum 2009;60:1305–1310.
85 Kollias G, Douni E, Kassiotis G, Kontoyiannis D: On the role of tumor necrosis factor and receptors in models of multiorgan failure, rheumatoid arthritis, multiple sclerosis and inflammatory bowel disease. Immunol Rev 1999;169:175–194.
86 Armaka M, Apostolaki M, Jacques P, Kontoyiannis DL, Elewaut D, Kollias G: Mesenchymal cell targeting by TNF as a common pathogenic principle in chronic inflammatory joint and intestinal diseases. J Exp Med 2008;205:331–337.
87 Xavier RJ, Podolsky DK: Unravelling the pathogenesis of inflammatory bowel disease. Nature 2007;448: 427–434.
88 Rioux JD, Xavier RJ, Taylor KD, Silverberg MS, Goyette P, Huett A, Green T, Kuballa P, Barmada MM, Datta LW, Shugart YY, Griffiths AM, Targan SR, Ippoliti AF, Bernard EJ, Mei L, Nicolae DL, Regueiro M, Schumm LP, Steinhart AH, Rotter JI, Duerr RH, Cho JH, Daly MJ, Brant SR: Genome-wide association study identifies new susceptibility loci for Crohn disease and implicates autophagy in disease pathogenesis. Nat Genet 2007;39:596–604.
89 Parkes M, Barrett JC, Prescott NJ, Tremelling M, Anderson CA, Fisher SA, Roberts RG, Nimmo ER, Cummings FR, Soars D, Drummond H, Lees CW, Khawaja SA, Bagnall R, Burke DA, Todhunter CE, Ahmad T, Onnie CM, McArdle W, Strachan D, Bethel G, Bryan C, Lewis CM, Deloukas P, Forbes A, Sanderson J, Jewell DP, Satsangi J, Mansfield JC, Cardon L, Mathew CG: Sequence variants in the autophagy gene IRGM and multiple other replicating loci contribute to Crohn's disease susceptibility. Nat Genet 2007;39:830–832.
90 Yamazaki K, McGovern D, Ragoussis J, Paolucci M, Butler H, Jewell D, Cardon L, Takazoe M, Tanaka T, Ichimori T, Saito S, Sekine A, Iida A, Takahashi A, Tsunoda T, Lathrop M, Nakamura Y: Single nucleotide polymorphisms in TNFSF15 confer susceptibility to Crohn's disease. Hum Mol Genet 2005; 14:3499–3506.
91 Duerr RH, Taylor KD, Brant SR, Rioux JD, Silverberg MS, Daly MJ, Steinhart AH, Abraham C, Regueiro M, Griffiths A, Dassopoulos T, Bitton A, Yang H, Targan S, Datta LW, Kistner EO, Schumm LP, Lee AT, Gregersen PK, Barmada MM, Rotter JI, Nicolae DL, Cho JH: A genome-wide association study identifies IL23R as an inflammatory bowel disease gene. Science 2006;314:1461–1463.
92 Strober W, Fuss IJ, Blumberg RS: The immunology of mucosal models of inflammation. Annu Rev Immunol 2002;20:495–549.
93 Rutgeerts P, Sandborn WJ, Feagan BG, Reinisch W, Olson A, Johanns J, Travers S, Rachmilewitz D, Hanauer SB, Lichtenstein GR, de Villiers WJ, Present D, Sands BE, Colombel JF: Infliximab for induction and maintenance therapy for ulcerative colitis. N Engl J Med 2005;353:2462–2476.
94 Kontoyiannis D, Boulougouris G, Manoloukos M, Armaka M, Apostolaki M, Pizarro T, Kotlyarov A, Forster I, Flavell R, Gaestel M, Tsichlis P, Cominelli F, Kollias G: Genetic dissection of the cellular pathways and signaling mechanisms in modeled tumor necrosis factor-induced Crohn's-like inflammatory bowel disease. J Exp Med 2002;196:1563–1574.
95 Powrie F, Leach MW, Mauze S, Menon S, Caddle LB, Coffman RL: Inhibition of Th1 responses prevents inflammatory bowel disease in scid mice reconstituted with CD45RBhi CD4+ T cells. Immunity 1994;1:553–562.
96 Neurath MF, Fuss I, Pasparakis M, Alexopoulou L, Haralambous S, Meyer zum Buschenfelde KH, Strober W, Kollias G: Predominant pathogenic role of tumor necrosis factor in experimental colitis in mice. Eur J Immunol 1997;27:1743–1750.

97 Kontoyiannis D, Kotlyarov A, Carballo E, Alexopoulou L, Blackshear PJ, Gaestel M, Davis R, Flavell R, Kollias G: Interleukin-10 targets p38 MAPK to modulate ARE-dependent TNF mRNA translation and limit intestinal pathology. EMBO J 2001;20:3760–3770.
98 Marini M, Bamias G, Rivera-Nieves J, Moskaluk CA, Hoang SB, Ross WG, Pizarro TT, Cominelli F: TNF-alpha neutralization ameliorates the severity of murine Crohn's-like ileitis by abrogation of intestinal epithelial cell apoptosis. Proc Natl Acad Sci USA 2003;100:8366–8371.
99 Nenci A, Becker C, Wullaert A, Gareus R, van Loo G, Danese S, Huth M, Nikolaev A, Neufert C, Madison B, Gumucio D, Neurath MF, Pasparakis M: Epithelial NEMO links innate immunity to chronic intestinal inflammation. Nature 2007;446:557–561.
100 Naito Y, Takagi T, Handa O, Ishikawa T, Nakagawa S, Yamaguchi T, Yoshida N, Minami M, Kita M, Imanishi J, Yoshikawa T: Enhanced intestinal inflammation induced by dextran sulfate sodium in tumor necrosis factor-alpha deficient mice. J Gastroenterol Hepatol 2003;18:560–569.
101 Mizoguchi E, Hachiya Y, Kawada M, Nagatani K, Ogawa A, Sugimoto K, Mizoguchi A, Podolsky DK: TNF receptor type I-dependent activation of innate responses to reduce intestinal damage-associated mortality. Gastroenterology 2008;134:470–480.
102 Holtmann MH, Douni E, Schutz M, Zeller G, Mudter J, Lehr HA, Gerspach J, Scheurich P, Galle PR, Kollias G, Neurath MF: Tumor necrosis factor-receptor 2 is up-regulated on lamina propria T cells in Crohn's disease and promotes experimental colitis in vivo. Eur J Immunol 2002;32:3142–3151.
103 Apostolaki M, Manoloukos M, Roulis M, Wurbel MA, Muller W, Papadakis KA, Kontoyiannis DL, Malissen B, Kollias G: Role of beta7 integrin and the chemokine/chemokine receptor pair CCL25/CCR9 in modeled TNF-dependent Crohn's disease. Gastroenterology 2008;134:2025–2035.
104 Cheroutre H: IELs: enforcing law and order in the court of the intestinal epithelium. Immunol Rev 2005;206:114–131.
105 Poussier P, Ning T, Banerjee D, Julius M: A unique subset of self-specific intraintestinal T cells maintains gut integrity. J Exp Med 2002;195:1491–1497.
106 Das G, Augustine MM, Das J, Bottomly K, Ray P, Ray A: An important regulatory role for CD4+CD8 alpha alpha T cells in the intestinal epithelial layer in the prevention of inflammatory bowel disease. Proc Natl Acad Sci U S A 2003;100:5324–5329.
107 Inagaki-Ohara K, Chinen T, Matsuzaki G, Sasaki A, Sakamoto Y, Hiromatsu K, Nakamura-Uchiyama F, Nawa Y, Yoshimura A: Mucosal T cells bearing TCRgammadelta play a protective role in intestinal inflammation. J Immunol 2004;173:1390–1398.
108 Kunkel EJ, Campbell JJ, Haraldsen G, Pan J, Boisvert J, Roberts AI, Ebert EC, Vierra MA, Goodman SB, Genovese MC, Wardlaw AJ, Greenberg HB, Parker CM, Butcher EC, Andrew DP, Agace WW: Lymphocyte CC chemokine receptor 9 and epithelial thymus-expressed chemokine (TECK) expression distinguish the small intestinal immune compartment: epithelial expression of tissue-specific chemokines as an organizing principle in regional immunity. J Exp Med 2000;192:761–768.
109 Svensson M, Marsal J, Ericsson A, Carramolino L, Broden T, Marquez G, Agace WW: CCL25 mediates the localization of recently activated CD8alphabeta(+) lymphocytes to the small-intestinal mucosa. J Clin Invest 2002;110:1113–1121.
110 Bouma G, Strober W: The immunological and genetic basis of inflammatory bowel disease. Nat Rev Immunol 2003;3:521–533.
111 Korn T, Bettelli E, Oukka M, Kuchroo VK: IL-17 and Th17 Cells. Annu Rev Immunol 2009;27:485–517.
112 Oppmann B, Lesley R, Blom B, Timans JC, Xu Y, Hunte B, Vega F, Yu N, Wang J, Singh K, Zonin F, Vaisberg E, Churakova T, Liu M, Gorman D, Wagner J, Zurawski S, Liu Y, Abrams JS, Moore KW, Rennick D, de Waal-Malefyt R, Hannum C, Bazan JF, Kastelein RA: Novel p19 protein engages IL-12p40 to form a cytokine, IL-23, with biological activities similar as well as distinct from IL-12. Immunity 2000;13:715–725.
113 Breese E, Braegger CP, Corrigan CJ, Walker-Smith JA, MacDonald TT: Interleukin-2- and interferon-gamma-secreting T cells in normal and diseased human intestinal mucosa. Immunology 1993;78:127–131.
114 Annunziato F, Cosmi L, Santarlasci V, Maggi L, Liotta F, Mazzinghi B, Parente E, Fili L, Ferri S, Frosali F, Giudici F, Romagnani P, Parronchi P, Tonelli F, Maggi E, Romagnani S: Phenotypic and functional features of human Th17 cells. J Exp Med 2007;204:1849–1861.
115 Yen D, Cheung J, Scheerens H, Poulet F, McClanahan T, McKenzie B, Kleinschek MA, Owyang A, Mattson J, Blumenschein W, Murphy E, Sathe M, Cua DJ, Kastelein RA, Rennick D: IL-23 is essential for T cell-mediated colitis and promotes inflammation via IL-17 and IL-6. J Clin Invest 2006;116:1310–1316.
116 Kullberg MC, Jankovic D, Feng CG, Hue S, Gorelick PL, McKenzie BS, Cua DJ, Powrie F, Cheever AW, Maloy KJ, Sher A: IL-23 plays a key role in *Helicobacter hepaticus*-induced T cell-dependent colitis. J Exp Med 2006;203:2485–2494.

117 Yang XO, Chang SH, Park H, Nurieva R, Shah B, Acero L, Wang YH, Schluns KS, Broaddus RR, Zhu Z, Dong C: Regulation of inflammatory responses by IL-17F. J Exp Med 2008;205:1063–1075.
118 O'Connor W, Jr., Kamanaka M, Booth CJ, Town T, Nakae S, Iwakura Y, Kolls JK, Flavell RA: A protective function for interleukin 17A in T cell-mediated intestinal inflammation. Nat Immunol 2009;10:603–609.
119 Wiekowski MT, Leach MW, Evans EW, Sullivan L, Chen SC, Vassileva G, Bazan JF, Gorman DM, Kastelein RA, Narula S, Lira SA: Ubiquitous transgenic expression of the IL-23 subunit p19 induces multiorgan inflammation, runting, infertility, and premature death. J Immunol 2001;166:7563–7570.
120 Andoh A, Bamba S, Brittan M, Fujiyama Y, Wright NA: Role of intestinal subepithelial myofibroblasts in inflammation and regenerative response in the gut. Pharmacol Ther 2007;114:94–106.
121 Bailey CJ, Hembry RM, Alexander A, Irving MH, Grant ME, Shuttleworth CA: Distribution of the matrix metalloproteinases stromelysin, gelatinases A and B, and collagenase in Crohn's disease and normal intestine. J Clin Pathol 1994;47:113–116.
122 van Deventer SJ: Transmembrane TNF-alpha, induction of apoptosis, and the efficacy of TNF-targeting therapies in Crohn's disease. Gastroenterology 2001;121:1242–1246.
123 Di Sabatino A, Pender SL, Jackson CL, Prothero JD, Gordon JN, Picariello L, Rovedatti L, Docena G, Monteleone G, Rampton DS, Tonelli F, Corazza GR, MacDonald TT: Functional modulation of Crohn's disease myofibroblasts by anti-tumor necrosis factor antibodies. Gastroenterology 2007;133:137–149.
124 Khan MA: Update on spondyloarthropathies. Ann Intern Med 2002;136:896–907.
125 Redlich K, Gortz B, Hayer S, Zwerina J, Kollias G, Steiner G, Smolen JS, Schett G: Overexpression of tumor necrosis factor causes bilateral sacroiliitis. Arthritis Rheum 2004;50:1001–1005.
126 Forster R, Mattis AE, Kremmer E, Wolf E, Brem G, Lipp M: A putative chemokine receptor, BLR1, directs B cell migration to defined lymphoid organs and specific anatomic compartments of the spleen. Cell 1996;87:1037–1047.
127 Ansel KM, Ngo VN, Hyman PL, Luther SA, Forster R, Sedgwick JD, Browning JL, Lipp M, Cyster JG: A chemokine-driven positive feedback loop organizes lymphoid follicles. Nature 2000;406:309–314.
128 Ansel KM, Harris RB, Cyster JG: CXCL13 is required for B1 cell homing, natural antibody production, and body cavity immunity. Immunity 2002; 16:67–76.
129 Allen CD, Ansel KM, Low C, Lesley R, Tamamura H, Fujii N, Cyster JG: Germinal center dark and light zone organization is mediated by CXCR4 and CXCR5. Nat Immunol 2004;5:943–952.
130 MacLennan IC: Germinal centers. Annu Rev Immunol 1994;12:117–139.
131 Tew JG, Wu J, Fakher M, Szakal AK, Qin D: Follicular dendritic cells: beyond the necessity of T-cell help. Trends Immunol 2001;22:361–367.
132 Randall TD, Carragher DM, Rangel-Moreno J: Development of secondary lymphoid organs. Annu Rev Immunol 2008;26:627–650.
133 Matsumoto M, Mariathasan S, Nahm MH, Baranyay F, Peschon JJ, Chaplin DD: Role of lymphotoxin and the type I TNF receptor in the formation of germinal centers. Science 1996;271:1289–1291.
134 Pasparakis M, Alexopoulou L, Grell M, Pfizenmaier K, Bluethmann H, Kollias G: Peyer's patch organogenesis is intact yet formation of B lymphocyte follicles is defective in peripheral lymphoid organs of mice deficient for tumor necrosis factor and its 55-kDa receptor. Proc Natl Acad Sci U S A 1997;94: 6319–6323.
135 Endres R, Alimzhanov MB, Plitz T, Futterer A, Kosco-Vilbois MH, Nedospasov SA, Rajewsky K, Pfeffer K: Mature follicular dendritic cell networks depend on expression of lymphotoxin beta receptor by radioresistant stromal cells and of lymphotoxin beta and tumor necrosis factor by B cells. J Exp Med 1999;189:159–168.
136 Ngo VN, Korner H, Gunn MD, Schmidt KN, Riminton DS, Cooper MD, Browning JL, Sedgwick JD, Cyster JG: Lymphotoxin alpha/beta and tumor necrosis factor are required for stromal cell expression of homing chemokines in B and T cell areas of the spleen. J Exp Med 1999;189:403–412.
137 Alexopoulou L, Pasparakis M, Kollias G: Complementation of lymphotoxin alpha knockout mice with tumor necrosis factor-expressing transgenes rectifies defective splenic structure and function. J Exp Med 1998;188:745–754.
138 Matsumoto M, Fu YX, Molina H, Huang G, Kim J, Thomas DA, Nahm MH, Chaplin DD: Distinct roles of lymphotoxin alpha and the type I tumor necrosis factor (TNF) receptor in the establishment of follicular dendritic cells from non-bone marrow-derived cells. J Exp Med 1997;186:1997–2004.
139 Tkachuk M, Bolliger S, Ryffel B, Pluschke G, Banks TA, Herren S, Gisler RH, Kosco-Vilbois MH: Crucial role of tumor necrosis factor receptor 1 expression on nonhematopoietic cells for B cell localization within the splenic white pulp. J Exp Med 1998;187:469–477.

140 Victoratos P, Lagnel J, Tzima S, Alimzhanov MB, Rajewsky K, Pasparakis M, Kollias G: FDC-specific functions of p55TNFR and IKK2 in the development of FDC networks and of antibody responses. Immunity 2006;24:65–77.
141 Cohen SB, Emery P, Greenwald MW, Dougados M, Furie RA, Genovese MC, Keystone EC, Loveless JE, Burmester GR, Cravets MW, Hessey EW, Shaw T, Totoritis MC: Rituximab for rheumatoid arthritis refractory to anti-tumor necrosis factor therapy: Results of a multicenter, randomized, double-blind, placebo-controlled, phase III trial evaluating primary efficacy and safety at twenty-four weeks. Arthritis Rheum 2006;54:2793–2806.
142 Edwards JC, Cambridge G: B-cell targeting in rheumatoid arthritis and other autoimmune diseases. Nat Rev Immunol 2006;6:394–403.
143 Martin F, Chan AC: B cell immunobiology in disease: evolving concepts from the clinic. Annu Rev Immunol 2006;24:467–496.
144 Drayton DL, Liao S, Mounzer RH, Ruddle NH: Lymphoid organ development: from ontogeny to neogenesis. Nat Immunol 2006;7:344–353.
145 Aloisi F, Pujol-Borrell R: Lymphoid neogenesis in chronic inflammatory diseases. Nat Rev Immunol 2006;6:205–217.
146 Sims GP, Shiono H, Willcox N, Stott DI: Somatic hypermutation and selection of B cells in thymic germinal centers responding to acetylcholine receptor in myasthenia gravis. J Immunol 2001;167:1935–1944.
147 Armengol MP, Juan M, Lucas-Martin A, Fernandez-Figueras MT, Jaraquemada D, Gallart T, Pujol-Borrell R: Thyroid autoimmune disease: demonstration of thyroid antigen-specific B cells and recombination-activating gene expression in chemokine-containing active intrathyroidal germinal centers. Am J Pathol 2001;159:861–873.
148 Salomonsson S, Jonsson MV, Skarstein K, Brokstad KA, Hjelmstrom P, Wahren-Herlenius M, Jonsson R: Cellular basis of ectopic germinal center formation and autoantibody production in the target organ of patients with Sjogren's syndrome. Arthritis Rheum 2003;48:3187–3201.
149 Nacionales DC, Weinstein JS, Yan XJ, Albesiano E, Lee PY, Kelly-Scumpia KM, Lyons R, Satoh M, Chiorazzi N, Reeves WH: B cell proliferation, somatic hypermutation, class switch recombination, and autoantibody production in ectopic lymphoid tissue in murine lupus. J Immunol 2009;182:4226–4236.
150 Luzina IG, Atamas SP, Storrer CE, daSilva LC, Kelsoe G, Papadimitriou JC, Handwerger BS: Spontaneous formation of germinal centers in autoimmune mice. J Leukoc Biol 2001;70:578–584.
151 Mandik-Nayak L, Seo SJ, Sokol C, Potts KM, Bui A, Erikson J: MRL-lpr/lpr mice exhibit a defect in maintaining developmental arrest and follicular exclusion of anti-double-stranded DNA B cells. J Exp Med 1999;189:1799–1814.
152 Seo SJ, Fields ML, Buckler JL, Reed AJ, Mandik-Nayak L, Nish SA, Noelle RJ, Turka LA, Finkelman FD, Caton AJ, Erikson J: The impact of T helper and T regulatory cells on the regulation of anti-double-stranded DNA B cells. Immunity 2002;16:535–546.
153 Cappione A, 3rd, Anolik JH, Pugh-Bernard A, Barnard J, Dutcher P, Silverman G, Sanz I: Germinal center exclusion of autoreactive B cells is defective in human systemic lupus erythematosus. J Clin Invest 2005;115:3205–3216.
154 Grammer AC, Slota R, Fischer R, Gur H, Girschick H, Yarboro C, Illei GG, Lipsky PE: Abnormal germinal center reactions in systemic lupus erythematosus demonstrated by blockade of CD154-CD40 interactions. J Clin Invest 2003;112:1506–1520.
155 Kouskoff V, Korganow AS, Duchatelle V, Degott C, Benoist C, Mathis D: Organ-specific disease provoked by systemic autoimmunity. Cell 1996;87:811–822.
156 Korganow AS, Ji H, Mangialaio S, Duchatelle V, Pelanda R, Martin T, Degott C, Kikutani H, Rajewsky K, Pasquali JL, Benoist C, Mathis D: From systemic T cell self-reactivity to organ-specific autoimmune disease via immunoglobulins. Immunity 1999;10:451–461.
157 Matsumoto I, Staub A, Benoist C, Mathis D: Arthritis provoked by linked T and B cell recognition of a glycolytic enzyme. Science 1999;286:1732–1735.
158 Basu D, Horvath S, Matsumoto I, Fremont DH, Allen PM: Molecular basis for recognition of an arthritic peptide and a foreign epitope on distinct MHC molecules by a single TCR. J Immunol 2000; 164:5788–5796.
159 Matsumoto I, Maccioni M, Lee DM, Maurice M, Simmons B, Brenner M, Mathis D, Benoist C: How antibodies to a ubiquitous cytoplasmic enzyme may provoke joint-specific autoimmune disease. Nat Immunol 2002;3:360–365.
160 Wipke BT, Wang Z, Kim J, McCarthy TJ, Allen PM: Dynamic visualization of a joint-specific autoimmune response through positron emission tomography. Nat Immunol 2002;3:366–372.
161 Wipke BT, Allen PM: Essential role of neutrophils in the initiation and progression of a murine model of rheumatoid arthritis. J Immunol 2001;167:1601–1608.

162 Ji H, Ohmura K, Mahmood U, Lee DM, Hofhuis FM, Boackle SA, Takahashi K, Holers VM, Walport M, Gerard C, Ezekowitz A, Carroll MC, Brenner M, Weissleder R, Verbeek JS, Duchatelle V, Degott C, Benoist C, Mathis D: Arthritis critically dependent on innate immune system players. Immunity 2002; 16:157–168.
163 Corr M, Crain B: The role of FcgammaR signaling in the K/BxN serum transfer model of arthritis. J Immunol 2002;169:6604–6609.
164 Lee DM, Friend DS, Gurish MF, Benoist C, Mathis D, Brenner MB: Mast cells: a cellular link between autoantibodies and inflammatory arthritis. Science 2002;297:1689–1692.
165 Victoratos P, Kollias G: Induction of autoantibody-mediated spontaneous arthritis critically depends on follicular dendritic cells. Immunity 2009;30:130–142.
166 Linterman MA, Rigby RJ, Wong RK, Yu D, Brink R, Cannons JL, Schwartzberg PL, Cook MC, Walters GD, Vinuesa CG: Follicular helper T cells are required for systemic autoimmunity. J Exp Med 2009;206:561–576.

George Kollias, PhD
Institute of Immunology, Biomedical Sciences Research Center 'Alexander Fleming'
34 Al. Fleming Street
GR–16672 Vari (Greece)
Tel. +30 210 9656507, Fax +30 210 9656563, E-Mail kollias@fleming.gr

Kollias G, Sfikakis PP (eds): TNF Pathophysiology. Molecular and Cellular Mechanisms.
Curr Dir Autoimmun. Basel, Karger, 2010, vol 11, pp 27–60

Transcriptional Control of the TNF Gene

James V. Falvo · Alla V. Tsytsykova · Anne E. Goldfeld

Immune Disease Institute and Harvard Medical School, Boston, Mass., USA

Abstract

The cytokine TNF is a critical mediator of immune and inflammatory responses. The TNF gene is an immediate early gene, rapidly transcribed in a variety of cell types following exposure to a broad range of pathogens and signals of inflammation and stress. Regulation of TNF gene expression at the transcriptional level is cell type- and stimulus-specific, involving the recruitment of distinct sets of transcription factors to a compact and modular promoter region. In this review, we describe our current understanding of the mechanisms through which TNF transcription is specifically activated by a variety of extracellular stimuli in multiple cell types, including T cells, B cells, macrophages, mast cells, dendritic cells, and fibroblasts. We discuss the role of nuclear factor of activated T cells and other transcription factors and coactivators in enhanceosome formation, as well as the contradictory evidence for a role for nuclear factor κB as a classical activator of the TNF gene. We describe the impact of evolutionarily conserved *cis*-regulatory DNA motifs in the TNF locus upon TNF gene transcription, in contrast to the neutral effect of single nucleotide polymorphisms. We also assess the regulatory role of chromatin organization, epigenetic modifications, and long-range chromosomal interactions at the TNF locus.

TNF plays a critical role in the innate and adaptive immune response and in the normal function of lymphocytes, monocytes, macrophages, neutrophils, and dendritic cells [1, 2]. Although TNF was initially described as a product of macrophages [3], later studies demonstrated that the TNF gene is in fact expressed in a wide range of cell types, including T cells, B cells, NK cells, mast cells, dendritic cells, and fibroblasts [4–11]. Although the secretion of TNF as a mature protein is regulated at the transcriptional, posttranscriptional, translational, and posttranslational levels, this review will examine our current understanding of the mechanisms that control activation of TNF gene expression at the level of transcription, the first step in TNF production.

At the level of transcription, the TNF gene is activated in response to a diversity of specific stimuli that are characteristic of cellular activation, inflammation, infection, and stress. Among these stimuli are calcium signaling, such as calcium influx triggered by ionophores; pathogens, such as bacteria and viruses; mitogens, such as phorbol esters; chemical stress, such as osmotic stress, and radiation, such as UV light

Table 1. Inducers of TNF transcription. Certain stimuli (asterisk) require a costimulus in some cell types.

	Stimuli	Reference
PRR ligands		
TLR2	Peptidoglycan (Gram-positive bacteria)	[214]
	Atypical LPS (*P. gingivalis*)	[215]
TLR2/TLR6	Lipoteichoic acid (Gram-positive bacteria)	[216]
	Diacylated lipoproteins, e.g. MALP-2	[217]
	Zymosan	[218]
TLR3	Double-stranded RNA, e.g. poly (I:C)	[219]
TLR4	LPS (Gram-negative bacteria)	[220, 221]
	Synthetic lipid A	[222]
	Taxol	[223]
TLR7	Loxoribine	[224]
TLR7/TLR8	Single-stranded RNA, e.g. poly I, poly C	[225]
	Imidazoquinoline compounds, e.g. imiquimod	[226]
TLR9	Bacterial CpG-DNA	[225]
NOD2	Muramyl dipeptide	[227]
Antigen receptor ligands		
T cell receptor	Anti-CD3	[15]
	PHA	[4]
B cell receptor	Anti-IgG	[13]
Fc receptor ligands		
Mast cell receptor (FcεRI)	IgE + antigen	[10]
NK cell receptor (FcγRIIIA/CD16a)	Anti-CD16, immune complexes	[228]
Other stimuli		
Cytokines	Interleukin-1	[221]
	Interleukin-2	[229]
	IFN-γ*	[230]
	Granulocyte-macrophage colony stimulating factor (GM-CSF)	[231]
	TNF	[232]
Mitogens	Concanavalin A	[233]
	PMA*	[221]
Superantigens	Staphylococcal toxic shock syndrome toxin-1	[234]
	Staphylococcal enterotoxin B	[234]
Phosphatase inhibitors	Okadaic acid	[235, 236]
	Calyculin A	[235]

Table 1. Continued

	Stimuli	Reference
Calcium ionophore	Ionomycin*	[15]
Radiation	UV light	[237]
	X-rays	[238]
Osmotic stress	Raffinose	[45]
High glucose		[239]
Silica particles		[240]
Bacteria	*Listeria monocytogenes*	[241, 242]
	Staphylococcus aureus	[7]
	Mycobacterium tuberculosis	[243]
	Salmonella typhimurium	[242]
	Escherichia coli	[244]
Viruses	Sendai virus	[245]
	Human cytomegalovirus	[246]
	Vesicular stomatitis virus	[219]
	Herpes simplex virus type II	[219]
Protozoans	*Plasmodium falciparum*	[247]
	Trypanosoma cruzi	[248]
	Schistosoma mansoni	[249]

(table 1). Inducers of TNF gene transcription also include ligands for several classes of receptors, including antigen receptors, such as the T cell receptor; pattern recognition receptors, such as Toll-like receptors [12], and receptors for cytokines, including the two cognate receptors for TNF itself (table 1).

Notably, induction of TNF gene transcription after exposure to certain stimuli in specific cell types is paradigmatic of an immediate early gene. For example, after T and B cell activation or after lipopolysaccharide (LPS) stimulation of monocytes, TNF mRNA is transcribed within minutes and is independent of de novo protein synthesis [13–15]. In T cells in particular, TNF is one of the first genes expressed after cellular activation and is one of the few genes that can be induced by signaling through the T cell receptor in the absence of protein synthesis [15] and a CD28 costumulatory signal [15, 16]. Furthermore, the calcium influx component of T cell activation alone can induce TNF transcription [15].

Tight control of TNF expression in specific cell types and after specific stimuli is essential for cellular homeostasis and normal physiology in humans, as evidenced by the finding that dysregulated TNF levels are associated with multiple disease states, including asthma, rheumatoid arthritis, cardiovascular diseases, Crohn's disease, type

II diabetes, eczema, multiple sclerosis, psoriasis, systemic lupus erythematosus, septic shock, and several different forms of cancer [17, 18]. Dysregulation of TNF expression has also been linked to differential susceptibility to several major infectious diseases including tuberculosis and cerebral malaria, when too little or too much TNF is produced, respectively [19, 20]. Thus, the study of TNF gene regulation not only provides an outstanding model system for the study of cell type- and stimulus-specific eukaryotic gene regulation, but also has direct translational implications for understanding a variety of human diseases. The understanding of basic regulatory pathways and identification of mediators leading to TNF gene expression in particular cell types and tissues can provide targets for the design and development of clinically important therapeutic agents that modulate its expression.

Cell Type- and Stimulus-Specific Regulation of TNF Gene Transcription

TNF gene transcription is regulated by nucleoprotein complexes known as enhanceosomes [21–24]. Enhanceosomes consist of sets of transcription factors and coactivators that associate in a higher-order structure with enhancer or promoter regions of a gene and then function in synergy to drive transcription [25, 26]. Notably, studies of TNF gene regulation expanded our understanding of the role of enhanceosomes in transcriptional regulation in general with the novel demonstration that TNF enhanceosome assembly is cell type- and stimulus-specific, involving distinct sets of transcription factors and coactivators [21–24, 27].

The early observation that TNF was produced by multiple cell types led to experiments to define the sequences involved in transcriptional regulation of the gene in several cell types in response to different stimuli. The 5′ UTR of the TNF gene contains a proximal promoter region of approximately 200 nucleotides (nt) upstream of the mRNA cap site (fig. 1a) that is very highly conserved in mammals [28–31] and almost totally conserved in higher primates [32, 33]. The –200 nt proximal TNF promoter in both humans and mice is sufficient to drive transcription in response to multiple stimuli, including T cell and B cell activation [15, 34–36], calcium ionophore [15], bacterial LPS [22, 37–42], virus infection [37], TNF [43, 44], *Mycobacterium tuberculosis* (MTb) [24], and osmotic stress [45]. In some cell types, an even smaller core region of the TNF proximal promoter suffices for induction, such as by the mitogen phorbol 12-myristate 13-acetate (PMA) [14, 43, 46, 47].

The TNF promoter is both compact and modular in its organization. The relative spacing of DNA motifs in the promoter is critical for TNF gene transcription [23, 24], consistent with the specific architectural requirements of distinct sets of factors sitting on the same face of the DNA helix for enhanceosome formation [25, 26]. Moreover, a number of TNF activator binding motifs can be recognized by more than one class of transcription factor depending on the cell type and stimulus and the ambient concentration of different factors in the nucleus [21–24, 34, 35] (fig. 1b, c). This allows the

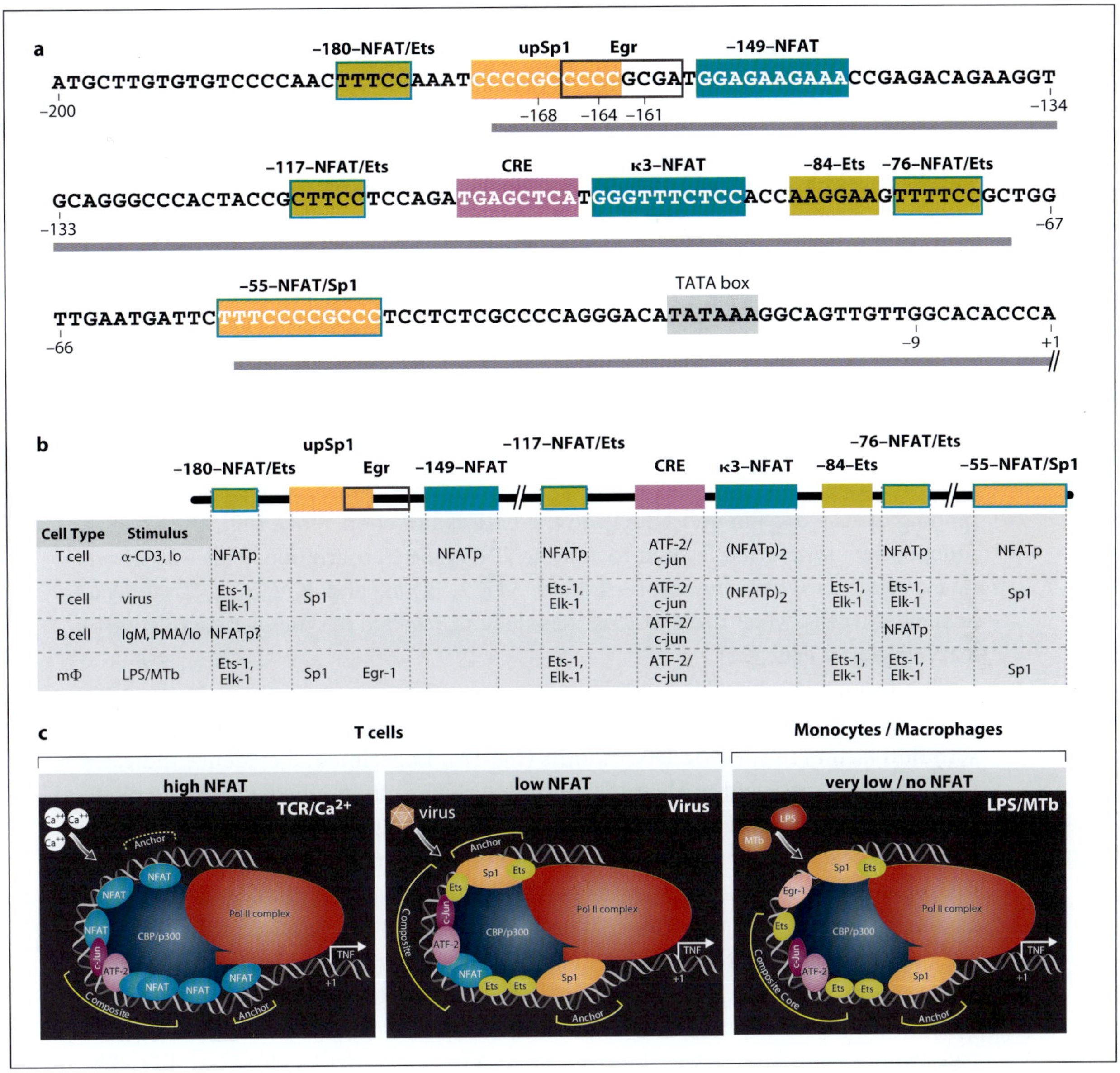

Fig. 1. Cell type- and stimulus-specific enhanceosome formation at the proximal TNF promoter. **a** Sequence of human proximal TNF promoter showing the positions of transcription factor binding sites (boxes), regions of highest sequence conservation in the primate lineage (bar) [33], and positions of fixed genetic differences in nonhuman primates (–9 G/T) and in orangutan and gibbon species (–168 G/A, –164 C/T, –161 C/T) [32, 33]. Adapted from Baena et al. [33]. **b** Transcription factors that bind and function at sites in the indicated cell types in response to the indicated stimuli [15, 21–24, 34, 35]. **c** Model of the role of ambient NFAT levels upon TNF enhanceosome formation, with proteins bound at the CRE/κ3/Ets composite site interacting with 'anchor' complexes to stabilize interactions with CBP/p300 and the RNA Pol II complex [23, 24]. Adapted from Barthel et al. [24].

TNF promoter a remarkable degree of flexibility to respond to a variety of stimuli in a specific manner through a short *cis*-regulatory region.

The proximal TNF promoter (fig. 1a) contains a TATA box and multiple *cis*-acting elements that are binding motifs for transcription factors. Six nuclear factor of activated T cells (NFAT) binding sites have been identified [15, 23, 35, 48], along with four Ets/Elk binding sites [22, 23, 49] and two Sp1 binding sites [14, 23, 50]. Three of the NFAT sites overlap Ets/Elk sites, while one NFAT site overlaps the downstream Sp1 site [22, 23]. There is also an Egr binding site [50], adjacent to the upstream Sp1 site [22], and a cyclic AMP response element (CRE) [46, 51].

This CRE site, which binds a heterodimer of the basic region-leucine zipper (bZIP) proteins ATF-2 and c-jun [21, 22, 24, 34, 35, 44, 45, 52–57], is a critical regulatory element of the TNF gene in all cell types and under all conditions so far tested [21–24, 34, 35, 39, 41, 42, 44, 45, 49, 51, 54–56]. The TNF CRE functions as part of a potent composite element together with the adjacent κ3-NFAT site when fused to a minimal TNF promoter or to a heterologous promoter [21, 22, 24, 34, 35, 39, 52]. Although the binding of ATF-2/c-jun and NFATp to the CRE is not cooperative [35], the proteins function in a synergistic fashion to activate TNF gene transcription. This was shown in experiments where overexpression of ATF-2, c-jun, and NFATp in *Drosophila* Schneider-2 cells, which are devoid of these factors, resulted in synergistic activation of a TNF reporter gene [21].

The formation of cell type- and stimulus-specific enhanceosomes at the TNF promoter involves mutually exclusive binding of NFAT, Sp1, and Ets/Elk transcription factors to specific DNA motifs (fig. 1b). Experimental evidence indicates that if there is sufficient nuclear localization of NFAT, then NFAT will outcompete Ets/Elk or Sp1 binding at these sequences. For example, the amount of NFATp that translocates to the nucleus in T cells upon ionomycin stimulation is higher than upon virus infection, consistent with higher levels of calcium influx induced by ionomycin relative to virus [21]. After ionomycin stimulation, all six NFAT sites are occupied by NFATp [21, 23, 34]. But after virus induction, with lower levels of nuclear NFATp, Sp1 can successfully compete for binding at both Sp1 sites in the proximal promoter [21, 23]. Similarly, after LPS stimulation of monocytes where no inducible translocation of NFATp can be detected, Sp1 and Ets/Elk proteins can successfully bind to sites that are occupied by NFATp in T cells [22, 23] (fig. 1b).

Cell type- and stimulus-specific assembly of TNF enhanceosomes is thus typified by differential occupancy of overlapping DNA motifs in the TNF promoter. These studies have led to a model in which the ambient level of NFAT in the nucleus determines the composition of the TNF enhanceosome, where a combination of core 'composite' complexes of ATF-2/c-jun-NFAT or ATF-2/c-jun-NFAT-Ets/Elk proteins, centered at the CRE/κ3-NFAT site, interact with other 'anchor' complexes to stabilize interactions with coactivators, such as CBP/p300, and the general transcription machinery (fig. 1c). This flexible configuration is consistent with activation of the

TNF gene in response to different transcription factors at different concentrations in the nucleus after a particular stimulus.

Role of Nuclear Factor of Activated T Cells in TNF Gene Regulation

Of all of the transcription factors involved in the regulation of TNF gene expression, NFAT has the clearest role. The NFAT family of proteins consists of five distinct transcription factors in vertebrates: NFATp (also known as NFATc2 or NFAT1), NFATc (NFATc1 or NFAT2), NFAT3 (NFATc4), NFAT4 (NFATc3 or NFATx), and NFAT5 (TonEBP or OREBP). NFATp, NFATc, NFAT3, and NFAT4 reside in the cytoplasm in a hyperphosphorylated state; upon activation they are dephosphorylated by the calcium-dependent phosphatase calcineurin and undergo rapid nuclear translocation, and are thus sensitive to the calcineurin inhibitors cyclosporin A (CsA) and FK506. Unlike the four calcineurin-dependent factors, NFAT5 is responsive to hypertonic stress and integrin activation in addition to T cell stimulation and is constitutively nuclear, save for a small proportion that translocates to the nucleus under hypertonic conditions. NFATp, NFATc, NFAT4, and NFAT5 are expressed in a wide range of cell types and tissues, while NFAT3 is restricted to nonlymphoid cells. NFAT5 binds DNA as an obligate dimer and is dimeric in solution, while the calcineurin-dependent NFAT proteins bind DNA as monomers or dimers; NFAT monomers bind cooperatively with other transcription factors at NFAT/AP-1 elements and other composite DNA motifs [58–61].

The first evidence that NFAT was involved in regulation of the TNF gene came from studies showing that TNF transcription in T cells in response to engagement of the T cell receptor with anti-CD3 or treatment with ionomycin was CsA-sensitive and dependent upon the κ3 site in the proximal promoter [15]. Expression of TNF mRNA was found to be independent of de novo protein synthesis, and a constitutive anti-CD3- and ionomycin-inducible factor that was blocked from the nucleus by CsA was found to bind the κ3 site [15]. This factor thus clearly exhibited the properties assigned to NFAT [62] prior to the cloning of the first NFAT family member, NFATp [63]. This was further supported by the observation that induction of TNF gene transcription was specifically dependent upon the phosphatase activity of calcineurin, both in anti-CD3- or ionomycin-stimulated T cells and in anti-IgG-stimulated B cells [64]. Later experiments confirmed that the observed CsA- and FK506-sensitive factor was NFATp, which bound to κ3 as a dimer [48]. NFATp was also later detected in the FK506-sensitive complex bound to κ3 following PMA and ionomycin treatment of a murine mast cell line [65].

The NFATp dimer was then shown to function in a cooperative fashion with ATF-2/c-jun at the CRE/κ3-NFAT composite element, also in a CsA-sensitive manner [34]. Unlike the canonical NFAT/AP-1 composite element of the interleukin-2 gene promoter, in which binding of NFAT monomer is required for recruitment of its bZIP partner [58, 59, 61], binding of NFATp dimer and ATF-2/c-jun to CRE/κ3-NFAT was shown to be noncooperative [35]. In addition to κ3-NFAT, which is located at –97 to –88, the five other NFAT binding sites identified in the human and murine TNF promoters through DNase footprinting with recombinant NFATp are designated by their positions at –180, –149, –117, –76, and –55 relative to the mRNA cap site (fig. 1a) [23, 35, 36]. NFATc, NFAT3, and NFAT4 bind to these same NFAT sites with slightly varying affinities [45].

A role for cell type-specific function of NFATp in activated T versus B cells was also demonstrated. For example, while binding of NFATp to the κ3-NFAT was shown to be dispensable for activation of the proximal TNF promoter in B cells (in response to ionomycin or PMA and ionomycin), the high-affinity –76-NFAT site was strictly required for TNF gene activation in T and B cells [34, 35]. All of the NFAT sites seem to play a role in T cell activation of TNF in that mutations in the –180, –149, –117, and –55-NFAT sites all reduce inducible levels of activation of the proximal TNF promoter in T cells [23, 34, 35]. Thus, DNA motifs with different affinities for NFAT and other factors, such as Ets/Elk and Sp1 proteins, are required for the regulation of TNF transcription and are involved in the precise modulation of gene expression in T cells and other cells.

Chromatin immunoprecipitation (ChIP) assays later confirmed that NFATp was indeed inducibly recruited to the TNF promoter in its native chromatin context upon treatment of T cells with ionophore or virus and upon treatment of B cells with PMA and ionomycin [21]. Similarly, NFATc was also shown by ChIP to be recruited to the TNF promoter in L929 fibroblasts, which lack NFATp, upon treatment with virus [21]. In ChIP assays with bone marrow-derived murine mast cells, both NFATp and NFATc were shown to be recruited to the TNF promoter upon treatment with ionomycin [66]. Taken together, these studies provided evidence that NFAT binding elements and NFAT proteins play a critical role in TNF gene regulation.

Detailed analysis of NFATp dimer binding at the CRE/κ3-NFAT composite element revealed that dimerization of NFATp is required for transcriptional activation mediated by the CRE/κ3-NFAT site. This was independent of the orientation of the κ3-NFAT site relative to the CRE site, consistent with the lack of physical interaction between ATF-2/c-jun and NFATp [52]. However, ATF-2/c-jun and NFATp function in a synergistic fashion at the CRE/κ3-NFAT element despite their lack of cooperative binding. This functional synergy suggested that the activation domains of NFATp played a role in TNF transcription. Indeed, two subdomains in the N- and C-terminal activation domains of NFATp are required for interaction with CBP and activation of transcription mediated by the CRE/κ3-NFAT site and the TNF promoter [52]. This is in agreement with a study in which the C-terminal activation domain of NFATp was shown to be required for TNF transcription and to function in isolation as a dominant-negative inhibitor of TNF transcription in the Jurkat human T cell line [67]. These studies demonstrated that in the context of the TNF promoter, the activation domains of NFATp specifically contribute to the activation of gene transcription.

NFAT3 has also been implicated in TNF gene regulation in cells of a nonlymphoid lineage. Activation of a TNF gene reporter and expression of endogenous TNF mRNA in response to UV radiation in murine embryonic fibroblasts is blocked by RNA interference (RNAi) of NFAT3; furthermore, in response to UV radiation, activation of the TNF reporter is strongly inhibited by mutation of the –180 and –76-NFAT sites, and NFAT3 is recruited to the endogenous TNF promoter [68]. RNAi of NFAT3 also inhibited TNF transcription in a murine epidermal cell line in response to silica or the carcinogen (+/–)-benzo[a]pyrene-7,8-diol-9,10-epoxide [69, 70]. A specific role for NFAT4 in the regulation of TNF gene transcription has not been characterized, although as noted above, recombinant NFAT4 can bind to all six NFAT binding sites in the TNF promoter [45].

By contrast, while recombinant NFATp, NFATc, NFAT3, and NFAT4 bind to the same set of sites in the TNF promoter in DNase footprinting assays, recombinant NFAT5 only binds to two of the NFAT sites in the TNF promoter: κ3-NFAT and –76-NFAT [45]. ChIP assays showed that NFAT5 associates with the TNF promoter in response to osmotic stress in Jurkat cells [71], while inhibition of NFAT5 by RNAi, transfection of a dominant-negative form of NFAT5, or mutation of the κ3-NFAT and –76-NFAT sites inhibits activation of the proximal TNF gene promoter in response to osmotic stress in L929 cells [45]. Thus, NFAT5 also plays a key role in TNF gene regulation in certain cell types and under certain conditions.

The critical role of NFAT in TNF gene regulation shown by these biochemical and cell-based assays has been supported by multiple studies using transgenic mice. In particular, these in vivo studies have provided evidence that NFATp is the NFAT protein primarily responsible for TNF gene regulation in T cells. NFATp-deficient mice display defects in T cell-derived TNF expression in response to anti-CD3 (or anti-CD3 and anti-CD28) antibodies and to superantigen [36, 72, 73]. This is in contrast to findings in NFATc-deficient murine T cells, which produce wild-type levels of TNF [74, 75]. Furthermore, TNF mRNA expression induced by treatment with ionomycin or IgE crosslinking in bone marrow-derived mast cells is inhibited by deficiency of NFATp but not NFAT4 [66], and TNF expression from murine T cells lacking both NFATp and NFAT4 is diminished at levels similar to those observed in NFATp-deficient mice [72, 76].

Strikingly, in vivo evidence for a stimulus-specific role for NFATp in TNF gene regulation was demonstrated in a murine model of TNF-mediated superantigen-induced lethal shock. NFATp-deficient mice were resistant to T cell-dependent septic shock as compared to wild-type control mice, but were susceptible to macrophage-driven LPS-induced septic shock [36]. These experiments thus provided definitive in vivo evidence that TNF gene expression is dependent upon distinct factors in T cells and in monocytic cells.

Residual TNF expression in NFATp-deficient cells appears to be driven by NFATc, as deletion of both NFATp and NFATc from murine lymphoid cells eliminates any detectable TNF expression in T cells upon primary or secondary stimulation under

Th1 and Th2 conditions [77]. Similarly, in bone marrow-derived mast cells, ionomycin-induced TNF mRNA expression is inhibited by RNAi of either NFATp or NFATc [66]. RNAi of NFATc in NFATp-deficient mice further inhibits ionomycin-induced TNF mRNA expression, and ectopic expression of NFATc rescues this expression [66]. Furthermore, transgenic mice expressing a constitutively nuclear form of NFATc display enhanced TNF transcription in unstimulated and anti-CD3-stimulated peripheral T lymphocytes [78], although expression of constitutively nuclear NFAT4 in transgenic mice also greatly enhances TNF expression in Th1 cells induced by PMA or PMA and ionomycin [79].

Notably, a parallel to these results from transgenic mice is found in studies of human T cells in which NFAT signaling is impaired. T cells from severe combined immune deficiency patients, which had nearly undetectable levels of NFAT binding to DNA in EMSAs, were unable to produce TNF [80]. Similarly, in Wiskott-Aldrich syndrome patients, TNF secretion was abrogated or strongly reduced in CD4+ and CD8+ T cells, which correlated with a reduction in nuclear localization of both NFATp and NFATc [81].

Nuclear Factor-κB-Independent TNF Gene Transcription

The nuclear factor-κB (NF-κB) proteins are a major family of transcription factors involved in control of the immune response. This family includes p50, p65 (RelA), c-Rel, p52, and RelB, which bind DNA as obligate homo- or heterodimers [82–84]. TNF is often described in the literature as one of the classical NF-κB-dependent proinflammatory cytokines, largely based upon early studies that concluded that there was a major role for NF-κB in TNF gene expression in response to LPS in cells of the monocyte/macrophage lineage. However, a direct role for NF-κB in the activation of TNF transcription was not clear from these initial studies.

Based on their similarity to known NF-κB binding sites, sequences in the 5′ flanking region of both the human and the murine TNF genes were initially designated as κB motifs [28, 37, 85, 86]. In the human TNF promoter, these sites were named κ1, κ2, and κ3 [37]. Later, three additional NF-κB-like sites were identified: κB1, κB2, and ζ [29, 87]. In the murine TNF promoter, these sites were initially called κB1, κB2, κB2b, κB3, and κB4 [28, 85, 86]; κB2b was later renamed κB2a [29]. Notably, all of these sites in the human and mouse genes, except the human κ3 site, lie outside of the proximal TNF promoter shown to be sufficient for induction of the gene by multiple stimuli and in diverse cell types (fig. 2a).

A role for NF-κB in LPS-mediated murine TNF gene expression was originally postulated based on the binding of inducible factors to sites in the distal murine TNF promoter, particularly κB3 (fig. 2a), a murine-specific κB site which exhibited the highest affinity for a factor with the properties of NF-κB and conferred LPS inducibility upon a heterologous promoter when present in two or three copies [28, 85, 86]. This view

a

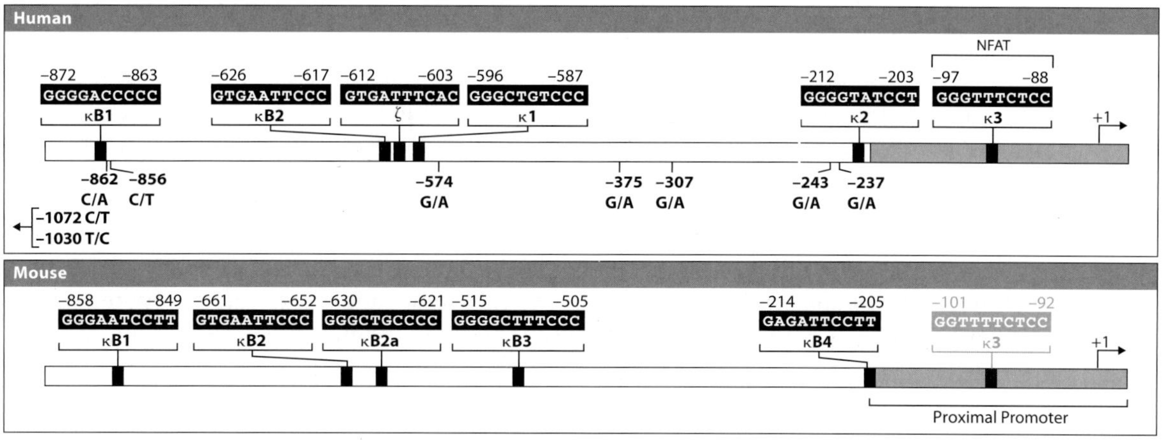

b

IP	Cell type	Stimulus	Amplified seq.	Ref.
p65, p50, c-Rel	BMDM	LPS (*E. coli*)	–752 to –547	[114]
p65, p50, RelB	THP-1 (monocytic)	LPS (*E. coli*)	–122 to +8	[123]
p65	THP-1	LPS (*E. coli*)	–161 to +62	[95]
p65	THP-1	LPS (*E. coli*)	–119 to +32 –345 to –195	[250]
p65, p50, c-Rel, p52, p53, RelB	U937 (monocytic)	LPS (*E. coli*)	–700 to +200	[251]
p65	U937	LPS (*E. coli*)	–420 to –49	[112]
p65	mur. cardiac monocyt.	LPS (*E. coli*)	–722 to –434	[94]
p65, p50	RBL-2H3 (basophil)	LPS (*E. coli*)	–563 to –404	[115]
p65	TEPM	LPS (*P. gingivalis*)	–683 to –529	[252]
p65, RelB	murine stromal cells	TNF	–545 to +20	[111]
p65, p50	U20S (osteosarcoma)	TNF	–167 to +42	[253]
p65	THP-1	TNF	–160 to –15	[116]
p65, p50	THP-1	high glucose	–453 to +33	[181]
p65	murine mast cells	FcεRI stimulation	–650 to –460	[254]
p65, RelB	murine stromal cells	LTβR stimulation	–545 to +20	[111]
p65	HeLa (endothelial)	TGF-β, *H. influenzae*	–420 to –49	[113]

c

Genotype	Cell type	Stimulus	Effect upon TNF	References
$p50^{-/-}$	BMDC,BMDM,TEPM FLDM,TEPM* spleen, liver	LPS LPS + IFN-γ LPS	none none elevated	[121,255] [121*,256] [257]
$c\text{-}Rel^{-/-}$	BMDC,BMDM,TEPM* FLDM, TEPM* splenic T cells*	LPS LPS + IFN-γ IL-2+mitogens	none none none/ slightly red.	[121*,129*,255] [256] [258]
$RelB^{-/-}$	BMDM* TEPM*	LPS LPS, LPS + IFN-γ	slightly red. reduced	[259] [98]
$p65^{-/-}$	BMDC BMDM TEPM, MEF FLDM	LPS LPS LPS LPS + IFN-γ	~none reduced strongly red. reduced, strongly red.*	[255] [255] [255] [256]
$p50^{-/-}c\text{-}Rel^{-/-}$	BMDC,BMDM FLDM, TEPM*	LPS LPS + IFN-γ	none none	[255] [256]
$p50^{-/-}p52^{-/-}$	TEPM peritoneal mϕ*	LPS LPS + IFN-γ	none none	[260] [261]
$p50^{-/-}p65^{+/-}$	spleen liver BMDM	LPS LPS LPS,TNF	elevated none none	[257] [257] [257]

Fig. 2. NF-κB and TNF gene regulation. **a** Putative NF-κB and NF-κB-like sites in the human and murine TNF promoters, along with the positions of SNPs in the human TNF promoter. **b** Stimulus-dependent association of NF-κB proteins (in at least one condition or time point studied) with the TNF promoter in ChIP assays. **c** Impact of deletions of NF-κB proteins in mice upon TNF gene transcription (asterisk denotes studies and cell types where only protein expression was assayed) in response to various stimuli). TEPM = Thioglycollate-elicited peritoneal macrophages; FLDM = fetal liver-derived macrophages; BMDM = bone marrow-derived macrophages; BMDC = bone marrow-derived dendritic cells; MEF = murine embryonic fibroblasts.

that NF-κB was involved in TNF gene regulation in monocytes was reinforced by indirect evidence from studies where NF-κB was inhibited by dexamethasone and pyrrolidine dithiocarbamate, and the correlation of these data with experiments showing that the same compounds also inhibited TNF mRNA expression in monocytes or macrophages [85, 88]. Subsequent studies reported that inhibition of NF-κB activity by dominant-negative versions of IκB-α and IκB kinase was also correlated with inhibition of LPS-induced TNF transcription in monocytic cell lines, murine dendritic cells, and primary human macrophages [29, 89–95].

However, deletion or site-directed mutation of κ1 or κ2 did not affect LPS induction of the human TNF promoter in gene reporter assays, and multiple copies of κ1, κ2, and κ3 did not confer LPS or virus inducibility on a heterologous minimal promoter [21, 22, 37, 39], as would be expected for a bona fide NF-κB binding site such as PRDII in the interferon-β promoter [96]. These observations led to the conclusion, even in 1990, that these sites were not typical NF-κB sites [37]. Although κB1 (fig. 2a) was shown to have the highest affinity for NF-κB in vitro of all of the distal human TNF promoter sites, its mutation also had little or no effect on TNF transcription [29, 87]. The human κ2 site has consistently been shown to have little or no affinity for NF-κB and to play no role in TNF gene transcription [29, 37–39, 87, 88, 97]. Furthermore, the κ2 and κB1 sites are not present in the murine promoter [28, 29], while the murine κB3 site, which has the highest affinity for NF-κB among the murine κB-like sites [29, 85, 86, 98], is not present in the TNF promoter in humans and other primates [28, 29, 32, 33]. In conclusion, all of these data argue against an important conserved regulatory role for these sites in murine or human TNF gene regulation.

Three NF-κB sites in the distal human TNF promoter, κB2, ζ, and κ1 (fig. 2a), are clustered in a region of higher evolutionary sequence conservation relative to the rest of the distal promoter, but only κB2 is completely conserved between mice and primates [29, 32, 33]. Although mutations in these low-affinity NF-κB sites were reported to reduce LPS inducibility of the human TNF gene by 40–50% in reporter assays [29, 87], these reports were hard to reconcile with studies showing that mutation of κ1 [37] and even deletion of the entire distal promoter [22, 37–39, 41] did not impact LPS induction of a human TNF reporter gene. Similarly, in gene reporter assays with the murine TNF promoter, deletion of the distal promoter region [42] and mutation of κB1 (and in some cases, κB3, κB2, and κB2a) had little to no effect on LPS-induced TNF transcription [29, 99]. Thus, there has been no clear functional role demonstrated for these NF-κB motifs in the activation of murine or human TNF gene transcription.

A number of studies have reported, using antibody supershift assays, the binding of NF-κB p50 and p65 to the most proximal of the human κB-like sites, κ3 [38, 39, 55, 97, 100–102]. However, the κ3 site (later renamed κ3-NFAT since it was shown to be an important functional NFAT site as described above) has little or no affinity for NF-κB, particularly in studies comparing the relative affinities of NF-κB (or

NF-κB-like) sites in EMSAs and quantitative DNase I footprinting assays [29, 87, 88, 103]. In fact, replacement of the κ3-NFAT site (5′-GGGTTTCTCC-3′) with the well-characterized functional NF-κB site from the interferon-β gene promoter (PRDII, 5′-GGGAAATTCC-3′) confers virus inducibility on the TNF proximal promoter in HeLa cells, a cell type in which the wild-type TNF proximal promoter is not virus inducible [103]. This is consistent with the DNA-binding properties of NF-κB, given that the κ3-NFAT site (5′-GGGTTTCTCC-3′) diverges from the canonical NF-κB (5′-GGGRNTTYCC-3′ or 5′-GGGRNNTYCC-3′) and p65 (5′-GGGRNTTTCC-3′ and 5′-NGGRNTTYCC-3′) consensus motifs at two key positions (underlined) and bears little resemblance to the noncanonical NF-κB consensus motif (5′-RGGAGATTG-3′) which preferentially binds RelB/p52 [95, 104–107]. Furthermore, the corresponding κ3 sequence in the murine proximal TNF promoter (5′-GGTTTTCTCC-3′) is an NFAT binding site [36] that has never been implicated as an NF-κB binding site. Notably, the G to T change at position three in mouse as compared to human (5′-GGGTTTCTCC-3′) would not favor NF-κB binding, but would be expected to make it an even stronger NFAT binding site.

Several studies employed the ChIP assay to provide evidence for the binding of NF-κB, typically p65, to the TNF promoter in response to a variety of stimuli in different cell types (fig. 2b). The region of the TNF promoter amplified in these ChIP assays varied considerably, with some encompassing the distal promoter and others the proximal (fig. 2b). The standard ChIP assay has a level of resolution of ~300 bp [108] and may disrupt native chromatin configuration [109]. In fact, global ChIP studies have identified multiple regions that are retained by immunoprecipitation of p50 or p65 but contain no NF-κB binding sites [95, 110]. Thus, it is important to correlate ChIP assays with other protein-DNA interaction assays and with functional data such as gene reporter assays before concluding that a protein detected by ChIP is functioning via a specific DNA site. While some of these ChIP-based studies employed EMSAs with consensus NF-κB motifs or NF-κB motifs from the distal TNF promoter or other gene promoters [111–116], none examined direct binding of NF-κB to the κ3-NFAT site, or to any other site in the proximal TNF promoter, and correlated such data with a functional role for the site.

Furthermore, in transgenic mice, deletion of the genes encoding the NF-κB proteins p50, c-Rel, p52, and RelB had little or no effect upon expression of TNF mRNA or protein, although in macrophages and embryonic fibroblasts lacking p65, TNF mRNA levels were strongly reduced (fig. 2c). Inhibition of NF-κB activity by genetic deletion of the upstream regulator NEMO/IκB kinase-γ or by the proteasome inhibitor lactacystin revealed that production of wild-type levels of TNF mRNA at late time points after treatment with LPS or virus was NF-κB-dependent, but induction of TNF transcription was not. This indicated a postinduction role for NF-κB, perhaps through control of signal transduction pathways or chromatin remodeling, rather than classical activation through NF-κB binding to the proximal promoter [103]. While the mechanistic details remain to be determined, this may at least partially reconcile the seemingly contradictory evidence regarding the role of NF-κB in TNF gene regulation in cells of the monocyte/macrophage lineage and explain how NF-κB inhibition could potentially inhibit TNF expression through a secondary effect.

Interestingly, interaction of NF-κB with the distal TNF promoter appears to have a role in another aspect of TNF transcription: LPS tolerance, the persistent state of repressed transcription following prolonged exposure to LPS [117]. In the murine TNF promoter, the κB3 site mediates this repression [118], and binding of p50 homodimer to the murine κB3 site and binding of p50 or p52 homodimer to the human κ1 site are upregulated during LPS tolerance [97, 118–121]. Moreover, abrogation or inhibition of p50 and RelB expression in primary macrophages or macrophage cell lines counteracts the repression of TNF transcription during LPS tolerance [121–123].

Other Transcription Factors Involved in Activating TNF Gene Transcription

A number of studies have implicated other transcription factors in TNF gene regulation, including proteins of the bZIP, signal transducer and activator of transcription (STAT), and interferon regulatory factor (IRF) families. Some of these results are relatively preliminary in nature. For example, a role for the bZIP protein C/EBPβ (also known as NF-IL6) in the regulation of TNF gene transcription in cells of the monocyte/macrophage lineage was initially postulated primarily from results with ectopic expression of full-length and dominant-negative forms of the protein and EMSAs with motifs overlapping the κ3-NFAT and –180-NFAT/Ets sites [124–126]. However, later studies contradicted a specific role for C/EBPβ in LPS-induced TNF gene transcription in macrophages [97, 127–129]. Another bZIP protein, Nrf1, has only been detected by EMSA in a complex that binds CRE/κ3-NFAT in stimulated mast cells [130, 131]. Similarly, ChIP assays showed phosphorylated STAT1 binding to the TNF promoter region upon IFN-γ stimulation of murine macrophages [132], although gene knockouts of STAT1 and STAT4 and mutation of a putative STAT3 binding site in the distal murine TNF promoter had only modest effects upon LPS-induced TNF expression and transcription [133, 134].

Another transcription factor implicated in TNF gene regulation is LPS-induced TNF-α factor (LITAF), also known as p53-induced gene 7 (PIG7) [135, 136]. LITAF binds a motif in the distal human TNF promoter (5′-CTCCC-3′, at –515 to –511) and forms a functional complex with STAT6(B) [137, 138]. Reduction in LITAF expression levels resulted in partial inhibition of endogenous TNF mRNA and protein expression [135, 136, 139], and ectopic expression of LITAF, or LITAF and STAT6(B), increased transcription mediated by the human TNF promoter and secretion of TNF protein [137, 138]. However, deletion of the LITAF binding site in the context of the human TNF promoter (including –991 nt) had no effect upon LPS-induced transcription in reporter assays [137]. Thus, it is possible that, like NF-κB, binding of LITAF to the distal TNF promoter plays a postinduction role in the maintenance of mRNA expression. Indeed, it has been suggested that the two factors act in an additive fashion via independent signal transduction pathways [139].

Recent studies have suggested a role for IRF proteins in TNF transcription. ChIP assays detected binding to the TNF promoter by IRF-1 and IRF-8 in response to IFN-γ stimulation of a murine macrophage cell line [140] and by IRF-3 in response to LPS and PMA under conditions of chronic ethanol exposure in a human macrophage cell line [141]. Furthermore, RNAi of IRF-3 in mouse embryonic fibroblasts inhibited LPS-induced TNF mRNA expression [142]. However, while some studies described putative IRF binding sites in the TNF promoter [140, 141], and IRF-3 in particular was implicated in a late-phase autocrine loop of TNF gene activation by LPS [142, 143], the binding of IRF proteins to a specific DNA site was not tested. Thus, a direct role of IRF proteins in TNF gene transcription remains to be elucidated.

Single Nucleotide Polymorphisms, Fixed Genetic Differences, and TNF Transcriptional Regulation

The TNF gene resides between the class I human leukocyte antigen (HLA)-B and class II HLA-DR loci in the most polymorphic region of the human genome: the major histocompatibility complex, or human histocompatibility locus (HLA), located on human chromosome 6 [144, 145]. Several single nucleotide polymorphisms (SNPs) have been identified in the vicinity of the human TNF locus, particularly in the distal TNF promoter, and several have been suggested to influence disease outcome through a direct impact upon the transcriptional regulation of TNF [146–150].

We note that initial studies incorrectly numbered the positions of SNPs relative to the TNF mRNA cap site, which should have proceeded 5′ from the adenine at the +1 position [151]. While the 'traditional' numbering system (e.g. –308 instead of –307) persists in some papers in the literature, the corrected numbering has been used in several studies and in the Cytokine Gene Polymorphism in Human Disease database [152] and is used throughout this review (fig. 2a). TNF promoter SNPs have been associated with a wide variety of disease conditions [146–150, 152]. Intriguingly, multiple SNPs in the TNF promoter (–1030 T/C, –862 C/A, –856 C/T, –307 G/A, and –243 G/A) are in linkage with HLA molecules in extended haplotypes [151, 153–155], which are conserved blocks of DNA sequences between HLA-B and HLA-DR [156, 157]. They in fact serve as markers of extended haplotypes [146–149, 152].

Notably, a number of TNF promoter SNPs, including several purported to impact TNF gene regulation, in fact serve as markers of ancestry in human populations (–1030 T/C, –862 C/A, –856 C/T, –574 G/A, –375 G/A, –307 G/A, –243 G/A, –237 G/A, +69 C/G, and +70 +C) and correspond to ancestral alleles in the primate lineage (–1072 C/T, –1030 T/C, –862 C/A, –375 G/A, –307 G/A, and –237 G/A) [30, 32, 33, 155]. For example, the –856 SNP was relatively common (13–14%) in individuals of Caucasian and Cambodian descent and unusually prevalent (30–45%) in Amerindian (Quechua and Paez) populations but was absent in Malawians [155]. Thus, TNF promoter SNPs are likely to be in linkage with other genes in the major histocompatibility complex locus, which may in turn impact resistance and susceptibility to, and severity of, disease independent of their effect, or lack of effect, on TNF gene expression. Consistent with this hypothesis that SNPs are markers of divergence, and not recently selected traits impacting disease susceptibility, is the finding that the rare allele of the –307 human SNP is present in Old World monkey species, indicating that the adenine at that position was fixed in Old World monkeys when they diverged from apes roughly 25 million years ago [32, 33].

Notably, all SNPs that have been detected upstream of the TNF gene in more than one individual lie outside of the proximal promoter (fig. 2a). Moreover, sequence analysis of the TNF promoter region (up to 1.2 kb upstream of the start site of transcription) in primates revealed areas of high conservation within the ~200 bp proximal promoter. Comparative sequence analysis led to the identification of 'phylogenetic footprints' between positions –131 to –63 and –53 to –45 [32]. Regions of very low accumulated sequence variation were also revealed by phylogenetic shadowing from –171 to –70 and from –54 to +29 [33]. Strikingly, these sequences overlap motifs essential for TNF gene regulation and for enhanceosome formation (fig. 1a).

Intriguingly, in great apes and in species representing the four gibbon genera, the proximal TNF promoter sequence is identical to that found in humans, except for a G to T transversion at position –9 common to all non-human primates and certain fixed genetic differences within or flanking the upstream Sp1 site at –172 to –163 [32, 33] (fig. 1a). In humans, African great apes, and Old World monkeys, this sequence is 5′-CCCCGCCCCCGCG-3′, while the sequence is 5′-CCCC<u>A</u>CCCCCGTG-3′ in orangutans *(Pongo pygmaeus* and *Pongo abelii)* and 5′-CCCC<u>A</u>CCC<u>T</u>CGCG-3′ in the northern white-cheeked gibbon *(Nomascus leucogenys leucogenys)*. The G to A transition at position –168 strongly inhibited the binding of Sp1 and Sp3 to the site in EMSAs. Strikingly, the orangutan and gibbon TNF promoters with this –168 G/A transition displayed decreased levels of transcriptional activation in response to LPS and MTb in monocytic cells but not in response to ionomycin in T cells [33]. This observation is consistent with the cell type- and inducer-specific role of Sp1 in the regulation of the TNF gene [21–24], and suggests that a distinct set of infectious disease pressures were involved in the evolution of the innate immune response and TNF in Asian versus African apes, which impacted the Sp1 site [33].

Despite the associations between TNF promoter SNPs and disease, a direct impact of TNF promoter polymorphisms upon TNF transcription has not been conclusively demonstrated [146–150]. For example, in the case of the most commonly studied TNF promoter SNP, –307 G/A, some reports concluded that the –307 A allele yielded an increase in transcription in reporter assays [158–162]. In some of these studies, the effect was only observed when the 3′ UTR was included in the reporter construct, and with only a small subset of cell types and stimuli [160, 162]. Other studies reported no differences between the alleles in comparable reporter assay systems [151, 163–165]. Contrasting results were obtained in EMSAs with nuclear extracts and probes spanning the –307 region of the TNF promoter [160, 161]. Furthermore, ChIP assays in human lymphoblastoid cell lines heterozygous at the –307 G/A SNP showed that there was no allele-specific difference in recruitment of RNA polymerase II to the TNF promoter [166]. Thus, taken together the data indicate that the –307 SNP, which appears as a fixed difference from humans in Old World monkeys, has no clear transcriptional role.

A number of studies have also reported that binding of the transcription factor Oct-1 to sites in the human TNF promoter is specific for the variant alleles of the –375 G/A, –856 C/T, and –862 C/A SNPs [167–170]. As was the case for the –307 G/A SNP, some groups reported a change in activity mediated by some or all of these rare alleles [154, 167, 171], while others reported no difference in activity [151, 165, 169, 172]. Since the binding of NF-κB p50 homodimer to κB1 (fig. 2a) is inhibited by the presence of an adenine at –862 [170, 171], the –862 C/A SNP might have an impact on LPS-mediated tolerance at the level of transcription; however, in reporter assays in human monocytes, the presence of the -862 A allele only resulted in an increase in LPS-induced transcription when the 3′ UTR was present [171]. Finally, like the –307 SNP and other SNPs, the –375 SNP falls outside of the evolutionarily conserved regions of the TNF promoter, consistent with the lack of a direct impact of the variant alleles upon activation of TNF gene transcription.

Epigenetic Regulation of TNF Transcription

Gene transcription involves not only the assembly of transcription factors and coactivators at gene promoter and enhancer regions, but also the regulation of accessibility to DNA in the context of chromatin. Transcriptionally active genes are marked by a number of covalent modifications of both histones and DNA [173–177]. For example, histone acetyltransferases, including general transcription factors, coactivators, and sequence-specific transcription factors, promote the formation of 'open' areas of chromatin by altering the net charge of nucleosomes through acetylation of lysine residues in the N-terminal tails of histones H3 and H4 [178–180]. TNF transcription is associated with multiple histone acetyltransferases, including ATF-2 [21, 22, 24, 34, 35, 44, 53–57], CBP/p300 [22, 24, 27, 181–183], p/CAF [181, 184], and GCN5 [184]. CBP, in particular, is specifically required for TNF gene transcription in response to T cell activation [27, 182]. Furthermore, a component of the SWI/SNF chromatin remodeling complex that interacts with acetylated histones, BRG1, is associated with the TNF promoter in unstimulated J774 monocytic cells, consistent with a poised preinduction open chromatin conformation [185].

In multiple studies, acetylation of histone H3 and histone H4 at the TNF promoter region has been correlated with TNF transcription in primary cells and cell lines of the monocyte/macrophage and T cell lineage. Increased histone H3 and H4 acetylation at the TNF promoter region (and, in some cases, at the third intron of TNF) in primary human monocytes or human monocytic cell lines (e.g. THP-1) has been associated with induction of TNF transcription by LPS [186, 187] and high glucose concentrations [181], maturation of monocytes to macrophages [188], diabetes [181], and systemic lupus erythematosus [189]. Moreover, IFN-γ treatment of primary human monocytes led to increased, persistent H4 acetylation along with recruitment of ATF-2 and RNA Pol II, yielding a 'poised' pretranscription state and, subsequently, elevated LPS-induced levels of both TNF transcription and histone H3 and H4 acetylation at the TNF promoter [187]. In Jurkat T cells, phytohemagglutinin (PHA)/PMA stimulation led to acetylation of histone H3 at the TNF promoter, correlating with the recruitment of both p/CAF and GCN5, while histone H4 was constitutively acetylated [184]. Notably, a transactivator of transcription (Tat) protein from HIV-1 subtype E (HIV-1_{TH64} Tat), which specifically inhibited TNF transcription, also inhibited p/CAF recruitment to the TNF promoter and decreased levels of histone H3 acetylation and GCN5 recruitment [184]. Enrichment of acetylated histone H3 and H4 was also reported in the vicinity of the TNF promoter in PMA/ionomycin-stimulated Jurkat cells [190].

The complex transcriptional regulatory code of histones includes not only acetylation, but also methylation, phosphorylation, and ubiquitination [174–177]. Histone modifications that are typically associated with gene derepression and transcription have been detected at the TNF promoter. Mono-, di-, and trimethylation of lysine 4 of histone H3 (H3K4) has been observed at the TNF promoter following LPS or TNF stimulation of THP-1 cells and PMA/ionomycin stimulation of Jurkat cells [116, 186, 190]. Phosphorylation of serine 10 of histone H3 (H3S10) has been observed at the TNF promoter in THP-1 cells, but not primary human dendritic cells, following LPS stimulation [123, 191]. Furthermore, levels of dimethylated H3K4, which often marks genes that are competent for transcription, have been shown to decrease at the TNF promoter following LPS induction, replaced by a peak (at the TNF promoter and, in some cases, TNF intron 3) of trimethylated H3K4, which typically marks active transcription [186]. By contrast, several kinds of non-TNF-expressing cells, including LPS-tolerant cells, display relatively lower levels of methylation at H3K4 and phosphorylation at H3S10 and higher levels of di- or trimethylation of H3K9, which is a marker of repressed genes that recruits heterochromatin protein 1 (HP1), which in turn promotes gene silencing [123, 186, 192]. The functional role of these modifications was indicated by the observation that inhibition of H3K4 methylation through RNAi of the histone methyltransferase SET7/9 or of components of the mixed-lineage leukemia histone methyltransferase complex reduced TNF transcription [116, 186], while inhibition of H3K9 methylation through RNAi of the histone methyltransferase G9a in LPS-tolerant cells decreased HP1 binding and restored TNF transcription

[192]. Thus, while the specific modifications involved vary with cell type and stimulus, histone methylation and phosphorylation influences TNF gene transcription.

In eukaryotic genomes, methylation of DNA at cytosine residues, typically at CpG or CpNpG sequences, generally represses gene transcription [174–176]. Initial studies demonstrated that in various primary cell types, including monocytes and lymphocytes, the TNF gene and proximal TNF promoter were unmethylated, while in non-TNF-expressing HeLa cells, they were highly methylated [193]. Moreover, fusion of murine 3T3 fibroblasts to RAW 264.7 macrophages resulted in hybrid cells which, like the 3T3 cells, had a highly methylated TNF locus and did not express TNF [194]. The TNF proximal promoter and first exon were also shown to be highly methylated in K562 cells and THP-1 clones that did not express TNF, but unmethylated in TNF-expressing THP-1 clones and HL60 promyelocytic leukemia cells. Similarly, demethylation of TNF locus correlates with both differentiation and competence to express TNF: the TNF proximal promoter and first exon were highly methylated in human embryonic stem cells and embryoid bodies, exon 1 was demethylated in hematopoietic stem cells and liver cells, and both the TNF proximal promoter and exon 1 were demethylated in primary monocytes and macrophages [186]. Inhibition of DNA methylation at the TNF locus can, in turn, enhance transcription of the TNF gene: 5-azacytidine enhanced LPS-mediated TNF production in THP-1 cells [186], while in LPS-tolerant THP-1 cells, RNAi of G9a inhibited recruitment of the DNA methyltransferase Dnmt3a/b, restoring TNF transcription [192]. Thus, DNA methylation is another epigenetic gene-regulatory mechanism that impacts TNF transcription.

Chromatin Remodeling at the TNF/Lymphotoxin Locus

'Open' regions in chromatin accessible to DNA-binding proteins yield DNase-hypersensitive sites (HSSs); many cytokine loci have been examined using DNase I hypersensitivity assays [173]. The TNF gene resides in a locus with the lymphotoxin-α and -β (LT-α and LT-β) genes, and numerous cell type-specific HSSs have been identified (fig. 3a). The first HSS identified in the TNF/LT locus was the proximal TNF promoter itself, consistent with its interaction with multiple transcription factors in the context of chromatin, as shown by in vivo genomic footprinting [195, 196]. The TNF promoter HSS was present in unstimulated and PMA-induced TNF-expressing human myelomonocytic cell lines (ML-1, U937, and KG-1), but not a non-TNF-expressing human erythroid cell line (HEL) [197]. In porcine peripheral blood mononuclear cells, constitutive HSSs at the TNF and LT-α promoter regions and the third intron of TNF were observed at the TNF/LT locus, while a macrophage-like porcine cell line, in which LT-α was not transcribed, had HSSs at the first and third introns of TNF (fig. 3a); HSSs were not observed in porcine fibroblasts or a porcine kidney cell line, in which TNF was not expressed [198]. Constitutive HSSs were detected in both

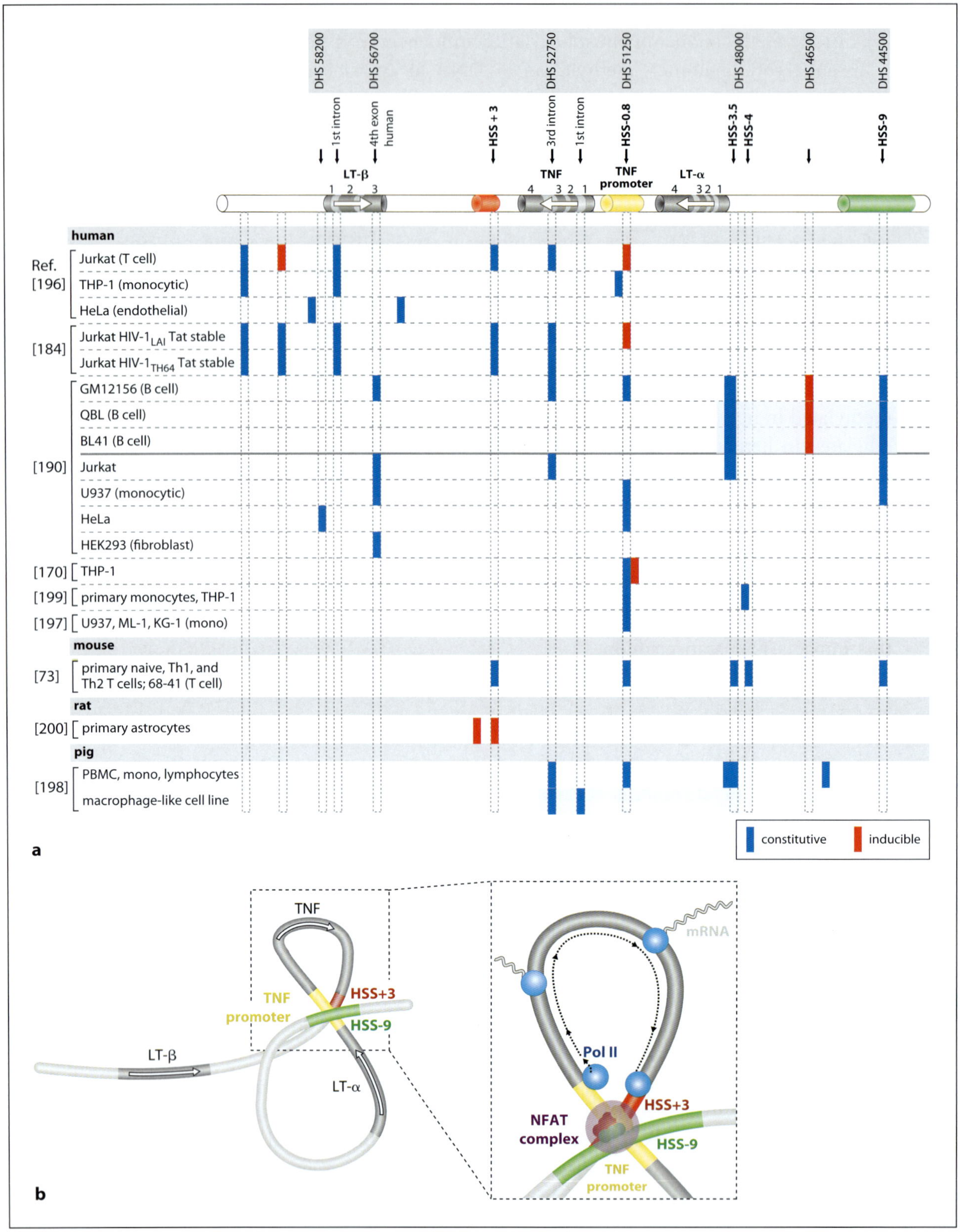
DHS 58200
DHS 56700
DHS 52750
DHS 51250
DHS 48000
DHS 46500
DHS 44500
1st intron
4th exon human
HSS + 3
3rd intron
1st intron
HSS-0.8
HSS-3.5
HSS-4
HSS-9
LT-β
TNF
TNF promoter
LT-α
human
Ref. [196]
Jurkat (T cell)
THP-1 (monocytic)
HeLa (endothelial)
[184]
Jurkat HIV-1LAI Tat stable
Jurkat HIV-1TH64 Tat stable
[190]
GM12156 (B cell)
QBL (B cell)
BL41 (B cell)
Jurkat
U937 (monocytic)
HeLa
HEK293 (fibroblast)
[170] THP-1
[199] primary monocytes, THP-1
[197] U937, ML-1, KG-1 (mono)
mouse
[73] primary naive, Th1, and Th2 T cells; 68-41 (T cell)
rat
[200] primary astrocytes
pig
[198]
PBMC, mono, lymphocytes
macrophage-like cell line
constitutive
inducible
a
TNF
TNF promoter
HSS+3
HSS-9
LT-β
LT-α
mRNA
Pol II
NFAT complex
HSS+3
HSS-9
TNF promoter
b

the TNF and LT-α promoters in THP-1 cells and human monocytes stimulated by superantigen and LPS, respectively [199]. Two TNF-inducible HSSs were identified ~3 kb downstream of the start site of TNF transcription in rat astrocytes; the stronger HSS bound NF-κB p50/p65 and was linked to TNF-induced TNF gene activation [200] (fig. 3a). In Jurkat T cells, a constitutive HSS in TNF intron 3 was functionally linked to T cell-specific TNF transcription in response to PMA/ionomycin [196]. Activation-dependent HSSs were also found in the TNF promoter and ~8 kb downstream of the start site of TNF transcription in Jurkat cells, with constitutive HSSs in the TNF 3′ UTR, the first intron of LT-β, and upstream of LT-β, while in unstimulated and MTb- or LPS-stimulated THP-1 cells, HSSs were present near the start site of TNF transcription, at the first LT-β intron, and upstream of LT-β (fig. 3a). By contrast, in non-TNF-expressing HeLa cells, constitutive HSSs were only found at the LT-β 5′ and 3′ UTRs (fig. 3a) [196]. In THP-1 cells, constitutive and LPS-inducible HSSs were also reported within the proximal and distal TNF promoter, respectively [170] (fig. 3a). Thus, cell type and stimulus appear to influence accessibility of the TNF locus to transcription factors and the general transcription machinery.

Another approach to epigenetic regulation at the TNF locus involves viral protein products. For example, the TNF promoter HSS was present upon PMA/PHA induction in Jurkat T cells stably expressing Tat protein derived from an HIV-1 subtype that promoted TNF transcription (HIV-1_{LAI} Tat) but not in cells expressing a Tat protein from a different HIV-1 subtype that inhibited p/CAF recruitment to the TNF promoter and suppressed TNF transcription (HIV-1_{TH64} Tat). The latter cells instead displayed increased cleavage at the HSS in the TNF 3′ UTR [184]. Thus, coinfection of T cells with HIV-1 and subsequent production of Tat proteins from different HIV subtypes results in epigenetic modification of the TNF chromatin environment, consistent with the important role chromatin remodeling has in TNF gene expression.

In primary naïve, Th1, and Th2 murine T cells and the murine T cell line 68-41, constitutive HSSs were observed at the TNF promoter (HSS-0.8), ~3 kb downstream of the TNF start site of transcription near the 3′ UTR (HSS+3), the LT-α promoter (HSS-3.5 and HSS-4), and ~9 kb upstream of the TNF gene and ~5 kb upstream of the LT-α gene (HSS-9; fig. 3a). HSS+3 and HSS-9 bind NFATp and exhibit increased acetylation of associated histone H3 and H4 upon stimulation with anti-CD3/CD28, and function as NFAT-dependent enhancers of TNF transcription [73]. Most recently,

Fig. 3. Cell type- and stimulus-specific chromatin organization of the TNF/LT locus. **a** Diagram of the genes (modeled after the murine locus, but note that human LT-β has one more exon than murine LT-β) and approximate positions of constitutive and inducible HSSs identified in the TNF/LT locus, with nomenclature from Tsytsykova et al. [73] (HSS) and Taylor et al. [190] (DHS) noted above. Shaded boxes indicate subsets of sites assayed within one study. **b** Model of the higher-order conformation of the TNF/LT locus in activated T cells. Activation-dependent intrachromosomal interactions would place NFAT-containing promoter/enhancer complexes into close proximity (TNF promoter-HSS+3, TNF promoter-HSS-9, and HSS+3-HSS-9) and circularize the TNF gene (TNF promoter-HSS+3) to facilitate reinitiation by the general transcription machinery. Adapted from Tsytsykova et al. [73].

a comprehensive DNase hypersensitivity profile of the TNF/LT locus was performed in a panel of human cell lines: the B cell line GM12156, Jurkat, U937, HeLa, and the embryonic kidney cell line HEK293T [190]. This confirmed the existence of HSS-9 and HSSs at the TNF promoter, LT-α promoter, and TNF third intron and identified a number of novel HSSs (fig. 3a), mainly in the region 10–12 kb upstream of LT-α [190]. In summary, consistent with a major impact upon regulation of TNF gene transcription, chromatin remodeling across the TNF/LT locus exposes highly conserved enhancer and promoter regions to transcription factors and other regulatory proteins in a cell type- and stimulus-specific fashion.

Higher-Order Intrachromosomal Interactions at the TNF/Lymphotoxin Locus

Yet another level of regulation that can impact gene transcription at native loci is provided by long-range intra- and interchromosomal interactions [173, 201–203]. Higher-order chromatin configurations that result from these interactions can bring widely separated gene regulatory regions into close proximity with gene promoters and physically organize gene loci into regions of high local concentrations of transcription factors, or 'transcription factories' [173, 202, 203]. These long-range interactions can be examined using chromatin conformation capture (3C) assays, which combine ChIP with ligation of DNA fragments that lie in close proximity in the native chromatin context [202, 203]. Using 3C assays, it was shown that upon activation of T cells, intrachromosomal interactions occur between the TNF promoter and HSS+3, between the TNF promoter and HSS-9, and between HSS+3 and HSS-9 [73]. These interactions would thus bring together distal enhancers and the TNF promoter, increasing the local concentration of enhancer complexes containing NFATp (fig. 3b). Interactions between the TNF promoter and HSS+3 would also circularize the TNF gene and thus facilitate the recycling of the general transcription machinery (fig. 3b), consistent with the rapid and early expression of high levels of TNF mRNA in T cells. This promotion of transcription through looping was previously proposed for yeast genes [204], but TNF is the first mammalian gene found to undergo activation-dependent looping. Given that certain transcription factors have been shown to be required for establishing higher-order intrachromosomal interactions [205–210], NFATp may be required for the interactions between the TNF promoter and distal enhancers at the TNF/LT locus [73, 211].

Higher-order configuration of gene loci can sequester certain genes into inactive chromatin regions [209, 212]. Thus, in addition to promoting transcription of TNF, intrachromosomal interactions at the TNF/LT locus may contribute to the distinct regulation of TNF, LT-α, and LT-β. For example, interactions between the TNF promoter and HSS-9 would place the LT-α gene in a loop that would sequester the transcriptional start site of the gene from enhancer complexes clustered at the TNF promoter (fig. 3b). Notably, the arrangement and orientation of the three genes has been preserved in mammalian evolution and is conserved between placentals and marsupials [31, 213]. It remains to be determined whether the TNF gene moves into an area where RNA polymerase II and actively transcribing genes are colocalized, that is, a bona fide transcription factory [202, 203], although it is interesting to note that both LPS stimulation and monocyte differentiation drives the TNF locus from heterochromatin to euchromatin, which correlates with the cells' ability to produce TNF [186]. Given the cell type-specific nature of HSSs in the TNF/LT locus, it will be fascinating to determine whether higher-order intra- and interchromosomal interactions involving the TNF/LT locus are also cell type- and stimulus-specific.

Conclusion

A multitude of potential input signals in the cell lead to the specific output of initiation of TNF gene transcription. The TNF gene thus presents a paradigm of cell type- and stimulus-specific gene regulation. This is reflected in the assembly of distinct enhanceosomes at the proximal TNF promoter and in the pattern of chromosomal organization of the TNF/LT locus. The functional importance of the TNF promoter and enhancer regions within the locus is underscored by the evolutionary conservation of sequences associated with transcription factor binding, epigenetic modifications, and higher-order chromosomal interactions. We have described the critical roles of NFAT, enhanceosome formation, coactivator recruitment, epigenetic modifications, chromatin remodeling, and intrachromosomal interactions in TNF gene regulation. Further characterization of the factors and elements involved in cell type- and stimulus-specific regulation of TNF gene transcription will continue to elucidate basic issues in eukaryotic gene regulation while defining potential therapeutic targets for precise manipulation of TNF expression in disease states.

Acknowledgements

We thank Renate Hellmiss for graphic artwork. Supported by NIH grant R01GM076685 to A.E.G.

References

1 Aggarwal BB, Samanta A, Feldmann M: TNFα; in Oppenheim J, Feldmann M (eds): Cytokine Reference. Orlando, Academic Press, 2000, pp 413–434.

2 Grivennikov SI, Kuprash DV, Liu ZG, Nedospasov SA: Intracellular signals and events activated by cytokines of the tumor necrosis factor superfamily: from simple paradigms to complex mechanisms. Int Rev Cytol 2006;252:129–161.

3 Beutler B, Cerami A: Cachectin and tumor necrosis factor as two sides of the same biological coin. Nature 1986;320:584.

4 Cuturi MC, Murphy M, Costa-Giomi MP, Weinmann R, Perussia B, Trinchieri G: Independent regulation of tumor necrosis factor and lymphotoxin production by human peripheral blood lymphocytes. J Exp Med 1987;165:1581–1594.

5 Steffen M, Ottmann OG, Moore MA: Simultaneous production of tumor necrosis factor-α and lymphotoxin by normal T cells after induction with IL-2 and anti-T3. J Immunol 1988;140:2621–2624.

6 Sung SS, Bjorndahl JM, Wang CY, Kao HT, Fu SM: Production of tumor necrosis factor/cachectin by human T cell lines and peripheral blood T lymphocytes stimulated by phorbol myristate acetate and anti-CD3 antibody. J Exp Med 1988;167:937–953.

7 Sung SS, Jung LK, Walters JA, Chen W, Wang CY, Fu SM: Production of tumor necrosis factor/cachectin by human B cell lines and tonsillar B cells. J Exp Med 1988;168:1539–1551.

8 Niitsu Y, Watanabe N, Neda H, Yamauchi N, Maeda M, Sone H, Kuriyama H: Induction of synthesis of tumor necrosis factor in human and murine cell lines by exogenous recombinant human tumor necrosis factor. Cancer Res 1988;48:5407–5410.

9 Goldfeld AE, Maniatis T: Coordinate viral induction of tumor necrosis factor α and interferon β in human B cells and monocytes. Proc Natl Acad Sci USA 1989;86:1490–1494.

10 Gordon JR, Galli SJ: Mast cells as a source of both preformed and immunologically inducible TNF-α/cachectin. Nature 1990;346:274–276.

11 Caux C, Massacrier C, Vanbervliet B, Dubois B, Van Kooten C, Durand I, Banchereau J: Activation of human dendritic cells through CD40 cross-linking. J Exp Med 1994;180:1263–1272.

12 Akira S, Takeda K: Toll-like receptor signalling. Nat Rev Immunol 2004;4:499–511.

13 Goldfeld AE, Flemington EK, Boussiotis VA, Theodos CM, Titus RG, Strominger JL, Speck SH: Transcription of the tumor necrosis factor α gene is rapidly induced by anti-immunoglobulin and blocked by cyclosporin A and FK506 in human B cells. Proc Natl Acad Sci USA 1992;89:12198–12201.

14 Goldfeld AE, Strominger JL, Doyle C: Human tumor necrosis factor α gene regulation in phorbol ester stimulated T and B cell lines. J Exp Med 1991; 174:73–81.

15 Goldfeld AE, McCaffrey PG, Strominger JL, Rao A: Identification of a novel cyclosporin-sensitive element in the human tumor necrosis factor α gene promoter. J Exp Med 1993;178:1365–1379.

16 Diehn M, Alizadeh AA, Rando OJ, Liu CL, Stankunas K, Botstein D, Crabtree GR, Brown PO: Genomic expression programs and the integration of the CD28 costimulatory signal in T cell activation. Proc Natl Acad Sci USA 2002;99:11796–11801.

17 Aggarwal BB, Shishodia S, Takada Y, Jackson-Bernitsas D, Ahn KS, Sethi G, Ichikawa H: TNF blockade: an inflammatory issue. Ernst Schering Res Found Workshop 2006:161–186.

18 Sethi G, Sung B, Aggarwal BB: TNF: A master switch for inflammation to cancer. Front Biosci 2008;13:5094–5107.

19 Lou J, Lucas R, Grau GE: Pathogenesis of cerebral malaria: recent experimental data and possible applications for humans. Clin Microbiol Rev 2001; 14:810–820.

20 Jacobs M, Samarina A, Grivennikov S, Botha T, Allie N, Fremond C, Togbe D, Vasseur V, Rose S, Erard F, Monteiro A, Quesniaux V, Ryffel B: Reactivation of tuberculosis by tumor necrosis factor neutralization. Eur Cytokine Netw 2007;18:5–13.

21 Falvo JV, Uglialoro AM, Brinkman BM, Merika M, Parekh BS, Tsai EY, King HC, Morielli AD, Peralta EG, Maniatis T, Thanos D, Goldfeld AE: Stimulus-specific assembly of enhancer complexes on the tumor necrosis factor alpha gene promoter. Mol Cell Biol 2000;20:2239–2247.

22 Tsai EY, Falvo JV, Tsytsykova AV, Barczak AK, Reimold AM, Glimcher LH, Fenton MJ, Gordon DC, Dunn IF, Goldfeld AE: A lipopolysaccharide-specific enhancer complex involving Ets, Elk-1, Sp1, and CREB binding protein and p300 is recruited to the tumor necrosis factor alpha promoter in vivo. Mol Cell Biol 2000;20:6084–6094.

23 Tsytsykova AV, Goldfeld AE: Inducer-specific enhanceosome formation controls tumor necrosis factor alpha gene expression in T lymphocytes. Mol Cell Biol 2002;22:2620–2631.

24 Barthel R, Tsytsykova AV, Barczak AK, Tsai EY, Dascher CC, Brenner MB, Goldfeld AE: Regulation of tumor necrosis factor alpha gene expression by mycobacteria involves the assembly of a unique enhanceosome dependent on the coactivator proteins CBP/p300. Mol Cell Biol 2003;23:526–533.

25 Carey M: The enhanceosome and transcriptional synergy. Cell 1998;92:5–8.

26 Merika M, Thanos D: Enhanceosomes. Curr Opin Genet Dev 2001;11:205–208.

27 Falvo JV, Brinkman BMN, Tsytsykova AV, Tsai EY, Yao T-P, Kung AL, Goldfeld AE: A stimulus-specific role for CREB-binding protein (CBP) in T cell receptor-activated tumor necrosis factor α gene expression. Proc Natl Acad Sci USA 2000;97:3925–3929.

28 Shakhov AN, Collart MA, Vassalli P, Nedospasov SA, Jongeneel CV: κB-type enhancers are involved in lipopolysaccharide-mediated transcriptional activation of the tumor necrosis factor α gene in primary macrophages. J Exp Med 1990;171:35–47.

29 Kuprash DV, Udalova IA, Turetskaya RL, Kwiatkowski D, Rice NR, Nedospasov SA: Similarities and differences between human and murine TNF promoters in their response to lipopolysaccharide. J Immunol 1999;162:4045–4052.

30 Goldfeld AE, Leung JY, Sawyer SA, Hartl DL: Post-genomics and the neutral theory: variation and conservation in the tumor necrosis factor-α promoter. Gene 2000;261:19–25.

31 Cross JG, Harrison GA, Coggill P, Sims S, Beck S, Deakin JE, Graves JA: Analysis of the genomic region containing the tammar wallaby *(Macropus eugenii)* orthologues of MHC class III genes. Cytogenet Genome Res 2005;111:110–117.

32 Leung JY, McKenzie FE, Uglialoro AM, Flores-Villanueva PO, Sorkin BC, Yunis EJ, Hartl DL, Goldfeld AE: Identification of phylogenetic footprints in primate tumor necrosis factor-α promoters. Proc Natl Acad Sci USA 2000;97:6614–6618.

33 Baena A, Mootnick AR, Falvo JV, Tsytskova AV, Ligeiro F, Diop OM, Brieva C, Gagneux P, O'Brien SJ, Ryder OA, Goldfeld AE: Primate TNF promoters reveal markers of phylogeny and evolution of innate immunity. PLoS One 2007;2:e621.

34 Tsai EY, Jain J, Pesavento PA, Rao A, Goldfeld AE: Tumor necrosis factor alpha gene regulation in activated T cells involves ATF-2/Jun and NFATp. Mol Cell Biol 1996;16:459–467.

35 Tsai EY, Yie J, Thanos D, Goldfeld AE: Cell-type-specific regulation of the human tumor necrosis factor alpha gene in B cells and T cells by NFATp and ATF-2/JUN. Mol Cell Biol 1996;16:5232–5244.

36 Tsytsykova AV, Goldfeld AE: Nuclear factor of activated T cells transcription factor NFATp controls superantigen-induced lethal shock. J Exp Med 2000; 192:581–586.

37 Goldfeld AE, Doyle C, Maniatis T: Human tumor necrosis factor α gene regulation by virus and lipopolysaccharide. Proc Natl Acad Sci USA 1990; 87:9769–9773.

38 Trede NS, Tsytsykova AV, Chatila T, Goldfeld AE, Geha RS: Transcriptional activation of the human TNF-α promoter by superantigen in human monocytic cells: role of NF-κB. J Immunol 1995;155:902–908.

39 Yao J, Mackman N, Edgington TS, Fan ST: Lipopolysaccharide induction of the tumor necrosis factor-α promoter in human monocytic cells. Regulation by Egr-1, c-Jun, and NF-κB transcription factors. J Biol Chem 1997;272:17795–17801.

40 Haudek SB, Natmeβnig BE, Redl H, Schlag G, Giroir BP: Genetic sequences and transcriptional regulation of the *TNFA* promoter: comparison of human and baboon. Immunogenetics 1998;48:202–207.

41 Means TK, Pavlovich RP, Roca D, Vermeulen MW, Fenton MJ: Activation of TNF-α transcription utilizes distinct MAP kinase pathways in different macrophage populations. J Leukoc Biol 2000;67:885–893.

42 O'Donnell PM, Taffet SM: The proximal promoter region is essential for lipopolysaccharide induction and cyclic AMP inhibition of mouse tumor necrosis factor-α. J Interferon Cytokine Res 2002;22:539–548.

43 Leitman DC, Mackow ER, Williams T, Baxter JD, West BL: The core promoter region of the tumor necrosis factor α gene confers phorbol ester responsiveness to upstream transcriptional activators. Mol Cell Biol 1992;12:1352–1356.

44 Brinkman BM, Telliez JB, Schievella AR, Lin LL, Goldfeld AE: Engagement of tumor necrosis factor (TNF) receptor 1 leads to ATF-2- and p38 mitogen-activated protein kinase-dependent TNF-α gene expression. J Biol Chem 1999;274:30882–30886.

45 Esensten JH, Tsytsykova AV, Lopez-Rodriguez C, Ligeiro FA, Rao A, Goldfeld AE: NFAT5 binds to the TNF promoter distinctly from NFATp, c, 3 and 4, and activates TNF transcription during hypertonic stress alone. Nucleic Acids Res 2005;33:3845–3854.

46 Economou JS, Rhoades K, Essner R, McBride WH, Gasson JC, Morton DL: Genetic analysis of the human tumor necrosis factor α/cachectin promoter region in a macrophage cell line. J Exp Med 1989; 170:321–326.

47 Rhoades KL, Golub SH, Economou JS: The regulation of the human tumor necrosis factor α promoter region in macrophage, T cell, and B cell lines. J Biol Chem 1992;267:22102–22107.

48 McCaffrey PG, Goldfeld AE, Rao A: The role of NFATp in cyclosporin A-sensitive tumor necrosis factor-α gene transcription. J Biol Chem 1994;269: 30445–30450.

49 Krämer B, Wiegmann K, Krönke M: Regulation of the human TNF promoter by the transcription factor Ets. J Biol Chem 1995;270:6577–6583.

50 Krämer B, Meichle A, Hensel G, Charnay P, Krönke M: Characterization of an Krox-24/Egr-1-responsive element in the human tumor necrosis factor promoter. Biochim Biophys Acta 1994;1219:413–421.

51 Leitman DC, Ribeiro RC, Mackow ER, Baxter JD, West BL: Identification of a tumor necrosis factor-responsive element in the tumor necrosis factor α gene. J Biol Chem 1991;266:9343–9346.

52 Falvo JV, Lin CH, Tsytsykova AV, Hwang PK, Thanos D, Goldfeld AE, Maniatis T: A dimer-specific function of the transcription factor NFATp. Proc Natl Acad Sci USA 2008.

53 Newell CL, Deisseroth AB, Lopez-Berestein G: Interaction of nuclear proteins with an AP-1/CRE-like promoter sequence in the human TNF-α gene. J Leukoc Biol 1994;56:27–35.

54 Diaz B, Lopez-Berestein G: A distinct element involved in lipopolysaccharide activation of the tumor necrosis factor-α promoter in monocytes. J Interferon Cytokine Res 2000;20:741–748.

55 Steer JH, Kroeger KM, Abraham LJ, Joyce DA: Glucocorticoids suppress tumor necrosis factor-α expression by human monocytic THP-1 cells by suppressing transactivation through adjacent NF-κB and c-Jun-activating transcription factor-2 binding sites in the promoter. J Biol Chem 2000;275:18432–18440.

56 Sato H, Watanabe A, Tanaka T, Koitabashi N, Arai M, Kurabayashi M, Yokoyama T: Regulation of the human tumor necrosis factor-α promoter by angiotensin II and lipopolysaccharide in cardiac fibroblasts: different *cis*-acting promoter sequences and transcriptional factors. J Mol Cell Cardiol 2003; 35:1197–1205.

57 Hayakawa J, Mittal S, Wang Y, Korkmaz KS, Adamson E, English C, Ohmichi M, McClelland M, Mercola D: Identification of promoters bound by c-Jun/ATF2 during rapid large-scale gene activation following genotoxic stress. Mol Cell 2004;16:521–535.

58 Hogan PG, Chen L, Nardone J, Rao A: Transcriptional regulation by calcium, calcineurin, and NFAT. Genes Dev 2003;17:2205–2232.

59 Macian F: NFAT proteins: Key regulators of T-cell development and function. Nat Rev Immunol 2005; 5:472–484.

60 Aramburu J, Drews Elger K, Estrada-Gelonch A, Minguillón J, Morancho B, Santiago V, López-Rodríguez C: Regulation of the hypertonic stress response and other cellular functions by the Rel-like transcription factor NFAT5. Biochem Pharmacol 2006;72:1597–1604.

61 Wu H, Peisley A, Graef IA, Crabtree GR: NFAT signaling and the invention of vertebrates. Trends Cell Biol 2007;17:251–260.

62 Ullman KS, Northrop JP, Verweij CL, Crabtree GR: Transmission of signals from the T lymphocyte antigen receptor to the genes responsible for cell proliferation and immune function: the missing link. Annu Rev Immunol 1990;8:421–452.

63 McCaffrey PG, Luo C, Kerppola TK, Jain J, Badalian TM, Ho AM, Burgeon E, Lane WS, Lambert JN, Curran T, Verdine GL, Rao A, Hogan PG: Isolation of the cyclosporin-sensitive T cell transcription factor NFATp. Science 1993;262:750–754.

64 Goldfeld AE, Tsai E, Kincaid R, Belshaw PJ, Schreiber SL, Strominger JL, Rao A: Calcineurin mediates human tumor necrosis factor α gene induction in stimulated T and B cells. J Exp Med 1994;180:763–768.

65 Prieschl EE, Gouilleux-Gruart V, Walker C, Harrer NE, Baumruker T: A nuclear factor of activated T cell-like transcription factor in mast cells is involved in IL-5 gene regulation after IgE plus antigen stimulation. J Immunol 1995;154:6112–6119.

66 Klein M, Klein-Hessling S, Palmetshofer A, Serfling E, Tertilt C, Bopp T, Heib V, Becker M, Taube C, Schild H, Schmitt E, Stassen M: Specific and redundant roles for NFAT transcription factors in the expression of mast cell-derived cytokines. J Immunol 2006;177:6667–6674.

67 Kaminuma O, Kitamura F, Kitamura N, Hiroi T, Miyoshi H, Miyawaki A, Miyatake S: Differential contribution of NFATc2 and NFATc1 to TNF-α gene expression in T cells. J Immunol 2008;180:319–326.

68 Song L, Li J, Ye J, Yu G, Ding J, Zhang D, Ouyang W, Dong Z, Kim SO, Huang C: p85α acts as a novel signal transducer for mediation of cellular apoptotic response to UV radiation. Mol Cell Biol 2007;27: 2713–2731.

69 Ke Q, Li J, Ding J, Ding M, Wang L, Liu B, Costa M, Huang C: Essential role of ROS-mediated NFAT activation in TNF-α induction by crystalline silica exposure. Am J Physiol Lung Cell Mol Physiol 2006; 291:L257–264.

70 Ouyang W, Hu Y, Li J, Ding M, Lu Y, Zhang D, Yan Y, Song L, Qu Q, Desai D, Amin S, Huang C: Direct evidence for the critical role of NFAT3 in benzo[a] pyrene diol-epoxide-induced cell transformation through mediation of inflammatory cytokine TNF induction in mouse epidermal Cl41 cells. Carcinogenesis 2007;28:2218–2226.

71 López-Rodriguez C, Aramburu J, Jin L, Rakeman AS, Michino M, Rao A: Bridging the NFAT and NF-κB families: NFAT5 dimerization regulates cytokine gene transcription in response to osmotic stress. Immunity 2001;15:47–58.

72 Hodge MR, Ranger AM, Charles de la Brousse F, Hoey T, Grusby MJ, Glimcher LH: Hyperproliferation and dysregulation of IL-4 expression in NF-ATp-deficient mice. Immunity 1996;4:397–405.

73 Tsytsykova AV, Rajsbaum R, Falvo JV, Ligeiro F, Neely SR, Goldfeld AE: Activation-dependent intrachromosomal interactions formed by the *TNF* gene promoter and two distal enhancers. Proc Natl Acad Sci USA 2007;104:16850–16855.

74 Ranger AM, Hodge MR, Gravallese EM, Oukka M, Davidson L, Alt FW, de la Brousse FC, Hoey T, Grusby M, Glimcher LH: Delayed lymphoid repopulation with defects in IL-4-driven responses produced by inactivation of NF-ATc. Immunity 1998; 8:125–134.

75 Yoshida H, Nishina H, Takimoto H, Marengere LE, Wakeham AC, Bouchard D, Kong YY, Ohteki T, Shahinian A, Bachmann M, Ohashi PS, Penninger JM, Crabtree GR, Mak TW: The transcription factor NF-ATc1 regulates lymphocyte proliferation and Th2 cytokine production. Immunity 1998;8:115–124.

76 Ranger AM, Oukka M, Rengarajan J, Glimcher LH: Inhibitory function of two NFAT family members in lymphoid homeostasis and Th2 development. Immunity 1998;9:627–635.

77 Peng SL, Gerth AJ, Ranger AM, Glimcher LH: NFATc1 and NFATc2 together control both T and B cell activation and differentiation. Immunity 2001; 14:13–20.

78 Pan M, Winslow MM, Chen L, Kuo A, Felsher D, Crabtree GR: Enhanced NFATc1 nuclear occupancy causes T cell activation independent of CD28 costimulation. J Immunol 2007;178:4315–4321.

79 Chen J, Amasaki Y, Kamogawa Y, Nagoya M, Arai N, Arai K, Miyatake S: Role of NFATx (NFAT4/NFATc3) in expression of immunoregulatory genes in murine peripheral $CD4^+$ T cells. J Immunol 2003; 170:3109–3117.

80 Feske S, Muller JM, Graf D, Kroczek RA, Drager R, Niemeyer C, Baeuerle PA, Peter HH, Schlesier M: Severe combined immunodeficiency due to defective binding of the nuclear factor of activated T cells in T lymphocytes of two male siblings. Eur J Immunol 1996;26:2119–2126.

81 Trifari S, Sitia G, Aiuti A, Scaramuzza S, Marangoni F, Guidotti LG, Martino S, Saracco P, Notarangelo LD, Roncarolo MG, Dupre L: Defective Th1 cytokine gene transcription in CD4+ and CD8+ T cells from Wiskott-Aldrich syndrome patients. J Immunol 2006;177:7451–7461.

82 Perkins ND: Integrating cell-signalling pathways with NF-κB and IKK function. Nat Rev Mol Cell Biol 2007;8:49–62.

83 Hayden MS, Ghosh S: Shared principles in NF-κB signaling. Cell 2008;132:344–362.

84 Vallabhapurapu S, Karin M: Regulation and function of NF-κB transcription factors in the immune system. Annu Rev Immunol 2009;27:693–733.

85 Collart MA, Baeuerle P, Vassalli P: Regulation of tumor necrosis factor alpha transcription in macrophages: involvement of four κB-like motifs and of constitutive and inducible forms of NF-κB. Mol Cell Biol 1990;10:1498–1506.

86 Drouet C, Shakhov AN, Jongeneel CV: Enhancers and transcription factors controlling the inducibility of the tumor necrosis factor-α promoter in primary macrophages. J Immunol 1991;147: 1694–1700.

87 Udalova IA, Knight JC, Vidal V, Nedospasov SA, Kwiatkowski D: Complex NF-κB interactions at the distal tumor necrosis factor promoter region in human monocytes. J Biol Chem 1998;273:21178–21186.

88 Ziegler-Heitbrock HW, Sternsdorf T, Liese J, Belohradsky B, Weber C, Wedel A, Schreck R, Bauerle P, Strobel M: Pyrrolidine dithiocarbamate inhibits NF-κB mobilization and TNF production in human monocytes. J Immunol 1993;151:6986–6993.

89 Foxwell B, Browne K, Bondeson J, Clarke C, de Martin R, Brennan F, Feldmann M: Efficient adenoviral infection with IκBα reveals that macrophage tumor necrosis factor α production in rheumatoid arthritis is NF-κB dependent. Proc Natl Acad Sci USA 1998;95:8211–8215.

90 O'Connell MA, Bennett BL, Mercurio F, Manning AM, Mackman N: Role of IKK1 and IKK2 in lipopolysaccharide signaling in human monocytic cells. J Biol Chem 1998;273:30410–30404.

91 Bondeson J, Browne KA, Brennan FM, Foxwell BMJ,Feldmann M: Selective regulation of cytokine induction by adenoviral gene transfer of IκBα into human macrophages: lipopolysaccharide-induced, but not zymosan-induced proinflammatory cytokines are inhibited, but IL-10 is nuclear factor-κB independent. J Immunol 1999;162:2939–2945.

92 Swantek JL, Christerson L, Cobb MH: Lipopolysaccharide-induced tumor necrosis factor-α promoter activity is inhibitor of nuclear factor-κB kinase-dependent. J Biol Chem 1999;274: 11667–11671.

93 Moore F, Buonocore S, Aksoy E, Ouled-Haddou N, Goriely S, Lazarova E, Paulart F, Heirman C, Vaeremans E, Thielemans K, Goldman M, Flamand V: An alternative pathway of NF-κB activation results in maturation and T cell priming activity of dendritic cells overexpressing a mutated IκBα. J Immunol 2007;178:1301–1311.

94 Hall G, Singh IS, Hester L, Hasday JD, Rogers TB: Inhibitor-κB kinase-β regulates LPS-induced TNF-α production in cardiac myocytes through modulation of NF-κB p65 subunit phosphorylation. Am J Physiol Heart Circ Physiol 2005;289:H2103–H2111.

95 Lim CA, Yao F, Wong JJ, George J, Xu H, Chiu KP, Sung WK, Lipovich L, Vega VB, Chen J, Shahab A, Zhao XD, Hibberd M, Wei CL, Lim B, Ng HH, Ruan Y, Chin KC: Genome-wide mapping of RELA(p65) binding identifies E2F1 as a transcriptional activator recruited by NF-κB upon TLR4 activation. Mol Cell 2007;27:622–635.

96 Maniatis T, Falvo JV, Kim TH, Kim TK, Lin CH, Parekh BS, Wathelet MG: Structure and function of the interferon-β enhanceosome. Cold Spring Harbor Symp Quant Biol 1998;63:609–620.

97 Liu H, Sidiropoulos P, Song G, Pagliari LJ, Birrer MJ, Stein B, Anrather J, Pope RM: TNF-α gene expression in macrophages: regulation by NF-κB is independent of c-Jun or C/EBPβ. J Immunol 2000; 164:4277–4285.
98 Weih F, Warr G, Yang H, Bravo R: Multifocal defects in immune responses in RelB-deficient mice. J Immunol 1997;158:5211–5218.
99 Ye J, Wang L, Zhang X, Tantishaiyakul V, Rojanasakul Y: Inhibition of TNF-α gene expression and bioactivity by site-specific transcription factor-binding oligonucleotides. Am J Physiol Lung Cell Mol Physiol 2003;284:L386–394.
100 Joyce DA, Gimblett G, Steer JH: Targets of glucocorticoid action on TNF-α release by macrophages. Inflamm Res 2001;50:337–340.
101 Xu Z, Dziarski R, Wang Q, Swartz K, Sakamoto KM, Gupta D: Bacterial peptidoglycan-induced *tnf*-α transcription is mediated through the transcription factors Egr-1, Elk-1, and NF-κB. J Immunol 2001; 167:6975–6982.
102 Tzagarakis-Foster C, Geleziunas R, Lomri A, An J, Leitman DC: Estradiol represses human T-cell leukemia virus type 1 Tax activation of tumor necrosis factor-α gene transcription. J Biol Chem 2002;277: 44772–44777.
103 Tsytsykova AV, Falvo JV, Schmidt-Supprian M, Courtois G, Thanos D, Goldfeld AE: Post-induction, stimulus-specific regulation of tumor necrosis factor mRNA expression. J Biol Chem 2007;282:11629–11638.
104 Kunsch C, Ruben SM, Rosen CA: Selection of optimal κB/Rel DNA-binding motifs: Interaction of both subunits of NF-κB with DNA is required for transcriptional activation. Mol Cell Biol 1992;12: 4412–4421.
105 Bonizzi G, Karin M: The two NF-κB activation pathways and their role in innate and adaptive immunity. Trends Immunol 2004;25:280–288.
106 Natoli G, Saccani S, Bosisio D, Marazzi I: Interactions of NF-κB with chromatin: the art of being at the right place at the right time. Nat Immunol 2005; 6:439–445.
107 Natoli G: Tuning up inflammation: how DNA sequence and chromatin organization control the induction of inflammatory genes by NF-κB. FEBS Lett 2006;580:2843–2849.
108 Orlando V: Mapping chromosomal proteins in vivo by formaldehyde-crosslinked-chromatin immunoprecipitation. Trends Biochem Sci 2000;25:99–104.
109 Litt MD, Simpson M, Gaszner M, Allis CD, Felsenfeld G: Correlation between histone lysine methylation and developmental changes at the chicken β-globin locus. Science 2001;293:2453–2455.
110 Martone R, Euskirchen G, Bertone P, Hartman S, Royce TE, Luscombe NM, Rinn JL, Nelson FK, Miller P, Gerstein M, Weissman S, Snyder M: Distribution of NF-κB-binding sites across human chromosome 22. Proc Natl Acad Sci USA 2003;100: 12247–12252.
111 Bonizzi G, Bebien M, Otero DC, Johnson-Vroom KE, Cao Y, Vu D, Jegga AG, Aronow BJ, Ghosh G, Rickert RC, Karin M: Activation of IKKα target genes depends on recognition of specific κB binding sites by RelB:P52 dimers. EMBO J 2004;23:4202–4210.
112 Taggart CC, Cryan SA, Weldon S, Gibbons A, Greene CM, Kelly E, Low TB, O'Neill S J, McElvaney NG: Secretory leucoprotease inhibitor binds to NF-κB binding sites in monocytes and inhibits p65 binding. J Exp Med 2005;202:1659–1668.
113 Ishinaga H, Jono H, Lim JH, Kweon SM, Xu H, Ha UH, Xu H, Koga T, Yan C, Feng XH, Chen LF, Li JD: TGF-β induces p65 acetylation to enhance bacteria-induced NF-κB activation. EMBO J 2007;26:1150–1162.
114 Carmody RJ, Ruan Q, Palmer S, Hilliard B, Chen YH: Negative regulation of toll-like receptor signaling by NF-κB p50 ubiquitination blockade. Science 2007;317:675–678.
115 El Gazzar MA, El Mezayen R, Nicolls MR, Dreskin SC: Thymoquinone attenuates proinflammatory responses in lipopolysaccharide-activated mast cells by modulating NF-kappaB nuclear transactivation. Biochim Biophys Acta 2007;1770:556–564.
116 Li Y, Reddy MA, Miao F, Shanmugam N, Yee JK, Hawkins D, Ren B, Natarajan R: Role of the histone H3 lysine 4 methyltransferase, SET7/9, in the regulation of NF-κB-dependent inflammatory genes. Relevance to diabetes and inflammation. J Biol Chem 2008;283:26771–26781.
117 Buckley JM, Wang JH, Redmond HP: Cellular reprogramming by Gram-positive bacterial components: a review. J Leukoc Biol 2006;80:731–741.
118 Baer M, Dillner A, Schwartz RC, Sedon C, Nedospasov S, Johnson PF: Tumor necrosis factor alpha transcription in macrophages is attenuated by an autocrine factor that preferentially induces NF-κB p50. Mol Cell Biol 1998;18:5678–5689.
119 Frankenberger M, Ziegler-Heitbrock HW: LPS tolerance in monocytes/macrophages: three 3′ cytosins are required in the DNA binding motif for detection of upregulated NF-κB p50 homodimers. Immunobiology 1997;198:81–90.
120 Wedel A, Frankenberger M, Sulski G, Petersmann I, Kuprash D, Nedospasov S, Ziegler-Heitbrock HW: Role of p52 (NF-κB2) in LPS tolerance in a human B cell line. Biol Chem 1999;380:1193–1199.

121 Bohuslav J, Kravchenko VV, Parry GC, Erlich JH, Gerondakis S, Mackman N, Ulevitch RJ: Regulation of an essential innate immune response by the p50 subunit of NF-κB. J Clin Invest 1998;102:1645–1652.
122 Wessells J, Baer M, Young HA, Claudio E, Brown K, Siebenlist U, Johnson PF: BCL-3 and NF-κB p50 attenuate lipopolysaccharide-induced inflammatory responses in macrophages. J Biol Chem 2004;279: 49995–50003.
123 El Gazzar M, Yoza BK, Hu JY, Cousart SL, McCall CE: Epigenetic silencing of tumor necrosis factor α during endotoxin tolerance. J Biol Chem 2007;282: 26857–26864.
124 Pope RM, Leutz A, Ness SA: C/EBPβ regulation of the tumor necrosis factor α gene. J Clin Invest 1994; 94:1449–1455.
125 Wedel A, Sulski G, Ziegler-Heitbrock HW: CCAAT/enhancer binding protein is involved in the expression of the tumour necrosis factor gene in human monocytes. Cytokine 1996;8:335–341.
126 Zagariya A, Mungre S, Lovis R, Birrer M, Ness S, Thimmapaya B, Pope R: Tumor necrosis factor alpha gene regulation: enhancement of C/EBPβ-induced activation by c-Jun. Mol Cell Biol 1998; 18:2815–2824.
127 Tanaka T, Akira S, Yoshida K, Umemoto M, Yoneda Y, Shirafuji N, Fujiwara H, Suematsu S, Yoshida N, Kishimoto T: Targeted disruption of the NF-IL6 gene discloses its essential role in bacteria killing and tumor cytotoxicity by macrophages. Cell 1995; 80:353–361.
128 Gorgoni B, Maritano D, Marthyn P, Righi M, Poli V: C/EBPβ gene inactivation causes both impaired and enhanced gene expression and inverse regulation of IL-12 p40 and p35 mRNAs in macrophages. J Immunol 2002;168:4055–4062.
129 Lu YC, Kim I, Lye E, Shen F, Suzuki N, Suzuki S, Gerondakis S, Akira S, Gaffen SL, Yeh WC, Ohashi PS: Differential role for c-Rel and C/EBPβ/δ in TLR-mediated induction of proinflammatory cytokines. J Immunol 2009;182:7212–7221.
130 Novotny V, Prieschl EE, Csonga R, Fabjani G, Baumruker T: Nrf1 in a complex with fosB, c-jun, junD and ATF2 forms the AP1 component at the TNFα promoter in stimulated mast cells. Nucleic Acids Res 1998;26:5480–5485.
131 Prieschl EE, Novotny V, Csonga R, Jaksche D, Elbe-Burger A, Thumb W, Auer M, Stingl G, Baumruker T: A novel splice variant of the transcription factor Nrf1 interacts with the TNFα promoter and stimulates transcription. Nucleic Acids Res 1998;26:2291–2297.
132 Takagi K, Takagi M, Kanangat S, Warrington KJ, Shigemitsu H, Postlethwaite AE: Modulation of TNF-α gene expression by IFN-γ and pamidronate in murine macrophages: regulation by STAT1-dependent pathways. J Immunol 2005;174:1801–1810.
133 Chappell VL, Le LX, LaGrone L, Mileski WJ: Stat proteins play a role in tumor necrosis factor ααgene expression. Shock 2000;14:400–402; discussion 402–403.
134 Kamezaki K, Shimoda K, Numata A, Matsuda T, Nakayama K, Harada M: The role of Tyk2, Stat1 and Stat4 in LPS-induced endotoxin signals. Int Immunol 2004;16:1173–1179.
135 Myokai F, Takashiba S, Lebo R, Amar S: A novel lipopolysaccharide-induced transcription factor regulating tumor necrosis factor α gene expression: molecular cloning, sequencing, characterization, and chromosomal assignment. Proc Natl Acad Sci USA 1999;96:4518–4523.
136 Bolcato-Bellemin A-L, Mattei M-G, Fenton M, Amar S: Molecular cloning and characterization of mouse LITAF cDNA: role in the regulation of tumor necrosis factor-α (TNF-α) gene expression. J Endotoxin Res 2004;10:15–23.
137 Tang X, Fenton MJ, Amar S: Identification and functional characterization of a novel binding site on TNF-α promoter. Proc Natl Acad Sci USA 2003; 100:4096–4101.
138 Tang X, Marciano DL, Leeman SE, Amar S: LPS induces the interaction of a transcription factor, LPS-induced TNF-α factor, and STAT6(B) with effects on multiple cytokines. Proc Natl Acad Sci USA 2005;102:5132–5137.
139 Tang X, Metzger D, Leeman S, Amar S: LPS-induced TNF-α factor (LITAF)-deficient mice express reduced LPS-induced cytokine: evidence for LITAF-dependent LPS signaling pathways. Proc Natl Acad Sci USA 2006;103:13777–13782.
140 Vila-del Sol V, Punzón C, Fresno M: IFN-γ-induced TNF-α expression is regulated by interferon regulatory factors 1 and 8 in mouse macrophages. J Immunol 2008;181:4461–4470.
141 Zhao XJ, Dong Q, Bindas J, Piganelli JD, Magill A, Reiser J, Kolls JK: TRIF and IRF-3 binding to the TNF promoter results in macrophage TNF dysregulation and steatosis induced by chronic ethanol. J Immunol 2008;181:3049–3056.
142 Covert MW, Leung TH, Gaston JE, Baltimore D: Achieving stability of lipopolysaccharide-induced NF-κB activation. Science 2005;309:1854–1857.
143 Werner SL, Barken D, Hoffmann A: Stimulus specificity of gene expression programs determined by temporal control of IKK activity. Science 2005;309: 1857–1861.
144 Mungall AJ, Palmer SA, Sims SK, et al: The DNA sequence and analysis of human chromosome 6. Nature 2003;425:805–811.
145 Horton R, Wilming L, Rand V, Lovering RC, Bruford EA, Khodiyar VK, Lush MJ, Povey S, Talbot CC Jr, Wright MW, Wain HM, Trowsdale J, Ziegler A, Beck S: Gene map of the extended human MHC. Nat Rev Genet 2004;5:889–899.

146 Allen RD: Polymorphism of the human TNF-α promoter – random variation or functional diversity? Mol Immunol 1999;36:1017–1027.
147 Hajeer AH, Hutchinson IV: TNF-α gene polymorphism: clinical and biological implications. Microsc Res Tech 2000;50:216–228.
148 Hajeer AH, Hutchinson IV: Influence of TNFα gene polymorphisms on TNFα production and disease. Hum Immunol 2001;62:1191–1199.
149 Bayley J-P, Ottenhoff TH, Verweij CL: Is there a future for TNF promoter polymorphisms? Genes Immun 2004;5:315–329.
150 Smith AJ, Humphries SE: Cytokine and cytokine receptor gene polymorphisms and their functionality. Cytokine Growth Factor Rev 2009;20:43–59.
151 Uglialoro AM, Turbay D, Pesavento PA, Delgado JC, McKenzie FE, Gribben JG, Hartl D, Yunis EJ, Goldfeld AE: Identification of three new single nucleotide polymorphisms in the human tumor necrosis factor-α gene promoter. Tissue Antigens 1998;52:359–367.
152 Hollegaard MV, Bidwell JL: Cytokine gene polymorphism in human disease: On-line databases, supplement 3. Genes Immun 2006;7:269–276.
153 Wilson AG, de Vries N, Pociot F, di Giovine FS, van der Putte LBA, Duff GW: An allelic polymorphism within the human tumor necrosis factor1α promoter region is strongly associated with HLA A1, B8, and DR3 alleles. J Exp Med 1993;177:557–560.
154 Higuchi T, Seki N, Kamizono S, Yamada A, Kimura A, Kato H, Itoh K: Polymorphism of the 5′-flanking region of the human tumor necrosis factor (TNF)-α gene in Japanese. Tissue Antigens 1998;51:605–612.
155 Baena A, Leung JY, Sullivan AD, Landires I, Vasquez-Luna N, Quinones-Berrocal J, Fraser PA, Uko GP, Delgado JC, Clavijo OP, Thim S, Meshnick SR, Nyirenda T, Yunis EJ, Goldfeld AE: TNF-α promoter single nucleotide polymorphisms are markers of human ancestry. Genes Immun 2002;3: 482–487.
156 Alper CA, Awdeh ZL, Raum DD, Yunis EJ: Extended major histocompatibility complex haplotypes in man: role of alleles analogous to murine τ mutants. Clin Immunol Immunopathol 1982;24:276–285.
157 Yunis EJ, Larsen CE, Fernandez-Vina M, Awdeh ZL, Romero T, Hansen JA, Alper CA: Inheritable variable sizes of DNA stretches in the human MHC: conserved extended haplotypes and their fragments or blocks. Tissue Antigens 2003;62:1–20.
158 Braun N, Michel U, Ernst BP, Metzner R, Bitsch A, Weber F, Rieckmann P: Gene polymorphism at position –308 of the tumor-necrosis-factor-α (TNF-α) in multiple sclerosis and it's influence on the regulation of TNF-α production. Neurosci Lett 1996; 215:75–78.
159 Wu WS, McClain KL: DNA polymorphisms and mutations of the tumor necrosis factor-α (TNF-α) promoter in Langerhans cell histiocytosis (LCH). J Interferon Cytokine Res 1997;17:631–635.
160 Kroeger KM, Carville KS, Abraham LJ: The –308 tumor necrosis factor-α promoter polymorphism effects transcription. Mol Immunol 1997;34:391–399.
161 Wilson AG, Symons JA, McDowell TL, McDevitt HO, Duff GW: Effects of a polymorphism in the human tumor necrosis factor α promoter on transcriptional activation. Proc Natl Acad Sci USA 1997; 94:3195–3199.
162 Kroeger KM, Steer JH, Joyce DA, Abraham LJ: Effects of stimulus and cell type on the expression of the -308 tumour necrosis factor promoter polymorphism. Cytokine 2000;12:110–119.
163 Brinkman BM, Zuijdeest D, Kaijzel EL, Breedveld FC, Verweij CL: Relevance of the tumor necrosis factor alpha (TNF α) –308 promoter polymorphism in TNF-α gene regulation. J Inflamm 1996;46:32–41.
164 Stuber F, Udalova IA, Book M, Drutskaya LN, Kuprash DV, Turetskaya RL, Schade FU, Nedospasov SA: –308 tumor necrosis factor (TNF) polymorphism is not associated with survival in severe sepsis and is unrelated to lipopolysaccharide inducibility of the human TNF promoter. J Inflamm 1995;46:42–50.
165 Bayley J-P, de Rooij H, van den Elsen PJ, Huizinga TW, Verweij CL: Functional analysis of linker-scan mutants spanning the –376, –308, –244, and –238 polymorphic sites of the TNF-α promoter. Cytokine 2001;14:316–323.
166 Knight JC, Keating BJ, Rockett KA, Kwiatkowski DP: In vivo characterization of regulatory polymorphisms by allele-specific quantification of RNA polymerase loading. Nat Genet 2003;33:469–475.
167 Knight JC, Udalova I, Hill AV, Greenwood BM, Peshu N, Marsh K, Kwiatkowski D: A polymorphism that affects OCT-1 binding to the TNF promoter region is associated with severe malaria. Nat Genet 1999;22:145–150.
168 Hohjoh H, Tokunaga K: Allele-specific binding of the ubiquitous transcription factor OCT-1 to the functional single nucleotide polymorphism (SNP) sites in the tumor necrosis factor-alpha gene (TNFA) promoter. Genes Immun 2001;2:105–109.
169 van Heel DA, Udalova IA, De Silva AP, McGovern DP, Kinouchi Y, Hull J, Lench NJ, Cardon LR, Carey AH, Jewell DP, Kwiatkowski D: Inflammatory bowel disease is associated with a TNF polymorphism that affects an interaction between the OCT1 and NF-κB transcription factors. Hum Mol Genet 2002;11:1281–1289.
170 Skoog T, Hamsten A, Eriksson P: Allele-specific chromatin remodeling of the tumor necrosis factor-α promoter. Biochem Biophys Res Commun 2006;351:777–783.

171 Udalova IA, Richardson A, Denys A, Smith C, Ackerman H, Foxwell B, Kwiatkowski D: Functional consequences of a polymorphism affecting NF-κB p50-p50 binding to the TNF promoter region. Mol Cell Biol 2000;20:9113–9119.
172 Kaijzel EL, van Krugten MV, Brinkman BM, Huizinga TW, van der Straaten T, Hazes JM, Ziegler-Heitbrock HW, Nedospasov SA, Breedveld FC, Verweij CL: Functional analysis of a human tumor necrosis factor α (TNF-α) promoter polymorphism related to joint damage in rheumatoid arthritis. Mol Med 1998;4:724–733.
173 Lee GR, Kim ST, Spilianakis CG, Fields PE, Flavell RA: T helper cell differentiation: regulation by *cis* elements and epigenetics. Immunity 2006;24:369–379.
174 Berger SL: The complex language of chromatin regulation during transcription. Nature 2007;447:407–412.
175 Mellor J, Dudek P, Clynes D: A glimpse into the epigenetic landscape of gene regulation. Curr Opin Genet Dev 2008;18:116–122.
176 Cedar H, Bergman Y: Linking DNA methylation and histone modification: patterns and paradigms. Nat Rev Genet 2009;10:295–304.
177 Li B, Carey M, Workman JL: The role of chromatin during transcription. Cell 2007;128:707–719.
178 Marmorstein R, Roth SY: Histone acetyltransferases: function, structure, and catalysis. Curr Opin Genet Dev 2001;11:155–161.
179 Shahbazian MD, Grunstein M: Functions of site-specific histone acetylation and deacetylation. Annu Rev Biochem 2007;76:75–100.
180 Lee KK, Workman JL: Histone acetyltransferase complexes: one size doesn't fit all. Nat Rev Mol Cell Biol 2007;8:284–295.
181 Miao F, Gonzalo IG, Lanting L, Natarajan R: In vivo chromatin remodeling events leading to inflammatory gene transcription under diabetic conditions. J Biol Chem 2004;279:18091–18097.
182 Kasper LH, Fukuyama T, Biesen MA, Boussouar F, Tong C, de Pauw A, Murray PJ, van Deursen JM, Brindle PK: Conditional knockout mice reveal distinct functions for the global transcriptional coactivators CBP and p300 in T-cell development. Mol Cell Biol 2006;26:789–809.
183 Xu W, Fukuyama T, Ney PA, Wang D, Rehg J, Boyd K, van Deursen JM, Brindle PK: Global transcriptional coactivators CREB-binding protein and p300 are highly essential collectively but not individually in peripheral B cells. Blood 2006;107:4407–4416.
184 Ranjbar S, Rajsbaum R, Goldfeld AE: Transactivator of transcription from HIV type 1 subtype E selectively inhibits TNF gene expression via interference with chromatin remodeling of the TNF locus. J Immunol 2006;176:4182–4190.
185 Ramirez-Carrozzi VR, Nazarian AA, Li CC, Gore SL, Sridharan R, Imbalzano AN, Smale ST: Selective and antagonistic functions of SWI/SNF and Mi-2β nucleosome remodeling complexes during an inflammatory response. Genes Dev 2006;20:282–296.
186 Sullivan KE, Reddy AB, Dietzmann K, Suriano AR, Kocieda VP, Stewart M, Bhatia M: Epigenetic regulation of tumor necrosis factor alpha. Mol Cell Biol 2007;27:5147–5160.
187 Garrett S, Dietzmann-Maurer K, Song L, Sullivan KE: Polarization of primary human monocytes by IFN-γ induces chromatin changes and recruits RNA Pol II to the TNF-α promoter. J Immunol 2008;180:5257–5266.
188 Lee JY, Kim NA, Sanford A, Sullivan KE: Histone acetylation and chromatin conformation are regulated separately at the TNF-α promoter in monocytes and macrophages. J Leukoc Biol 2003;73:862–871.
189 Sullivan KE, Suriano A, Dietzmann K, Lin J, Goldman D, Petri MA: The TNFα locus is altered in monocytes from patients with systemic lupus erythematosus. Clin Immunol 2007;123:74–81.
190 Taylor JM, Wicks K, Vandiedonck C, Knight JC: Chromatin profiling across the human tumour necrosis factor gene locus reveals a complex, cell type-specific landscape with novel regulatory elements. Nucleic Acids Res 2008;36:4845–4862.
191 Saccani S, Pantano S, Natoli G: p38-dependent marking of inflammatory genes for increased NF-κB recruitment. Nat Immunol 2002;3:69–75.
192 El Gazzar M, Yoza BK, Chen X, Hu J, Hawkins GA, McCall CE: G9a and HP1 couple histone and DNA methylation to TNFα transcription silencing during endotoxin tolerance. J Biol Chem 2008;283:32198–32208.
193 Kochanek S, Radbruch A, Tesch H, Renz D, Doerfler W: DNA methylation profiles in the human genes for tumor necrosis factors α and β in subpopulations of leukocytes and in leukemias. Proc Natl Acad Sci USA 1991;88:5759–5763.
194 Kruys V, Thompson P, Beutler B: Extinction of the tumor necrosis factor locus, and of genes encoding the lipopolysaccharide signaling pathway. J Exp Med 1993;177:1383–1390.
195 Becker C, Barbulescu K, Wirtz S, Meyer zum Buschenfelde KH, Pettersson S, Neurath MF: Constitutive and inducible in vivo protein-DNA interactions at the tumor necrosis factor-α promoter in primary human T lymphocytes. Gene Expr 1999;8:115–127.
196 Barthel R, Goldfeld AE: T cell-specific expression of the human TNF-α gene involves a functional and highly conserved chromatin signature in intron 3. J Immunol 2003;171:3612–3619.

197 Sung SJ, Walters JA, Hudson J, Gimble JM: Tumor necrosis factor-α mRNA accumulation in human myelomonocytic cell lines. Role of transcriptional regulation by DNA sequence motifs and mRNA stabilization. J Immunol 1991;147:2047–2054.
198 Kuhnert P, Peterhans E, Pauli U: Chromatin structure and DNase I hypersensitivity in the transcriptionally active and inactive porcine tumor necrosis factor gene locus. Nucleic Acids Res 1992;20:1943–1948.
199 Espel E, Garcia-Sanz JA, Aubert V, Menoud V, Sperisen P, Fernandez N, Spertini F: Transcriptional and translational control of TNF-α gene expression in human monocytes by major histocompatibility complex class II ligands. Eur J Immunol 1996;26:2417–2424.
200 Kwon J, Lee SJ, Benveniste EN: A 3′ *cis*-acting element is involved in tumor necrosis factor-α gene expression in astrocytes. J Biol Chem 1996;271:22383–22390.
201 Dekker J: A closer look at long-range chromosomal interactions. Trends Biochem Sci 2003;28:277–280.
202 de Laat W: Long-range DNA contacts: romance in the nucleus? Curr Opin Cell Biol 2007;19:317–320.
203 Carter DR, Eskiw C, Cook PR: Transcription factories. Biochem Soc Trans 2008;36:585–589.
204 O'Sullivan JM, Tan-Wong SM, Morillon A, Lee B, Coles J, Mellor J, Proudfoot NJ: Gene loops juxtapose promoters and terminators in yeast. Nat Genet 2004;36:1014–1018.
205 Drissen R, Palstra RJ, Gillemans N, Splinter E, Grosveld F, Philipsen S, de Laat W: The active spatial organization of the β-globin locus requires the transcription factor EKLF. Genes Dev 2004;18:2485–2490.
206 Spilianakis CG, Lalioti MD, Town T, Lee GR, Flavell RA: Interchromosomal associations between alternatively expressed loci. Nature 2005;435:637–645.
207 Vakoc CR, Letting DL, Gheldof N, Sawado T, Bender MA, Groudine M, Weiss MJ, Dekker J, Blobel GA: Proximity among distant regulatory elements at the β-globin locus requires GATA-1 and FOG-1. Mol Cell 2005;17:453–462.
208 Splinter E, Heath H, Kooren J, Palstra RJ, Klous P, Grosveld F, Galjart N, de Laat W: CTCF mediates long-range chromatin looping and local histone modification in the β-globin locus. Genes Dev 2006;20:2349–2354.
209 Kurukuti S, Tiwari VK, Tavoosidana G, Pugacheva E, Murrell A, Zhao Z, Lobanenkov V, Reik W, Ohlsson R: CTCF binding at the H19 imprinting control region mediates maternally inherited higher-order chromatin conformation to restrict enhancer access to Igf2. Proc Natl Acad Sci USA 2006;103:10684–10689.
210 Ling JQ, Li T, Hu JF, Vu TH, Chen HL, Qiu XW, Cherry AM, Hoffman AR: CTCF mediates interchromosomal colocalization between Igf2/H19 and Wsb1/Nf1. Science 2006;312:269–272.
211 Cockerill PN: NFAT is well placed to direct both enhancer looping and domain-wide models of enhancer function. Sci Signal 2008;1:pe15.
212 Murrell A, Heeson S, Reik W: Interaction between differentially methylated regions partitions the imprinted genes *Igf2* and *H19* into parent-specific chromatin loops. Nat Genet 2004;36:889–893.
213 Deakin JE, Papenfuss AT, Belov K, Cross JG, Coggill P, Palmer S, Sims S, Speed TP, Beck S, Graves JA: Evolution and comparative analysis of the MHC class III inflammatory region. BMC Genomics 2006;7:281.
214 Gupta D, Jin YP, Dziarski R: Peptidoglycan induces transcription and secretion of TNF-α and activation of lyn, extracellular signal-regulated kinase, and rsk signal transduction proteins in mouse macrophages. J Immunol 1995;155:2620–2630.
215 Shapira L, Takashiba S, Amar S, Van Dyke TE: *Porphyromonas gingivalis* lipopolysaccharide stimulation of human monocytes: dependence on serum and CD14 receptor. Oral Microbiol Immunol 1994;9:112–117.
216 English BK, Patrick CC, Orlicek SL, McCordic R, Shenep JL: Lipoteichoic acid from viridans streptococci induces the production of tumor necrosis factor and nitric oxide by murine macrophages. J Infect Dis 1996;174:1348–1351.
217 Sato S, Nomura F, Kawai T, Takeuchi O, Muhlradt PF, Takeda K, Akira S: Synergy and cross-tolerance between toll-like receptor (TLR) 2- and TLR4-mediated signaling pathways. J Immunol 2000;165:7096–7101.
218 Bondeson J, Sundler R: Auranofin inhibits the induction of interleukin 1 beta and tumor necrosis factor α mRNA in macrophages. Biochem Pharmacol 1995;50:1753–1759.
219 Wong GH, Goeddel DV: Tumour necrosis factors α and β inhibit virus replication and synergize with interferons. Nature 1986;323:819–822.
220 Beutler B, Krochin N, Milsark IW, Luedke C, Cerami A: Control of cachectin (tumor necrosis factor) synthesis: mechanisms of endotoxin resistance. Science 1986;232:977–980.
221 Goeddel DV, Aggarwal BB, Gray PW, Leung DW, Nedwin GE, Palladino MA, Patton JS, Pennica D, Shepard HM, Sugarman BJ, Wong GHW: Tumor necrosis factors: gene structure and biological activities. Cold Spring Harbor Symp Quant Biol 1986;60:597–609.

222 Perera PY, Manthey CL, Stutz PL, Hildebrandt J, Vogel SN: Induction of early gene expression in murine macrophages by synthetic lipid A analogs with differing endotoxic potentials. Infect Immun 1993;61:2015–2023.
223 Manthey CL, Brandes ME, Perera PY, Vogel SN: Taxol increases steady-state levels of lipopolysaccharide-inducible genes and protein-tyrosine phosphorylation in murine macrophages. J Immunol 1992;149:2459–2465.
224 Pope BL, MacIntyre JP, Kimball E, Lee S, Zhou L, Taylor GR, Goodman MG: The immunostimulatory compound 7-allyl-8-oxoguanosine (loxoribine) induces a distinct subset of murine cytokines. Cell Immunol 1995;162:333–339.
225 Stacey KJ, Sweet MJ, Hume DA: Macrophages ingest and are activated by bacterial DNA. J Immunol 1996;157:2116–2122.
226 Megyeri K, Au WC, Rosztoczy I, Raj NB, Miller RL, Tomai MA, Pitha PM: Stimulation of interferon and cytokine gene expression by imiquimod and stimulation by Sendai virus utilize similar signal transduction pathways. Mol Cell Biol 1995;15: 2207–2218.
227 Le Contel C, Temime N, Charron DJ, Parant MA: Modulation of lipopolysaccharide-induced cytokine gene expression in mouse bone marrow-derived macrophages by muramyl dipeptide. J Immunol 1993;150:4541–4549.
228 Anegón I, Cuturi MC, Trinchieri G, Perussia B: Interaction of Fc receptor (CD16) ligands induces transcription of interleukin 2 receptor (CD25) and lymphokine genes and expression of their products in human natural killer cells. J Exp Med 1988; 167:452–472.
229 Kovacs EJ, Beckner SK, Longo DL, Varesio L, Young HA: Cytokine gene expression during the generation of human lymphokine-activated killer cells: early induction of interleukin 1β by interleukin 2. Cancer Res 1989;49:940–944.
230 Collart MA, Belin D, Vassalli JD, de Kossodo S, Vassalli P: γ-Interferon enhances macrophage transcription of the tumor necrosis factor/cachectin, interleukin 1, and urokinase genes, which are controlled by short-lived repressors. J Exp Med 1986; 164:2113–2118.
231 Cannistra SA, Rambaldi A, Spriggs DR, Herrmann F, Kufe D, Griffin JD: Human granulocyte-macrophage colony-stimulating factor induces expression of the tumor necrosis factor gene by the U937 cell line and by normal human monocytes. J Clin Invest 1987;79:1720–1728.
232 Spriggs D, Imamura K, Rodriguez C, Horiguchi J, Kufe DW: Induction of tumor necrosis factor expression and resistance in a human breast tumor cell line. Proc Natl Acad Sci USA 1987;84:6563–6566.
233 Collart MA, Belin D, Vassalli JD, Vassalli P: Modulations of functional activity in differentiated macrophages are accompanied by early and transient increase or decrease in *c-fos* gene transcription. J Immunol 1987;139:949–955.
234 Trede NS, Geha RS, Chatila T: Transcriptional activation of IL-1β and tumor necrosis factor-α genes by MHC class II ligands. J Immunol 1991;146:2310–2315.
235 Sung SJ, Walters JA, Fu SM: Stimulation of tumor necrosis factor-α production in human monocytes by inhibitors of protein phosphatase 1 and 2A. J Exp Med 1992;176:897–901.
236 Rieckmann P, Thevenin C, Kehrl JH: Okadaic acid is a potent inducer of AP-1, NF-κB, and tumor necrosis factor-α in human B lymphocytes. Biochem Biophys Res Commun 1992;187:51–57.
237 Kock A, Schwarz T, Kirnbauer R, Urbanski A, Perry P, Ansel JC, Luger TA: Human keratinocytes are a source for tumor necrosis factor α: evidence for synthesis and release upon stimulation with endotoxin or ultraviolet light. J Exp Med 1990;172:1609–1614.
238 Hallahan DE, Spriggs DR, Beckett MA, Kufe DW, Weichselbaum RR: Increased tumor necrosis factor-α mRNA after cellular exposure to ionizing radiation. Proc Natl Acad Sci USA 1989;86:10104–10107.
239 Guha M, Bai W, Nadler JL, Natarajan R: Molecular mechanisms of tumor necrosis factor α gene expression in monocytic cells via hyperglycemia-induced oxidant stress-dependent and -independent pathways. J Biol Chem 2000;275:17728–17739.
240 Piguet PF, Collart MA, Grau GE, Sappino AP, Vassalli P: Requirement of tumour necrosis factor for development of silica-induced pulmonary fibrosis. Nature 1990;344:245–247.
241 Ehlers S, Mielke MEA, Blankenstein T, Hahn H: Kinetic analysis of cytokine gene expression in the livers of naive and immune mice infected with *Listeria monocytogenes*. J Immunol 1992;149:3016–3022.
242 Poston RM, Kurlander RJ: Cytokine expression in vivo during murine listeriosis. Infection with live, virulent bacteria is required for monokine and lymphokine messenger RNA accumulation in the spleen. J Immunol 1992;149:3040–3044.
243 Barnes PF, Fong SJ, Brennan PJ, Twomey PE, Mazumder A, Modlin RL: Local production of tumor necrosis factor and IFN-γ in tuberculous pleuritis. J Immunol 1990;145:149–154.
244 Rugo HS, O'Hanley P, Bishop AG, Pearce MK, Abrams JS, Howard M, O'Garra A: Local cytokine production in a murine model of *Escherichia coli* pyelonephritis. J Clin Invest 1992;89:1032–1039.

245 Berent SL, Torczynski RM, Bollon AP: Sendai virus induces high levels of tumor necrosis factor mRNA in human peripheral blood leukocytes. Nucleic Acids Res 1986;14:8997–9015.
246 Dudding L, Haskill S, Clark BD, Auron PE, Sporn S, Huang ES: Cytomegalovirus infection stimulates expression of monocyte-associated mediator genes. J Immunol 1989;143:3343–3352.
247 Goodier MR, Lundqvist C, Hammarstrom ML, Troye-Blomberg M, Langhorne J: Cytokine profiles for human Vγ9+ T cells stimulated by *Plasmodium falciparum*. Parasite Immunol 1995;17:413–423.
248 Van Voorhis WC: Coculture of human peripheral blood mononuclear cells with *Trypanosoma cruzi* leads to proliferation of lymphocytes and cytokine production. J Immunol 1992;148:239–248.
249 Wynn TA, Oswald IP, Eltoum IA, Caspar P, Lowenstein CJ, Lewis FA, James SL, Sher A: Elevated expression of Th1 cytokines and nitric oxide synthase in the lungs of vaccinated mice after challenge infection with *Schistosoma mansoni*. J Immunol 1994;153:5200–5209.
250 Suriano AR, Sanford AN, Kim N, Oh M, Kennedy S, Henderson MJ, Dietzmann K, Sullivan KE: GCF2/LRRFIP1 represses tumor necrosis factor alpha expression. Mol Cell Biol 2005;25:9073–9081.
251 Schreiber J, Jenner RG, Murray HL, Gerber GK, Gifford DK, Young RA: Coordinated binding of NF-κB family members in the response of human cells to lipopolysaccharide. Proc Natl Acad Sci USA 2006;103:5899–5904.
252 Amar S, Zhou Q, Shaik-Dasthagirisaheb Y, Leeman S: Diet-induced obesity in mice causes changes in immune responses and bone loss manifested by bacterial challenge. Proc Natl Acad Sci USA 2007; 104:20466–20471.
253 Cvoro A, Tzagarakis-Foster C, Tatomer D, Paruthiyil S, Fox MS, Leitman DC: Distinct roles of unliganded and liganded estrogen receptors in transcriptional repression. Mol Cell 2006;21:555–564.
254 Yang YJ, Chen W, Carrigan SO, Chen WM, Roth K, Akiyama T, Inoue J, Marshall JS, Berman JN, Lin TJ: TRAF6 specifically contributes to FcεRI-mediated cytokine production but not mast cell degranulation. J Biol Chem 2008;283:32110–32118.
255 Wang J, Wang X, Hussain S, Zheng Y, Sanjabi S, Ouaaz F, Beg AA: Distinct roles of different NF-κB subunits in regulating inflammatory and T cell stimulatory gene expression in dendritic cells. J Immunol 2007;178:6777–6788.
256 Sanjabi S, Hoffmann A, Liou HC, Baltimore D, Smale ST: Selective requirement for c-Rel during IL-12 P40 gene induction in macrophages. Proc Natl Acad Sci USA 2000;97:12705–12710.
257 Gadjeva M, Tomczak MF, Zhang M, Wang YY, Dull K, Rogers AB, Erdman SE, Fox JG, Carroll M, Horwitz BH: A role for NF-κB subunits p50 and p65 in the inhibition of lipopolysaccharide-induced shock. J Immunol 2004;173:5786–5793.
258 Gerondakis S, Strasser A, Metcalf D, Grigoriadis G, Scheerlinck JY, Grumont RJ: Rel-deficient T cells exhibit defects in production of interleukin 3 and granulocyte-macrophage colony-stimulating factor. Proc Natl Acad Sci USA 1996;93:3405–3409.
259 Caamaño J, Alexander J, Craig L, Bravo R, Hunter CA: The NF-κB family member RelB is required for innate and adaptive immunity to *Toxoplasma gondii*. J Immunol 1999;163:4453–4461.
260 Franzoso G, Carlson L, Xing L, Poljak L, Shores EW, Brown KD, Leonardi A, Tran T, Boyce BF, Siebenlist U: Requirement for NF-κB in osteoclast and B-cell development. Genes Dev 1997;11:3482–3496.
261 Iotsova V, Caamaño J, Loy J, Yang Y, Lewin A, Bravo R: Osteopetrosis in mice lacking NF-κB1 and NF-κB2. Nat Med 1997;3:1285–1289.

Dr. James V. Falvo and Dr. Anne E. Goldfeld
Immune Disease Institute and Harvard Medical School
200 Longwood Avenue
Boston, MA 02115 (USA)
Tel. +1 617 713 8778, Fax +1 617 713 8788, E-Mail falvo@idi.harvard.edu and goldfeld@idi.harvard.edu

Kollias G, Sfikakis PP (eds): TNF Pathophysiology. Molecular and Cellular Mechanisms.
Curr Dir Autoimmun. Basel, Karger, 2010, vol 11, pp 61–79

Posttranscriptional Regulation of TNF mRNA: A Paradigm of Signal-Dependent mRNA Utilization and Its Relevance to Pathology

Panagiota Stamou · Dimitris L. Kontoyiannis

Institute of Immunology, Biomedical Sciences Research Center 'Alexander Fleming', Vari, Greece

Abstract

The relationship between TNF and immune pathology forced an intense research into the regulation of its biosynthesis that extends to multiple mechanisms controlling the utilization of its mRNA. These posttranscriptional mechanisms gradually and variably impose a series of flexible rate-limiting controls to modify the abundance of the TNF mRNA and the rate of its translation in response to environmental signals. Mechanistically, these controls consist of signaling networks converging to RNA-binding proteins and microRNAs, which in turn target a code of secondary or tertiary ribonucleotide structures located on the TNF mRNA. The outcome of these interactions is the stringent control of this mRNA's maturation, localization, turnover and translation. A wealth of molecular and genetic data highlighted that if these posttranscriptional interactions fail, they perturb cellular responses to provide the impetus for TNF-mediated inflammatory disease. Here, we highlight the parameters guiding the posttranscriptional regulation of TNF mRNA and their relevance to homeostasis and pathology.

The biosynthesis of the TNF protein is under the control of multiple regulatory mechanisms which facilitate its rapid production during inflammation whilst maintaining its levels under homeostatic thresholds. The flexibility of TNF's response relies heavily on biosynthetic processes taking place between the transcription of the TNF gene and the translational cleavage of the TNF protein. Posttranscriptional mechanisms (i.e. TNF mRNA maturation, shuttling, stability and translation) control the time and the extent of intracellular utilization of the TNF mRNA. Such evidence was initially provided during the late 1980s where the TNF mRNA was identified as a labile mRNA and its accumulation profile appeared discordant to the TNF protein in LPS-challenged macrophages [1]. Subsequent studies demonstrated that the TNF mRNA is hypoadenylated and that the length of its poly (A)

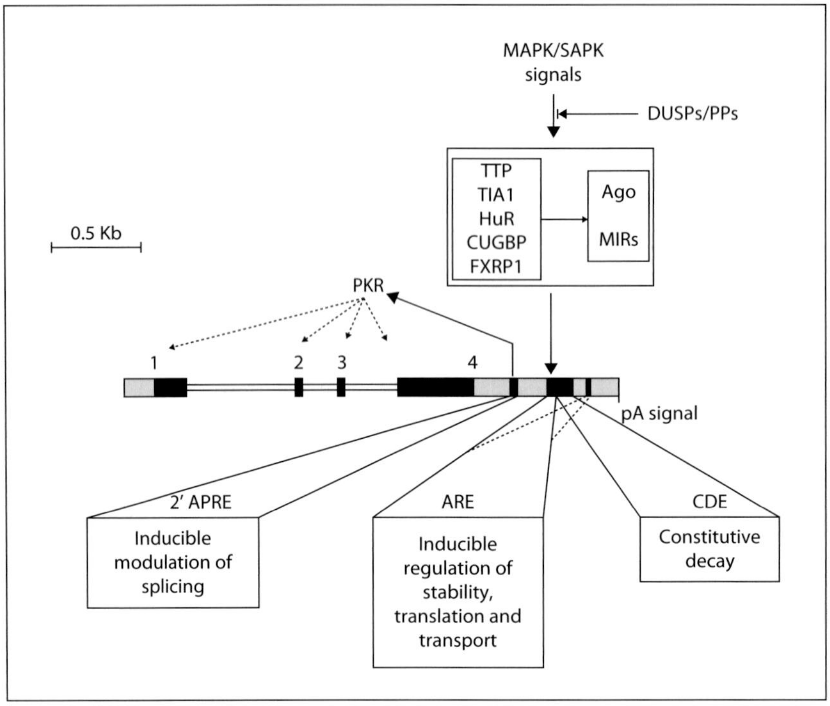

Fig. 1. Diagrammatic representation of the cis-elements (2′ APRE, AREs and CDE) residing in the 3′ UTR encoded by the 4th exon of the human TNF gene. Shown are their connections to signaling cascades like those inducing PKR for modulation of splicing and those transduced by MAPK and SAPK towards interfacing RNA interactors for the TNF 3′ ARE and opposed by protein phosphatases (DUSPs/PPs).

tail could be altered following stimulation, suggesting a complex pattern of TNF mRNA turnover and translation [2–4]. Most early studies concluded that the post-transcriptional control of TNF mRNA is guided by cis-elements residing in the 3′ UTR [5, 6]. In pathological terms, this was exemplified in TNF-transgenic mice; irrespectively of the promoters used, the replacement of this 3′ UTR with that from the β-globin mRNA, was sufficient to drive TNF overexpression supporting the development of systemic and organ-specific inflammation [see 15]. These early findings paved the way for a large amount of data collected over the last 15 years on the functional and molecular parameters involved in the posttranscriptional activation and suppression of TNF mRNA. The TNF 3′ UTR contains several elements that act as an RNA code for intracellular recognition by RNA-binding proteins (RBPs) and microRNAs (miRNAs) in response to a multitude of signals (fig. 1). Below, we focus on these elements, their interactions and their relevance to TNF-induced pathologies.

Cis-Elements and Their Associated Signals

Elements Controlling TNF mRNA Maturation

The first indications for a regulatory network affecting TNF pre-RNA splicing stemmed from the analysis of TNF biosynthesis in human lymphoid cells treated with 2-aminopurine (2-AP). This adenosine analogue induced a blockade in the splicing of the TNF precursor transcripts and reduced the presence of mature TNF mRNA [7]. The effect of 2-AP was attributed to its inhibitory effect on the RNA-inducible protein kinase, PKR [8]. In the context of viral infection, PKR is induced by inflammatory cytokines and activated by viral and particularly double-stranded RNA. The human TNF 3′ UTR contains a 104-nt-long sequence assuming a 5′ proximal stem-loop conformation that is required for sensitivity to 2-AP (2′ AP response element; 2′ APRE) and activates PKR (fig. 1) [8]. Although the mechanistic details remain to be elucidated, this 'RNA sensor system' may act to increase the efficiency of TNF pre-RNA splicing following detection by PKR, thus providing a positive feedback loop that maximizes the output of TNF-mRNA during infection [9]. Although, the current knowledge on 2′ APRE has been restricted to human peripheral blood mononuclear cells and lymphoid cells, it can also apply to other TNF-expressing cells. Recently, the herb-derived naphthoquinone shikonin was found to block TNF pre-RNA splicing and PKR activation in LPS-stimulated human monocytic cells [10]. However, despite the phylogenetic conservation of 2′ APRE in several species, its presence is not always apparent like in the case of the murine TNF gene. Still, the splicing of the murine TNF mRNA is also under signal-dependent controls, as has been exemplified in TCR-stimulated naïve murine lymphocytes [11], and the biosynthesis of TNF is compromised in Toll-like receptor (TLR)-stimulated/PKR-deficient macrophages [12], suggesting that similar structures remain to be identified in the murine TNF mRNA. Interestingly, the 2′ APRE does not interfere with the cytoplasmic translation of the TNF mRNA; this is puzzling since PKR can phosphorylate the eukaryotic initiation factor 2 to limit translational initiation [9]. The location of the 2′ APRE upstream of elements targeting TNF mRNA turnover and translation may provide an explanation since factors binding to these elements in the cytoplasm may mask the 2′ APRE from cytoplasmic PKR, so that they can modulate mRNA turnover/translation in response to different signals.

Adenine/Uracyl-Rich Elements Coordinating TNF mRNA Export, Turnover and Translation

Downstream of the 2′ APRE, the TNF 3′ UTR contains a stretch of 53–69 nt overlapping adenine/uracyl-rich motifs named AU-rich elements (AREs), whose history parallels that of TNF biology (fig. 1). Historically, the TNF 3′ ARE was initially described as a regulator of the inducible expression of TNF in innate cells [13].

Subsequently, AREs derived from TNF, GM-CSF and interferons were found to promote mRNA decay and translational inhibition when they were placed downstream of stable reporter mRNAs [14, 15]. At the level of the primary sequence, these phylogenetically conserved elements [16] are composed of a variable number of copies of the AUUUA or UUAUUUAUU nonamers and were originally classified into three separable functional subgroups based on their efficacy to induce mRNA degradation (class I-discontinuous nonamers; class II-continuous/overlapping nonamers; class II, undefined AU-rich structures). In subsequent years, bioinformatic approaches revealed that AREs can be found in 8% of human mRNAs [17]. Most intriguingly, the majority of ARE-containing mRNAs encode for factors involved in developmental, immune and cancer-related processes rendering the AREs as 'pathophysiologically relevant' RNA signatures. Based on the data assembled in the form of the ARED-database [18], the ARE classification scheme is now expanded to include 5 additional ARE subsets in class II AREs (clusters I–V relating to the number of nonamers). Biochemical studies on class II AREs, which includes the TNF 3′ ARE as a prototype, highlighted their importance in mediating an inducible form of mRNA decay, known as ARE-mediated decay (AMD). Current concepts support that AREs promote either the shortening of the poly-A tail (deadenylation) and/or removal of the 5′ cap (decapping) [19, 20]. Degradation then may proceed through the recruitment of 5′-3′ exonucleases or through the sequestration of a complex of 3′-5′ exonucleases called the exosome [17, 21, 22]. It appears that these events do not proceed in a linear fashion but are rather imposed by associations with RBPs and/or micro-RNA subsets that direct ARE-containing mRNAs in cytoplasmic foci containing decapping enzymes like the processing bodies (P-bodies). Similarly, stability enhancement and translational silencing of such mRNAs may proceed via a 'triage' step where ARE-containing mRNAs can be sorted either for AMD or translation, as has been suggested for the stress-induced granules (stress granules, SGs).

Concomitantly to the molecular and biochemical analyses of the TNF 3′ ARE, evidence started to come forth on its importance in inflammation and autoimmunity. Genetic analysis of autoimmunity-prone mouse strains revealed that a spontaneous dinucleotide insertion in the TNF 3′ ARE correlated with reduced levels of TNF protein and contributed to the development of lupus autoimmunity [23]. However, gene targeting in embryonic stem ES cells was meant to provide the definitive proof of the TNF 3′ ARE function. The genetic deletion of a 69-bp sequence containing the ARE from the mouse TNF locus, resulted in severe polyarthritic and inflammatory bowel disease phenotypes with resemblance to the human conditions of Crohn's disease, rheumatoid arthritis, spondyloarthritis and sacroiliitis [24–26]. In the absence of the ARE, the induction of TNF protein from mouse macrophages and lymphocytes was excessively prolonged, supporting a continuous state of innate activation as well as hypersensitivity to inflammatory agonists [25–27]. In addition, this small deletion was sufficient to drive the ectopic expression of TNF in cells of mesenchymal origin that do not normally produce this cytokine (e.g. synovial fibroblasts), indicating that the AREs modulate both temporal and spatial parameters of mRNA expression [25]. In molecular terms, these phenomena were attributed to the increased stability of the TNF mRNA and the absence of translational silencing mechanisms [25–27].

One important point that these studies revealed was that the AREs can respond to both immune-activating and immune-suppressive signals. For example, the absence of the TNF 3′ ARE rendered this cytokine partially or totally unresponsive to anti-inflammatory inhibition [25, 27]. Interestingly, the deletion of the TNF 3′ ARE alleviated also the inhibitory effects of nonsteroid pyridinyl imidazole compounds known to hinder the phosphorylation activity of the Ser/Thr stress-activated protein kinase (SAPK), p38 [28]. The use of such small molecule inhibitors and genetic systems deficient in molecules involved in p38/SAPK signaling revealed that p38α/β isoforms can modulate the translation and the stability of ARE-containing cytokine mRNAs. For the TNF mRNA, this effect is partially mediated through the p38-activated mitogen-activated protein kinase (MAPK)-activated protein kinase 2 (MAPKAPK-2, or MK2) and to a lesser extent by MAPKAP-3 (MK3) [29]. Mutant mice with a deficiency in MK2 produce low levels of TNF protein and are resistant to models of systemic and organ-specific inflammation, whilst remaining sensitive to infection, reciprocating states of TNF/TNFR deficiency [30–32]. In contrast, the absence of the TNF 3′ ARE in MK2-deficient cells restored TNF levels, proving that MK2 targets ARE-mediated processes [29–31, 33]. It is now clear that the p38/MK2 pathway blocks ARE-mediated destabilization in activated macrophages [29]. In contrast, the effect of p38 signals on the translational activation of the TNF mRNA in the same cells remains elusive. Studies on T cell lines suggested that this effect may proceed through the activation of MAP kinase signal-integrating kinases (Mnks), which are activated by the p38 and ERK pathways [34]. Mnks are known to affect the translational initiation via their interaction with the cap-binding initiation factor complex. In addition, they can phosphorylate selective ARE-interacting factors binding to TNF mRNA [34, 35]. However, the pharmacological inhibition of Mnks in mouse macrophages reduces the accumulation of TNF mRNA, obscuring the assessment of translational parameters [36]. Detailed analysis of TNF expression in mice lacking Mnks is thus required to conclude on their role towards TNF mRNA translation.

The JNK pathway has also been implicated in the posttranscriptional regulation of the TNF mRNA. The inhibition of JNKs via kinase-defective mutants, glucocorticoids or pharmacological inhibitors can block the translation of TNF mRNA in LPS-stimulated macrophages [37, 38]. This effect is alleviated in the absence of the TNF 3′ ARE [25]. Additional inflammatory signals converging to the TNF ARE include those proceeding via the MAP3 kinase Tpl2/Cot. In unstimulated cells, this Ser/Thr protein kinase remains inactive and stable due to its interaction with NF-κB1/p105, but in the presence of proinflammatory signals, it is activated by IKKβ and mediates the activation of ERK, JNK and NF-κB signals [39, 40]. TLR4-stimulated, Tpl2-deficient macrophages show a defect in the activation of the Erk1/2 kinases, as do those incubated with pharmacological inhibitors of MEK1, and produce lower amounts of several inflammatory mediators, including TNF [41]. The effect of the Tpl2/ERK pathway on TNF production may occur at several levels including post-translational processing [42]; however, it appears to have a predominant effect on the posttranscriptional regulation since TNF levels are restored in $Tpl2^{-/-}$ macrophages when the TNF mRNA lacks its ARE [41, 43]. Intriguingly, the Tpl2/ERK axis does not appear to target mRNA stability or translation, but it rather affects the nucleocytoplasmic localization of the TNF mRNA. Subsequent in vitro studies indicated that this process may proceed via the tethering of exon junction complex proteins like TAP and NxT1 towards the spliced portions of TNF mRNA, which facilitate nuclear export in an ERK/ARE-dependent manner [43]. Although the ARE interacting factors contributing to this process remain to be determined, the findings point towards an additional posttranscriptional level of control for TNF mRNA and connect its AREs to the regulation by nucleocytoplasmic shuttling.

On the antipode of the positive signals modulating ARE-dependent posttranscriptional controls, lie those that enforce destabilization and translational repression. Logically, such signals may proceed through the abrogation of 'activator' signals, e.g. via the dephosphorylation of MAPKs by dual-specificity phosphatases. A most likely candidate is MKP1 (DUSP1) which can dephosphorylate ERKs, p38α and JNK1 following TLR engagement. Mice with deficiencies in MKP1 show increased susceptibility endotoxemia, whereas their corresponding macrophages show enhanced cytokine outputs – including TNF – in response to a variety of TLR ligands [44–46]. Thus, it is logical to conceptualize that MKP1 may also affect ARE-mediated processes by means of MAPK/SAPK deactivation, but direct evidence is currently lacking. The involvement of protein phosphatases in ARE-dependent control has also been indicated by the effects of the protein phosphatase 2A inhibition on TNF mRNA destabilization via the dephosphorylation of ARE-interacting factors [47]. With regard to the active suppression of TNF production by anti-inflammatory signals, several pieces of evidence indicate counteracting ARE-related signaling cross-talks emanating from the concomitant engagement of inflammatory and anti- or coinflammatory receptors (TLR to MAPK modules versus IL-10/IFNs to JAK/STAT/SOCS modules); whether these effects are directly targeting specific ARE-mediated processes or are interfering with the p38/MK2 pathway is currently unclear, although evidence for both has been provided in the case of IL-10-associated signals [27, 46, 48]. Furthermore, whether IL-10 targets TNF mRNA translation, stability or both is still a matter of debate, although signal and species-specific variations may be responsible for measurement discrepancies. Still, the inability of anti-inflammatory cytokines to modulate TNF biosynthesis due to the absence of TNF 3′ ARE appears as a key deterministic parameter for the development of inflammatory disease, as was exemplified in the case of IL-10/TNF axis in inflammatory bowel disease. It is conceivable that other anti-inflammatory cues may also be affected, as has been previously suggested for TGFβ-, IL-4- and IL-13-mediated suppression of TNF biosynthesis [49].

Elements Controlling Basal Stability

The deletion of the main TNF 3′ ARE may have increased the stability of the TNF mRNA, but these mutant 3′ UTRs were still capable of stimulating decay, suggesting that alternative elements can induce TNF mRNA degradation independently of AMD. A detailed mutation analysis of the sequences downstream of the ARE in the TNF 3′ UTR identified the presence of a 99-nt sequence which contained a 15-nt core capable of eliciting decay. Interestingly, this phylogenetically conserved element was not responding to known signals controlling AMD, indicating that it acts in a constitutive fashion – hence it was named as constitutive decay element (CDE) (fig. 1) [50]. The molecular details of CDE's interactions with RBPs as well as its importance in TNF expression in vivo remain to be determined. Still, it appears that the CDE has evolved as a fail-safe mechanism to keep the basal levels of TNF mRNA at an essential minimum.

Indications for Additional Cis-Elements and Signals

Several pieces of evidence support the presence of additional elements controlling TNF mRNA utilization. Downstream of the CDE and 147 nt of the main ARE lies a second phylogenetically conserved ARE-like sequence, but its function remains to be determined (fig. 1). It appears that this element could synergize with the main ARE, since several protein complexes can bind to both and their combined absence abolishes the translational inducibility of reporter mRNAs by LPS. However, the study focusing on CDE functions revealed that downstream sequences failed to affect the stability of reporter mRNAs. Nevertheless, additional mechanisms affecting the translational response of TNF mRNA, in the absence of the main ARE, have been reported in the literature. For example, the engineered immunomodulatory peptide RDP58 (Allotrap 1258) that resembles HLA class I peptides, has been identified as a potent inhibitor of TNF protein synthesis. Interestingly, the inhibitory action of the peptide on TNF mRNA required the presence of an intact 3′ UTR but not of the main TNF 3′ ARE. Thus, and despite the dominant role of the main AREs, the inducible modulation of TNF mRNA translation may be fine tuned by the additional contributions of other cis-elements.

Determinants of TNF mRNA Utilization

A collection of cis-elements residing in an mRNA may act as a utilization code by posttranscriptional 'machines'; as such, these elements do not determine the fate of an mRNA per se, but rather indicate the possible routes that an mRNA may follow. The operational decision to activate or repress a posttranscriptional process depends on the collection of factors interacting with these elements at any given time. Such factors include small RNA populations and RBPs. RBPs represent one of the largest conserved protein families, which reflects the pleiotropy of their functions. It appears that the inducible determination of mRNA's fate is the collective sum of signal-induced assemblies of numerous RBPs with their target RNAs in ribonucleoprotein particles (RNPs). The dynamic associations between RBPs and RNAs in RNPs are complex and difficult to monitor in the context of a complex cellular response, e.g the inflammatory response; however, they can be inferred by the effects of individual RBP components on their mRNA targets.

The functional properties of the TNF 3′ UTR and its relevance to TNF-related pathologies prompted an intense investigation of factors that bind to this region. Most of the factors identified to date appear to interact with the TNF 3′ ARE. In general, numerous RBPs have been identified as ARE-binding proteins (AREBPs) and either contain RNA recognition motifs, positively charged KH domains or zinc finger motifs [51]. The variability in the RNA recognition motifs of the AREBPs suggests that the AREs are 'modular' in nature and consist of different binding sites. Most of the AREBPs have been ascribed as negative determinants of mRNA's fate being either destabilization or translational silencing, whereas only one family – the Elavl/Hu – has been ascribed as a positive regulator. Below, we will focus on those ARE interactors that have a proven involvement in regulation of TNF mRNA, as has been suggested by biochemical, molecular and genetic studies.

Tristetraprolins

Tristetraprolins are a family of RBPs with a tandem CCCH zinc finger motif which includes four members: TTP (ZFP36, TIS11), BRF1 (ZFP36L1, TIS11b), BRF2 (ZFP36L2, TIS11d) and the mouse placenta-restricted ZFP36L3 [52]. Genetic association studies indicated a possible relationship between the aberrant expression of these proteins and autoimmune disease [53–55]. The prototype member, TTP, emerged as a dominant regulator of TNF mRNA stability. TTP-deficient mice develop an inflammatory syndrome, characterized by polyarticular arthritis, myeloid hyperplasia and cachexia [56] due to the increased production of TNF and GM-CSF. Subsequent studies on innate cells demonstrated that TTP binds to the TNF ARE and induces the destabilization of the TNF mRNA [57]. TTP can interact with various components of the basic RNA decay machinery responsible for deadenylation, decapping and exonucleolytic activity in order to promote ARE-mediated mRNA decay [58]. In addition, TTP cooperates or antagonizes the functions of other ARE-BPs in RNPs controlling stability, translation [59] or micro-RNA-mediated silencing [60]. Although the molecular activities of TTP are still under investigation, the current data are sufficient to propose a model towards its effects on TNF mRNA turnover. According to this model, in unstimulated cells low levels of unphosphorylated TTPs are able to bind and destabilize the TNF mRNA. TLR agonists, TNF-receptor signals and γ-IFNs induce the transcription, stabilization and translation of the TTP mRNA, whilst the TTP protein is phosphorylated at multiple serine, threonine and a few tyrosine residues [61, 62]. In relation to p38/MK2 signals, TTP is directly phosphorylated by MK2 on two serine residues (Ser 52 and Ser 178) [63]. This phosphorylation event promotes the cytoplasmic localization of TTP and allows its interaction with 14-3-3 proteins [64, 65]. The TTP/14-3-3 interaction (a) stabilizes the TTP protein by protecting it from proteosomal degradation [66, 65], and (b) inhibits TTP's destabilizing activity by excluding its sequestration towards its target mRNAs like the TNF mRNA [65]. As a consequence, the TNF mRNA is protected from degradation and can be used for translation. This has been exemplified in mice rendered deficient for both MK2 and TTP in which the low TNF levels seen in $MK2^{-/-}$ mice are restored to the high levels that are detected in $TTP^{-/-}$ mice. On the other end, and when the expression of TNF mRNA needs to cease, protein phosphatases induced by both inflammatory and anti-inflammatory signals (e.g. IL-10): (a) deactivate the p38/MK2 signal (e.g. MKP1), or (b) compete with 14-3–3 for binding to TTP and proceed with its dephosphorylation and consequent activation of destabilizing activities. Evidence for this latter mechanism has been provided via the pharmacological inhibition of protein phosphatase 2A which increases TTP phosphorylation and enhances the interaction with 14-3–3 [47]. At the same time, anti-inflammatory cytokines, like IL-4 and TGFβ, induce further the transcription of the TTP gene via STAT6 [67] and SMAD3/SMAD4 [68], respectively. This late burst of activation in TTP activity subsequently limits the accumulation of TNF mRNA which then returns to baseline values.

A similar mode of function has been recently ascribed for the second family member, Zfp36L1/BRF1. Unfortunately, mice deficient for this molecule, as well as for BRF2, die at early embryonic stages, thus preventing the assessment of their functions in inflammation [69, 70, 71]. Like TTP, BRF1 can promote AMD [72], and is regulated by phosphorylation and interaction with 14-3-3 proteins; however, BRF1 responds to the PI3-kinase pathway and is phosphorylated by PKB/Akt. Both TTP and BRF1 can facilitate the localization of ARE-mRNAs in P-bodies [73]. In vitro studies have demonstrated BRF1's capacity to bind to the TNF ARE and suppress the accumulation of reporter transcripts in cell-free systems [74]. Thus, it is highly likely that these proteins play similar roles either in other cellular compartments (e.g. lymphocytes) or in response to additional inflammatory signaling cascades, which remain to be revealed.

T Cell Intracellular Antigen-1 and -R

T cell intracellular antigen-1 and -R (TIA-1 and TIAR) are closely related members of the RRM family of RBPs which can bind numerous mRNAs containing a U-rich motif. Both proteins inhibit the translation of TNF transcripts in macrophages. TIA-

1-deficient mice are phenotypically normal in mixed genetic configuration, but in the inbred C57Bl/6 background they display mild symptoms of arthritis [59, 75]. These mice also appear very sensitive to mouse models of endotoxemia [59]. TIA-1-deficient macrophages overproduce TNF protein, and the percentage of TNF transcripts found in polysomes is significantly increased, suggesting that TIA-1 functions as a translational silencer.

The molecular mechanism of TIA's action has been revealed in cellular systems of oxidative stress, where the eukaryotic initiation factor 2α is phosphorylated to prevent the initiation of protein synthesis [76]. Under these conditions, TIA-1 and TIAR assemble with components of translation-initiation machinery which is directed to SGs [77]. SGs are also rich in RNP complexes and thus are regarded as 'triage' sites communicating with both P-bodies and the ribosomal machinery where mRNAs are stalled prior to decision for their destruction or translation. The demonstration that phosphorylated TTP is excluded from SGs whereas its non-phosphorylated form is present conforms to this notion and suggests that translational inhibition is indirectly linked to destabilization [65].

Heterologous Nuclear Ribonucleoproteins

Heterologous nuclear ribonucleoproteins (hnRNPs) are a large family of RBPs known to bind initially to RNA transcripts concomitant to transcription and form RNPs essential for posttranscriptional events that range from mRNA packaging and transport to splicing and silencing. Several hnRNPs have been identified to interact with AREs. The prototype in this category is hnRNPD/AUF1 which is an RRM containing RBP that exists as four different alternatively spliced isoforms (p37, p40, p42 and p45) [78]. AUF1 shuttles between the nucleus and the cytoplasm and was originally described as an inducer of mRNA decay [78–81]; however, subsequent studies revealed that it can also enhance mRNA stability and promote translation [82–85]. The signature motif for AUF1 has been recently identified to be <80% AU rich and to overlap with those from TIA1 and HuR [85]. Mechanistically, AUF1 can associate with heat shock proteins (e.g. hsp70), the translation initiation factor eIF4G and the poly(A)-binding protein PABP [86]. The decay of AUF1-associated transcripts requires the release of AUF1 from eIF4G and proteasome-mediated destruction or recruitment to the exosome. Similarly to TTP, AUF1 is phosphorylated in vivo [51], indicating a more prominent response to proliferating and stress signals. Evidence for AUF1's involvement in the posttranscriptional control of TNF mRNA was recently exemplified in AUF-1-deficient mice. These mice appear sensitized to endotoxemia and develop a form of skin inflammation relating to increases in the stability of both proinflammatory mRNAs like the TNF mRNA and Th2 cytokines [52]. Although there is no clear molecular evidence on the role of AUF-1 towards the posttranscriptional control of the TNF mRNA, these findings suggest that AUF-1 may be involved in the suppression of TNF biosynthesis. Additional hnRNPs have been inferred to be able to target the TNF 3′ ARE via their interaction with signaling cascades targeting this element. Such is the case for hnRNPA0 that is directly phosphorylated by MK2 in LPS-stimulated macrophages [87] or for hnRNPA1 that is phosphorylated by Mnks in response to T cell activation [88].

Elavl1/HuR Family

HuR is the prototypical member of the *Elavl/Hu* family of RRM containing RBPs which was originally identified as tumor antigens in paraneoplastic syndromes [89]. As in the case of AUF1, HuR shuttles between the nucleus and the cytoplasm acting as an RNA adaptor. With respect to inflammation, studies on macrophage cell lines suggested that innate sensitizers increase the cytoplasmic binding of HuR to TNF 3′ ARE supporting the stabilization of the TNF mRNA [90–92]. Furthermore, in the murine NZB/W model of spontaneous lupus nephritis, the disease-contributing reduction in TNF levels correlates with a mutation hindering HuR binding to the TNF 3′ ARE [93]. These observations suggested that HuR could act as positive regulator of TNF biosynthesis; however, conditional transgenic systems challenged this idea. Transgenic mice overexpressing HuR in macrophages in an inducible fashion displayed reduced inflammatory responses in modeled endotoxemia and in inflammatory hepatitis [94]. The anti-inflammatory effect of the transgenic HuR correlated with the reduced production of TNF protein. Although the overexpression of HuR increased the stability of the TNF mRNA, it also imposed a strong translational block, limiting the production of TNF protein. The genetic elimination of TTP or TIA-1 in the context of HuR overexpression revealed that HuR required the functions of TIA-1 to inhibit the translation of TNF mRNA, and this effect was occurring even in the absence of TTP deficiency. Although HuR-deficient embryos have a number of developmental defects [95], a T-cell-restricted mutant has been generated and indicated that indeed HuR-deficient T cells overproduce TNF protein [96]. By combining the current data on HuR, TTP and TIA-1, these studies postulate a functional network towards the modulation of TNF 3′ ARE, where HuR is tethering the TNF mRNA towards TIA-1-mediated translational inhibition, which in turn tethers towards destabilization by TTP. This is also compatible with the inclusion of all three RBPs in SGs and with positive correlation between HuR and TIA1 expression in rheumatoid arthritis patients [97, 98]. Currently a number of signal-dependent posttranslational modifications have been identified to modulate HuR's shuttling and binding [99], but the inflammatory signals controlling HuR functions are not clear. However, recent evidence suggests that HuR is methylated in response to LPS [100] and can be phosphorylated by p38 [101], whereas its absence may interfere with inflammatory signals stemming from TNFR superfamily members [96]. Thus, it is quite likely that inflammatory signals activate HuR TNF 3′ ARE interaction, which will consequently drive the posttranscriptional utilization of this mRNA.

CUG Triplet Repeat Binding Protein 1

CUG triplet repeat binding protein (CUG BP1) belongs to the CELF family of proteins involved in mRNA splicing, editing, translation and posttranscriptional regulation. The role of CUG BP1 in the regulation of TNF production by muscle cells was recently revealed via the analysis of the type I myotonic dystrophy disorder. This disorder is characterized by excess TNF production, muscle wasting, insulin resistance and cardiac dysfunction and is caused by an expansion of a CTG repeat within the 3′ UTR of the dystrophia myotonia protein kinase gene, which induces increased expression and stability of CUG BP1. The aberrant expression of CUG BP1 affects the expression of muscle-specific genes and cytokines like TNF. CUG BP1 can interact with GU-rich sequences that flank the TNF 3′ ARE and the PARN deadenylase in a PKC-dependent manner favoring TNF mRNA decay [102, 103]. Whether CUGBP1 affects the regulation of TNF mRNA in immune cells and interferes with the functions of AREBPs remains an open question.

Micro-RNAs, FXRMP1 and the RNA-Induced Silencing Complex

Recent evidence suggests that RNA-mediated gene silencing is involved in the posttranscriptional regulation of TNF mRNA. RNA silencing relies on sequence-specific interactions between target RNAs and small RNAs – such as the miRNAs (miRs). miRs are transcribed as precursor molecules and are first processed to a length of 60 nt by the RNaseIII type endonuclease Drosha; subsequently, they are exported to the nucleoplasm to undergo second cleavage step by the endonunclease Dicer to yield the final 22-nt-long miRs. In that form, miRs are loaded to the multisubunit RNA-induced silencing complex in P-bodies and associate with the Argonaute proteins possessing endonuclease activities against miR target mRNAs. miRNA-mediated silencing can involve both inhibition of translation as well as mRNA degradation via deadenylation and decapping [104].

The role of the RNA-induced silencing complex in posttranscriptional regulation of TNF mRNA was first indicated in an RNA interference-based screen in *insect* cells and then confirmed in mammalian cell lines, where it was shown that Dicer1, Ago-1 and Ago-2 are required for the rapid decay of the TNF mRNA [60]. In this work, miR16 was shown to induce rapid degradation of ARE-containing mRNAs via an indirect interaction with TTP. TTP was proposed to interact with Ago proteins, thus tethering mir16 to the ARE-containing mRNA. Similarly, the RBP FXR1P was recently shown to interfere with the translation of the TNF mRNA via its association with Ago2 and miR369-3. The involvement of FXR1P in the ARE-dependent modulation of the TNF mRNA was subsequently highlighted in FXR1P mouse mutants that overexpress TNF and develop muscle wasting, have a decreased growth rate and increased neonatal mortality [105, 106]. However, a recent report suggested that the association of FXR1P with Ago2 can induce translation activation rather that repression, necessitating the need for further in vivo experimentation to understand the dynamics of Ago interactions with other RBPs interacting with the TNF mRNAs [105, 107].

miRNAs miR155 and miR125b have also been implicated in the regulation of TNF production in immune cells. miR155 unregulated in monocytes upon LPS stimuli [108, 109]. Mice deficient in miR155 show impaired B cell responses [110, 111] linked to reduced TNF production [112]. Similarly, transgenic mice that overexpress miR-155 on B cells show elevated serum TNF and increased susceptibility to endotoxemia [113]. Contrastingly, miR-125b is downregulated in LPS-stimulated macrophages and can target the 3′ UTR of TNF, indicating its role in posttranscriptional repression of TNF transcripts.

Collectively, and although the collection of miRs targeting TNF biosynthesis remains to be elucidated, it is becoming apparent that RBPs targeting TNF mRNA cis-elements combine their functions with the miR machinery towards both activation and suppression.

Conclusions

From the above, it is clear that the posttranscriptional regulation of the TNF mRNA involves a series of complex interactions between RNA elements, RBPs and signaling modules. Given the pleiotropy of TNF activities, it is not surprising that evolution has imposed such an extensive array of regulating mechanisms ensuring the prudent production of TNF protein. What is surprising is how much we have learnt by studying the relatively small TNF mRNA. In molecular terms, the regulation of this mRNA highlighted that immune posttranscriptional control is governed by signaling/RBP modules that recognize a code of cis-elements and in turn determine the outcome of a biosynthetic response. In genetic and pathological terms, we have learnt that mutations targeting elements regulating RNA utilization, RBPs or associated signalosomes can contribute to disease development. Whilst the intricacies of posttranscriptional regulators are being unveiled, current and future research holds promise for a novel class of biological therapeutics against posttranscriptional processes in immune diseases.

Acknowledgements

We wish to thank the members of the Kontoyiannis lab for their comments. Dr. Stamou is supported by funding under the 7th Research Programme of the European Union, Integrated Project Masterswitch No. 223404.

References

1 Beutler B, Krochin N, Milsark IW, Luedke C, Cerami A: Control of cachectin (tumor necrosis factor) synthesis: mechanisms of endotoxin resistance. Science 1986;232:977–980.

2 Mijatovic T, Houzet L, Defrance P, Droogmans L, Huez G, Kruys V: Tumor necrosis factor-alpha mRNA remains unstable and hypoadenylated upon stimulation of macrophages by lipopolysaccharides. Eur J Biochem 2000;267:6004–6012.

3 Lieberman AP, Pitha PM, Shin ML: Poly(A) removal is the kinase-regulated step in tumor necrosis factor mRNA decay. J Biol Chem 1992;267:2123–2126.

4 Crawford EK, Ensor JE, Kalvakolanu I, Hasday JD:The role of 3′ poly(A) tail metabolism in tumor necrosis factor-alpha regulation. J Biol Chem 1997; 272:21120–21127.

5 Han J, Huez G, Beutler B: Interactive effects of the tumor necrosis factor promoter and 3′-untranslated regions. J Immunol 1991;146:1843–1848.

6 Han JH, Beutler B, Huez G: Complex regulation of tumor necrosis factor mRNA turnover in lipopolysaccharide-activated macrophages. Biochim Biophys Acta 1991;1090:22–28.

7 Jarrous N, Osman F, Kaempfer R: 2-Aminopurine selectively inhibits splicing of tumor necrosis factor alpha mRNA. Mol Cell Biol 1996;16:2814–2822.

8 Osman F, Jarrous N, Ben-Asouli Y, Kaempfer R: A cis-acting element in the 3′-untranslated region of human TNF-alpha mRNA renders splicing dependent on the activation of protein kinase PKR. Genes Dev 1999;13:3280–3293.

9 Kaempfer R: RNA sensors: novel regulators of gene expression. EMBO Rep 2003;4:1043–1047.

10 Chiu SC, Yang NS: Inhibition of tumor necrosis factor-alpha through selective blockade of Pre-mRNA splicing by shikonin. Mol Pharmacol 2007; 71:1640–1645.

11 Yang Y, Chang JF, Parnes JR, Fathman CG: T cell receptor (TCR) engagement leads to activation-induced splicing of tumor necrosis factor (TNF) nuclear pre-mRNA. J Exp Med 1998;188:247–254.

12 Cabanski M, Steinmuller M, Marsh LM, Surdziel E, Seeger W, Lohmeyer J: PKR regulates TLR2/TLR4-dependent signaling in murine alveolar macrophages. Am J Respir Cell Mol Biol 2008;38:26–31.

13 Caput D, Beutler B, Hartog K, Thayer R, Brown-Shimer S, Cerami A: Identification of a common nucleotide sequence in the 3′-untranslated region of mRNA molecules specifying inflammatory mediators. Proc Natl Acad Sci USA 1986;83:1670–1674.

14 Shaw G, Kamen R: A conserved AU sequence from the 3′ untranslated region of GM-CSF mRNA mediates selective mRNA degradation. Cell 1986;46:659–667.

15 Kruys V, Marinx O, Shaw G, Deschamps J, Huez G: Translational blockade imposed by cytokine-derived UA-rich sequences. Science 1989;245:852–855.

16 Asson-Batres MA, Spurgeon SL, Diaz J, DeLoughery TG, Bagby GC Jr: Evolutionary conservation of the AU-rich 3′ untranslated region of messenger RNA. Proc Natl Acad Sci USA 1994;91:1318–1322.

17 Chen CY, Shyu AB: AU-rich elements: characterization and importance in mRNA degradation. Trends Biochem Sci 1995;20:465–470.

18 Bakheet T, Williams BR, Khabar KS: ARED 3.0:the large and diverse AU-rich transcriptome. Nucleic Acids Res 2006;34:D111–D114.

19 Wilusz CJ, Gao M, Jones CL, Wilusz J, Peltz SW: Poly(A)-binding proteins regulate both mRNA deadenylation and decapping in yeast cytoplasmic extracts. RNA 2001;7:1416–1424.

20 Xu N, Chen CY, Shyu AB: Modulation of the fate of cytoplasmic mRNA by AU-rich elements: key sequence features controlling mRNA deadenylation and decay. Mol Cell Biol 1997;17:4611–4621.

21 Mukherjee D, Gao M, O'Connor JP, Raijmakers R, Pruijn G, Lutz CS, Wilusz J: The mammalian exosome mediates the efficient degradation of mRNAs that contain AU-rich elements. EMBO J 2002;21: 165–174.

22 Stoecklin G, Mayo T, Anderson P: ARE-mRNA degradation requires the 5′-3′ decay pathway. EMBO Rep 2006;7:72–77.

23 Jacob CO, Tashman NB: Disruption in the AU motif of the mouse TNF-alpha 3′ UTR correlates with reduced TNF production by macrophages in vitro. Nucleic Acids Res 1993;21:2761–2766.

24 Armaka M, Apostolaki M, Jacques P, Kontoyiannis DL, Elewaut D, Kollias G: Mesenchymal cell targeting by TNF as a common pathogenic principle in chronic inflammatory joint and intestinal diseases. J Exp Med 2008;205:331–337.

25 Kontoyiannis D, Pasparakis M, Pizarro TT, Cominelli F, Kollias G: Impaired on/off regulation of TNF biosynthesis in mice lacking TNF AU-rich elements: implications for joint and gut-associated immunopathologies. Immunity 1999;10:387–398.

26 Kontoyiannis D, Boulougouris G, Manoloukos M, Armaka M, Apostolaki M, Pizarro T, Kotlyarov A, Forster I, Flavell R, Gaestel M, Tsichlis P, Cominelli F, Kollias G: Genetic dissection of the cellular pathways and signaling mechanisms in modeled tumor necrosis factor-induced Crohn's-like inflammatory bowel disease. J Exp Med 2002;196:1563–1574.

27 Kontoyiannis D, Kotlyarov A, Carballo E, Alexopoulou L, Blackshear PJ, Gaestel M, Davis R, Flavell R, Kollias G: Interleukin-10 targets p38 MAPK to modulate ARE-dependent TNF mRNA translation and limit intestinal pathology. EMBO J 2001;20:3760–3770.

28 Lee JC, Young PR: Role of CSB/p38/RK stress response kinase in LPS and cytokine signaling mechanisms. J Leukoc Biol 1996;59:152–157.

29 Gaestel M: MAPKAP kinases – MKs – two's company, three's a crowd. Nat Rev Mol Cell Biol 2006;7: 120–130.

30 Lehner MD, Schwoebel F, Kotlyarov A, Leist M, Gaestel M, Hartung T: Mitogen-activated protein kinase-activated protein kinase 2-deficient mice show increased susceptibility to *Listeria monocytogenes* infection. J Immunol 2002;168:4667–4673.

31 Kotlyarov A, Neininger A, Schubert C, Eckert R, Birchmeier C, Volk HD, Gaestel M: MAPKAP kinase 2 is essential for LPS-induced TNF-alpha biosynthesis. Nat Cell Biol 1999;1:94–97.

32 Hegen M, Gaestel M, Nickerson-Nutter CL, Lin LL, Telliez JB: MAPKAP kinase 2-deficient mice are resistant to collagen-induced arthritis. J Immunol 2006;177:1913–1917.

33 Neininger A, Kontoyiannis D, Kotlyarov A, Winzen R, Eckert R, Volk HD, Holtmann H, Kollias G, Gaestel M: MK2 targets AU-rich elements and regulates biosynthesis of tumor necrosis factor and interleukin-6 independently at different post-transcriptional levels. J Biol Chem 2002;277:3065–3068.

34 Buxade M, Parra JL, Rousseau S, Shpiro N, Marquez R, Morrice N, Bain J, Espel E, Proud CG: The Mnks are novel components in the control of TNF alpha biosynthesis and phosphorylate and regulate hnRNP A1. Immunity 2005;23:177–189.

35 Buxade M, Morrice N, Krebs DL, Proud CG: The PSF.p54nrb complex is a novel Mnk substrate that binds the mRNA for tumor necrosis factor alpha. J Biol Chem 2008;283:57–65.

36 Rowlett RM, Chrestensen CA, Nyce M, Harp MG, Pelo JW, Cominelli F, Ernst PB, Pizarro TT, Sturgill TW, Worthington MT: MNK kinases regulate multiple TLR pathways and innate proinflammatory cytokines in macrophages. Am J Physiol Gastrointest Liver Physiol 2008;294:G452–G459.

37 Bennett BL, Sasaki DT, Murray BW, O'Leary EC, Sakata ST, Xu W, Leisten JC, Motiwala A, Pierce S, Satoh Y, Bhagwat SS, Manning AM, Anderson DW: SP600125, an anthrapyrazolone inhibitor of Jun N-terminal kinase. Proc Natl Acad Sci USA 2001;98: 13681–13686.

38 Swantek JL, Cobb MH, Geppert TD: Jun N-terminal kinase/stress-activated protein kinase (JNK/SAPK) is required for lipopolysaccharide stimulation of tumor necrosis factor alpha (TNF-alpha) translation: glucocorticoids inhibit TNF-alpha translation by blocking JNK/SAPK. Mol Cell Biol 1997;17:6274–6282.

39 Das S, Cho J, Lambertz I, Kelliher MA, Eliopoulos AG, Du K, Tsichlis PN: Tpl2/cot signals activate ERK, JNK, and NF-kappaB in a cell-type and stimulus-specific manner. J Biol Chem 2005;280:23748–23757.

40 Waterfield MR, Zhang M, Norman LP, Sun SC: NF-kappaB1/p105 regulates lipopolysaccharide-stimulated MAP kinase signaling by governing the stability and function of the Tpl2 kinase. Mol Cell 2003;11:685–694.

41 Dumitru CD, Ceci JD, Tsatsanis C, Kontoyiannis D, Stamatakis K, Lin JH, Patriotis C, Jenkins NA, Copeland NG, Kollias G, Tsichlis PN: TNF-alpha induction by LPS is regulated posttranscriptionally via a Tpl2/ERK-dependent pathway. Cell 2000;103: 1071–1083.

42 Rousseau S, Papoutsopoulou M, Symons A, Cook D, Lucocq JM, Prescott AR, O'Garra A, Ley SC, Cohen P: TPL2-mediated activation of ERK1 and ERK2 regulates the processing of pre-TNF alpha in LPS-stimulated macrophages. J Cell Sci 2008;121: 149–154.

43 Skinner SJ, Deleault KM, Fecteau R, Brooks SA: Extracellular signal-regulated kinase regulation of tumor necrosis factor-alpha mRNA nucleocytoplasmic transport requires TAP-NxT1 binding and the AU-rich element. J Biol Chem 2008;283:3191–3199.

44 Zhao Q, Wang X, Nelin LD, Yao Y, Matta R, Manson ME, Baliga RS, Meng X, Smith CV, Bauer JA, Chang CH, Liu Y: MAP kinase phosphatase 1 controls innate immune responses and suppresses endotoxic shock. J Exp Med 2006;203:131–140.

45 Salojin KV, Owusu IB, Millerchip KA, Potter M, Platt KA, Oravecz T: Essential role of MAPK phosphatase-1 in the negative control of innate immune responses. J Immunol 2006;176:1899–1907.

46 Zhao Q, Shepherd EG, Manson ME, Nelin LD, Sorokin A, Liu Y: The role of mitogen-activated protein kinase phosphatase-1 in the response of alveolar macrophages to lipopolysaccharide: attenuation of proinflammatory cytokine biosynthesis via feedback control of p38. J Biol Chem 2005;280:8101–8108.

47 Sun L, Stoecklin G, Van Way S, Hinkovska-Galcheva V, Guo RF, Anderson P, Shanley TP: Tristetraprolin (TTP)-14-3-3 complex formation protects TTP from dephosphorylation by protein phosphatase 2a and stabilizes tumor necrosis factor-alpha mRNA. J Biol Chem 2007;282:3766–3777.
48 Rajasingh J, Bord E, Luedemann C, Asai J, Hamada H, Thorne T, Qin G, Goukassian D, Zhu Y, Losordo DW, Kishore R: IL-10-induced TNF-alpha mRNA destabilization is mediated via IL-10 suppression of p38 MAP kinase activation and inhibition of HuR expression. FASEB J 2006;20:2112–2114.
49 Mijatovic T, Kruys V, Caput D, Defrance P, Huez G: Interleukin-4 and -13 inhibit tumor necrosis factor-alpha mRNA translational activation in lipopolysaccharide-induced mouse macrophages. J Biol Chem 1997;272:14394–14398.
50 Stoecklin G, Lu M, Rattenbacher B, Moroni C: A constitutive decay element promotes tumor necrosis factor alpha mRNA degradation via an AU-rich element-independent pathway. Mol Cell Biol 2003; 23:3506–3515.
51 Barreau C, Paillard L, Osborne HB: AU-rich elements and associated factors: are there unifying principles? Nucleic Acids Res 2006;33:7138–7150.
52 Blackshear PJ: Tristetraprolin and other CCCH tandem zinc-finger proteins in the regulation of mRNA turnover. Biochem Soc Trans 2002;30:945–952.
53 Suzuki T, Tsutsumi A, Suzuki H, Suzuki E, Sugihara M, Muraki Y, Hayashi T, Chino Y, Goto D, Matsumoto I, Ito S, Miyazawa K, Sumida T: Tristetraprolin (TTP) gene polymorphisms in patients with rheumatoid arthritis and healthy individuals. Mod Rheumatol 2008;18:472–479.
54 Carrick DM, Chulada P, Donn R, Fabris M, McNicholl J, Whitworth W, Blackshear PJ: Genetic variations in ZFP36 and their possible relationship to autoimmune diseases. J Autoimmun 2006;26:182–196.
55 Blackshear PJ, Phillips RS, Vazquez-Matias J, Mohrenweiser H: Polymorphisms in the genes encoding members of the tristetraprolin family of human tandem CCCH zinc finger proteins. Prog Nucleic Acid Res Mol Biol 2003;75:43–68.
56 Taylor GA, Carballo E, Lee DM, Lai WS, Thompson MJ, Patel DD, Schenkman DI, Gilkeson GS, Broxmeyer HE, Haynes BF, Blackshear PJ: A pathogenetic role for TNF alpha in the syndrome of cachexia, arthritis, and autoimmunity resulting from tristetraprolin (TTP) deficiency. Immunity 1996;4:445–454.
57 Carballo E, Gilkeson GS, Blackshear PJ: Bone marrow transplantation reproduces the tristetraprolin-deficiency syndrome in recombination activating gene-$2^{-/-}$ mice. Evidence that monocyte/macrophage progenitors may be responsible for TNFalpha overproduction. J Clin Invest 1997;100: 986–995.
58 Sandler H, Stoecklin G: Control of mRNA decay by phosphorylation of tristetraprolin. Biochem Soc Trans 2008;36:491–496.
59 Piecyk M, Wax S, Beck AR, Kedersha N, Gupta M, Maritim B, Chen S, Gueydan C, Kruys V, Streuli M, Anderson P: TIA-1 is a translational silencer that selectively regulates the expression of TNF-alpha. EMBO J 2000;19:4154–4163.
60 Jing Q, Huang S, Guth S, Zarubin T, Motoyama A, Chen J, Di Padova F, Lin SC, Gram H, Han J: Involvement of microRNA in AU-rich element-mediated mRNA instability. Cell 2005;120:623–634.
61 Cao H, Tuttle JS, Blackshear PJ: Immunological characterization of tristetraprolin as a low abundance, inducible, stable cytosolic protein. J Biol Chem 2004;279:21489–21499.
62 Cao H, Deterding LJ, Venable JD, Kennington EA, Yates JR 3rd, Tomer KB, Blackshear PJ: Identification of the anti-inflammatory protein tristetraprolin as a hyperphosphorylated protein by mass spectrometry and site-directed mutagenesis. Biochem J 2006; 394:285–297.
63 Chrestensen CA, Schroeder MJ, Shabanowitz J, Hunt DF, Pelo JW, Worthington MT, Sturgill TW: MAPKAP kinase 2 phosphorylates tristetraprolin on in vivo sites including Ser178, a site required for 14–3–3 binding. J Biol Chem 2004;279:10176–10184.
64 Johnson BA, Stehn JR, Yaffe MB, Blackwell TK: Cytoplasmic localization of tristetraprolin involves 14-3–3-dependent and -independent mechanisms. J Biol Chem 2002;277:18029–18036.
65 Stoecklin G, Stubbs T, Kedersha N, Wax S, Rigby WF, Blackwell TK, Anderson P: MK2-induced tristetraprolin:14-3–3 complexes prevent stress granule association and ARE-mRNA decay. EMBO J 2004;23:1313–1324.
66 Brook M, Tchen CR, Santalucia T, McIlrath J, Arthur JS, Saklatvala J, Clark AR: Posttranslational regulation of tristetraprolin subcellular localization and protein stability by p38 mitogen-activated protein kinase and extracellular signal-regulated kinase pathways. Mol Cell Biol 2006;26:2408–2418.

67 Suzuki K, Nakajima H, Ikeda K, Maezawa Y, Suto A, Takatori H, Saito Y, Iwamoto I: IL-4-Stat6 signaling induces tristetraprolin expression and inhibits TNF-alpha production in mast cells. J Exp Med 2003;198: 1717–1727.
68 Ogawa K, Chen F, Kim YJ, Chen Y: Transcriptional regulation of tristetraprolin by transforming growth factor-beta in human T cells. J Biol Chem 2003;278: 30373–30381.
69 Bell SE, Sanchez MJ, Spasic-Boskovic O, Santalucia T, Gambardella L, Burton GJ, Murphy JJ, Norton JD, Clark AR, Turner M: The RNA binding protein Zfp36l1 is required for normal vascularisation and post-transcriptionally regulates VEGF expression. Dev Dyn 2006;235:3144–3155.
70 Stumpo DJ, Byrd NA, Phillips RS, Ghosh S, Maronpot RR, Castranio T, Meyers EN, Mishina Y, Blackshear PJ: Chorioallantoic fusion defects and embryonic lethality resulting from disruption of Zfp36L1, a gene encoding a CCCH tandem zinc finger protein of the tristetraprolin family. Mol Cell Biol 2004;24:6445–6455.
71 Ramos SB, Stumpo DJ, Kennington EA, Phillips RS, Bock CB, Ribeiro-Neto F, Blackshear PJ: The CCCH tandem zinc-finger protein Zfp36l2 is crucial for female fertility and early embryonic development. Development 2004;131:4883–4893.
72 Stoecklin G, Colombi M, Raineri I, Leuenberger S, Mallaun M, Schmidlin M, Gross B, Lu M, Kitamura T, Moroni C: Functional cloning of BRF1, a regulator of ARE-dependent mRNA turnover. EMBO J 2002;21:4709–4718.
73 Eulalio A, Behm-Ansmant I, Izaurralde E: P bodies: at the crossroads of post-transcriptional pathways. Nat Rev Mol Cell Biol 2007;8:9–22.
74 Lai WS, Carballo E, Thorn JM, Kennington EA, Blackshear PJ: Interactions of CCCH zinc finger proteins with mRNA. Binding of tristetraprolin-related zinc finger proteins to Au-rich elements and destabilization of mRNA. J Biol Chem 2000;275: 17827–17837.
75 Phillips K, Kedersha N, Shen L, Blackshear PJ, Anderson P: Arthritis suppressor genes TIA-1 and TTP dampen the expression of tumor necrosis factor alpha, cyclooxygenase 2, and inflammatory arthritis. Proc Natl Acad Sci USA 2004;101:2011–2016.
76 Kedersha NL, Gupta M, Li W, Miller I, Anderson P: RNA-binding proteins TIA-1 and TIAR link the phosphorylation of eIF-2 alpha to the assembly of mammalian stress granules. J Cell Biol 1999;147: 1431–1442.
77 Gilks N, Kedersha N, Ayodele M, Shen L, Stoecklin G, Dember LM, Anderson P: Stress granule assembly is mediated by prion-like aggregation of TIA-1. Mol Biol Cell 2004;15:5383–5398.
78 Raineri I, Wegmueller D, Gross B, Certa U, Moroni C: Roles of AUF1 isoforms, HuR and BRF1 in ARE-dependent mRNA turnover studied by RNA interference. Nucleic Acids Res 2004;32:1279–1288.
79 Brewer G: An A + U-rich element RNA-binding factor regulates c-myc mRNA stability in vitro. Mol Cell Biol 1991;11:2460–2466.
80 Fialcowitz EJ, Brewer BY, Keenan BP, Wilson GM: A hairpin-like structure within an AU-rich mRNA-destabilizing element regulates trans-factor binding selectivity and mRNA decay kinetics. J Biol Chem 2005;280:22406–22417.
81 Loflin P, Chen CY, Shyu AB: Unraveling a cytoplasmic role for hnRNP D in the in vivo mRNA destabilization directed by the AU-rich element. Genes Dev 1999;13:1884–1897.
82 Sela-Brown A, Silver J, Brewer G, Naveh-Many T: Identification of AUF1 as a parathyroid hormone mRNA 3′-untranslated region-binding protein that determines parathyroid hormone mRNA stability. J Biol Chem 2000;275:7424–7429.
83 Xu N, Chen CY, Shyu AB: Versatile role for hnRNP D isoforms in the differential regulation of cytoplasmic mRNA turnover. Mol Cell Biol 2001;21:6960–6971.
84 Liao B, Hu Y, Brewer G: Competitive binding of AUF1 and TIAR to MYC mRNA controls its translation. Nat Struct Mol Biol 2007;14:511–518.
85 Mazan-Mamczarz K, Kuwano Y, Zhan M, White EJ, Martindale JL, Lal A, Gorospe M: Identification of a signature motif in target mRNAs of RNA-binding protein AUF1. Nucleic Acids Res 2009;37:204–214.
86 Laroia G, Cuesta R, Brewer G, Schneider RJ: Control of mRNA decay by heat shock-ubiquitin-proteasome pathway. Science 1999;284:499–502.
87 Rousseau S, Morrice N, Peggie M, Campbell DG, Gaestel M, Cohen P: Inhibition of SAPK2a/p38 prevents hnRNP A0 phosphorylation by MAPKAP-K2 and its interaction with cytokine mRNAs. EMBO J 2002;21:6505–6514.
88 Buxade M, Parra JL, Rousseau S, Shpiro N, Marquez R, Morrice N, Bain J, Espel E, Proud CG: The Mnks are novel components in the control of TNF alpha biosynthesis and phosphorylate and regulate hnRNP A1. Immunity 2005;23:177–189.
89 Manley GT, Smitt PS, Dalmau J, Posner JB: Hu antigens: reactivity with Hu antibodies, tumor expression, and major immunogenic sites. Ann Neurol 1995;38:102–110.
90 Brennan CM, Steitz JA: HuR and mRNA stability. Cell Mol Life Sci 2001;58:266–277.

91 McMullen MR, Cocuzzi E, Hatzoglou M, Nagy LE: Chronic ethanol exposure increases the binding of HuR to the TNFalpha 3′-untranslated region in macrophages. J Biol Chem 2003;278:38333–38341.
92 Dean JL, Wait R, Mahtani KR, Sully G, Clark AR, Saklatvala J: The 3′ untranslated region of tumor necrosis factor alpha mRNA is a target of the mRNA-stabilizing factor HuR. Mol Cell Biol 2001; 21:721–730.
93 Di Marco S, Hel Z, Lachance C, Furneaux H, Radzioch D: Polymorphism in the 3′-untranslated region of TNFalpha mRNA impairs binding of the post-transcriptional regulatory protein HuR to TNFalpha mRNA. Nucleic Acids Res 2001;29:863–871.
94 Katsanou V, Papadaki O, Milatos S, Blackshear PJ, Anderson P, Kollias G, Kontoyiannis DL: HuR as a negative posttranscriptional modulator in inflammation. Mol Cell 2005;19:777–789.
95 Katsanou V, Milatos S, Yiakouvaki A, Sgantzis N, Kotsoni A, Alexiou M, Harokopos V, Aidinis V, Hemberger M, Kontoyiannis DL: The RNA-binding protein Elavl1/HuR is essential for placental branching morphogenesis and embryonic development. Mol Cell Biol 2009;29:2762–2776.
96 Papadaki O, Milatos S, Grammenoudi S, Mukherjee N, Keene JD, Kontoyiannis DL: Control of thymic T cell maturation, deletion and egress by the RNA-binding protein HuR. J Immunol 2009;182:6779–6788.
97 Suzuki E, Tsutsumi A, Sugihara M, Mamura M, Goto D, Matsumoto I, Ito S, Ikeda K, Ochiai N, Sato Y, Sumida T: Expression of TNF-alpha, tristetraprolin, T-cell intracellular antigen-1 and Hu antigen R genes in synovium of patients with rheumatoid arthritis. Int J Mol Med 2006;18:273–278.
98 Sugihara M, Tsutsumi A, Suzuki E, Wakamatsu E, Suzuki T, Ogishima H, Hayashi T, Chino Y, Ishii W, Mamura M, Goto D, Matsumoto I, Ito S, Sumida T: Effects of infliximab therapy on gene expression levels of tumor necrosis factor alpha, tristetraprolin, T cell intracellular antigen 1, and Hu antigen R in patients with rheumatoid arthritis. Arthritis Rheum 2007;56:2160–2169.
99 Doller A, Pfeilschifter J, Eberhardt W: Signalling pathways regulating nucleo-cytoplasmic shuttling of the mRNA-binding protein HuR. Cell Signal 2008;20:2165–2173.
100 Li H, Park S, Kilburn B, Jelinek MA, Henschen-Edman A, Aswad DW, Stallcup MR, Laird-Offringa IA: Lipopolysaccharide-induced methylation of HuR, an mRNA-stabilizing protein, by CARM1. Coactivator-associated arginine methyltransferase. J Biol Chem 2002;277:44623–44630.
101 Lafarga V, Cuadrado A, Lopez de Silanes I, Bengoechea R, Fernandez-Capetillo O, Nebreda AR: p38 Mitogen-activated protein kinase- and HuR-dependent stabilization of p21(Cip1) mRNA mediates the G(1)/S checkpoint. Mol Cell Biol 2009; 29:4341–4351.
102 Moraes KC, Wilusz CJ, Wilusz J: CUG-BP binds to RNA substrates and recruits PARN deadenylase. RNA 2006;12:1084–1091.
103 Zhang L, Lee JE, Wilusz J, Wilusz CJ: The RNA-binding protein CUGBP1 regulates stability of tumor necrosis factor mRNA in muscle cells: implications for myotonic dystrophy. J Biol Chem 2008; 283:22457–22463.
104 Eulalio A, Huntzinger E, Izaurralde E: Getting to the root of miRNA-mediated gene silencing. Cell 2008;132:9–14.
105 Garnon J, Lachance C, Di Marco S, Hel Z, Marion D, Ruiz MC, Newkirk MM, Khandjian EW, Radzioch D: Fragile X-related protein FXR1P regulates proinflammatory cytokine tumor necrosis factor expression at the post-transcriptional level. J Biol Chem 2005;280:5750–5763.
106 Mientjes EJ, Willemsen R, Kirkpatrick LL, Nieuwenhuizen IM, Hoogeveen-Westerveld M, Verweij M, Reis S, Bardoni B, Hoogeveen AT, Oostra BA, Nelson DL: Fxr1 knockout mice show a striated muscle phenotype: implications for Fxr1p function in vivo. Hum Mol Genet 2004;13:1291–1302.
107 Vasudevan S, Steitz JA: AU-rich-element-mediated upregulation of translation by FXR1 and Argonaute 2. Cell 2007;128:1105–1118.
108 O'Connell RM, Taganov KD, Boldin MP, Cheng G, Baltimore D: MicroRNA-155 is induced during the macrophage inflammatory response. Proc Natl Acad Sci USA 2007;104:1604–1609.
109 Taganov KD, Boldin MP, Chang KJ, Baltimore D: NF-kappaB-dependent induction of microRNA miR-146, an inhibitor targeted to signaling proteins of innate immune responses. Proc Natl Acad Sci USA 2006;103:12481–12486.
110 Rodriguez A, Vigorito E, Clare S, Warren MV, Couttet P, Soond DR, van Dongen S, Grocock RJ, Das PP, Miska EA, Vetrie D, Okkenhaug K, Enright AJ, Dougan G, Turner M, Bradley A: Requirement of bic/microRNA-155 for normal immune function. Science 2007;316:608–611.
111 Vigorito E, Perks KL, breu-Goodger C, Bunting S, Xiang Z, Kohlhaas S, Das PP, Miska EA, Rodriguez A, Bradley A, Smith KG, Rada C, Enright AJ, Toellner KM, Maclennan IC, Turner M: microRNA-155 regulates the generation of immunoglobulin class-switched plasma cells. Immunity 2007;27: 847–859.

112 Thai TH, Calado DP, Casola S, Ansel KM, Xiao C, Xue Y, Murphy A, Frendewey D, Valenzuela D, Kutok JL, Schmidt-Supprian M, Rajewsky N, Yancopoulos G, Rao A, Rajewsky K: Regulation of the germinal center response by microRNA-155. Science 2007;316:604–608.

113 Tili E, Michaille JJ, Cimino A, Costinean S, Dumitru CD, Adair B, Fabbri M, Alder H, Liu CG, Calin GA, Croce CM: Modulation of miR-155 and miR-125b levels following lipopolysaccharide/TNF-alpha stimulation and their possible roles in regulating the response to endotoxin shock. J Immunol 2007;179: 5082–5089.

Dr. Dimitris L. Kontoyiannis
Institute of Immunology, Biomedical Sciences Research Center 'Alexander Fleming'
34 Al. Fleming Street
GR–16672 Vari (Greece)
Tel. +30 210 9654335, Fax +30 210 9656563, E-Mail kontoyiannis@fleming.gr

Kollias G, Sfikakis PP (eds): TNF Pathophysiology. Molecular and Cellular Mechanisms.
Curr Dir Autoimmun. Basel, Karger, 2010, vol 11, pp 80–93

Role of TNF in Pathologies Induced by Nuclear Factor κB Deficiency

Katerina Vlantis · Manolis Pasparakis

Centre for Molecular Medicine and Cologne Excellence Cluster on Cellular Stress Responses in Aging-Associated Diseases, Institute of Genetics, University of Cologne, Cologne, Germany

Abstract

TNF is a potent cytokine with an important role in the regulation of a multitude of cellular responses and in coordinating immune and inflammatory reactions. TNF exerts its effects by binding to the TNFR1- and TNFR2-specific cell surface receptors, which activate a number of intracellular signaling cascades including the nuclear factor κB (NF-κB) and mitogen-activated protein kinase pathways. Activation of NF-κB mediates many of the functions of TNF by transmitting information from the cell surface TNF receptors to the nucleus, where it coordinates a gene expression program that allows the cell to survive and elicit its responses. The intimate interplay of TNF with the NF-κB signaling pathway is highlighted by results obtained in transgenic and knockout mice with defects in NF-κB signaling components, where TNF has been shown to contribute to different pathologies observed in these mice. This chapter focuses on the function of TNF in pathologies induced by NF-κB deficiency and discusses the implications of these findings for our understanding of inflammatory diseases.

TNF and Its Receptors

TNF is a potent cytokine with a central role in the regulation of immune and inflammatory responses. TNF is expressed as a 26-kDa membrane-bound protein (mTNF) that forms trimers [1]. The 17-kDa soluble form of TNF is produced by the proteolytic cleavage of mTNF by the metalloproteinase TNF-α-converting enzyme [2]. The bioactive form of soluble TNF is also trimeric with a molecular weight of approximately 52 kDa. TNF is produced mainly by macrophages, although other immune cells such as dendritic cells, T cells, B cells and mast cells, but also epithelial and endothelial cells and fibroblasts can also be induced to express TNF.

TNF exerts a wide range of cellular responses by binding to two distinct cell surface receptors, termed TNF receptor 1 (TNFR1, CD120a, p55/60) and TNF receptor 2 (TNFR2, CD120b, p75/p80) [3]. TNFR1, a type 1 transmembrane glycoprotein, is constitutively expressed in most cell types including immune and nonimmune cells, whereas TNFR2 is more restricted to cells of the hematopoietic lineage and endothelial cells and its expression is regulated. The two TNF receptors do not possess endogenous catalytic activity and therefore depend on the recruitment of other signaling molecules in order to initiate intracellular signaling pathways.

Due to its potent effects, the production of TNF itself needs to be tightly controlled. Therefore, in addition to strict transcriptional control, the TNF mRNA stability is also tightly regulated. The 3′ untranslated region of the TNF transcript was shown to bear an AU-rich element (ARE), which is important for the regulation of TNF expression on a posttranscriptional level. Through the binding of different factors to the ARE sequence, the TNF mRNA stability and translational efficiency are modulated [4].

Ligation of the TNF receptors leads to various cellular responses, including programmed cell death, cell survival, cell differentiation, proliferation and activation of immune functions. The exact function of TNFR2 in TNF-mediated responses is less well established, but it is thought to play a role in TNF-induced proliferation of T lymphocytes. Most of the effects of TNF are mediated by TNFR1; therefore, in most cases research tends to focus on the more broadly involved functions of TNFR1 signaling.

TNF Receptor 1 Signaling

Upon binding of TNF to the extracellular cysteine-rich domains of TNFR1, the receptor molecules form trimers and trigger the activation of downstream signaling pathways [5]. TNFR1 induces the activation of multiple intracellular cascades including the transcription factor nuclear factor κB (NF-κB) and the mitogen-activated protein kinase (MAPK) pathways JNK, p38 and ERK. Activation of NF-κB and MAPK signaling induces the expression of a large number of genes that regulate immune and inflammatory responses including proinflammatory cytokines and chemokines and cell adhesion molecules. At the same time, TNF-mediated NF-κB activation induces the expression of several antiapoptotic molecules including cellular FLICE-like inhibitory protein and Bcl2 family members, and also a number of antioxidant enzymes such as superoxide dismutases (Cu/Zn SOD, MnSODs). In addition to activating proinflammatory signaling cascades, TNFR1 is also a potent inducer of apoptosis by recruiting and activating caspase 8. The outcome of TNFR1 stimulation is determined by the interplay between proapoptotic and antiapoptotic intracellular signaling cascades induced downstream of the receptor.

The cytoplasmic events that activate signaling downstream of TNFR1 have been extensively characterized. The intracellular part of TNFR1 contains a protein-protein interaction domain, called death domain (DD), which is essential for TNFR1-induced signaling. Upon TNF binding, the trimerized TNFR1 recruits a number of molecules, including the TNFR1-associated DD (TRADD) protein, the RIP1 kinase, and the ubiquitin ligases TNF receptor-associated factors TRAF2 and TRAF5 and cellular inhibitors of apoptosis cIAP1 and cIAP2, forming a receptor-proximal signaling complex, also termed complex I [6]. Dissecting the function of each of these molecules in TNFR1-induced signaling has been made possible with the generation of knockout mice with targeted inactivation of these proteins. Mouse embryonic fibroblasts (MEFs) lacking RIP1 cannot activate NF-κB and show increased death in response to TNF stimulation, showing that RIP1 is important for TNFR1-induced NF-κB activation but not for TNFR1-induced apoptosis. Cells with a combined deficiency of TRAF2 and TRAF5 also cannot activate NF-κB and show increased death in response to TNF stimulation, demonstrating that TRAF2 and TRAF5 are essential for NF-κB activation but not apoptosis induced by TNFR1. Recent studies on cells from TRADD-deficient mice demonstrated that TRADD is an essential adapter for TNFR1-induced signaling [7]. TNFR1-induced proinflammatory signaling to NF-κB and MAPK pathways was severely impaired but not completely blocked in the absence of TRADD. TNFR1-induced apoptosis was totally abrogated in MEFs or in hepatocytes from TRADD-deficient mice. Thus, TRADD is essential for TNFR1-induced apoptosis and very important but not completely indispensable for TNFR1-mediated proinflammatory signaling. Studies in TRADD-deficient cells also dissected the molecular function of TRADD in the TNFR1 signaling complex. TRADD is recruited to TNFR1 via a DD interaction and facilitates the recruitment of TRAF2 and TRAF5. RIP1 recruitment to the intracellular domain of TNFR1 was more efficient in cells expressing TRADD, although RIP1 could also associate with the ligated TNFR1 in the absence of TRADD. In cells expressing TRADD, TNF binding to TNFR1 induces the polyubiquitination of RIP1, which then serves as a platform for the recruitment and activation of downstream kinase complexes leading to the induction of NF-κB and MAPK pathways. RIP1 ubiquitination depends on the presence of TRAF2 and TRAF5, as TNFR1 stimulation cannot induce RIP1 ubiquitination in cells lacking TRAF2 and TRAF5. More recently, the cellular inhibitor of apoptosis proteins cIAP1 and cIAP2 were also shown to be important for TNFR1 signaling, although the molecular mechanisms by which IAPs control TNFR1 signaling have not been fully elucidated.

Extensive experimental evidence suggests that ubiquitination plays a crucial role in the activation of TNFR1-mediated signaling [8]. K63-linked polyubiquitination of RIP1 is thought to be essential for the activation of NF-κB and MAPK signaling upon binding of TNF to TNFR1. K63-linked polyubiquitin chains on RIP1 have been proposed to serve as a platform for the recruitment of the transforming growth factor-β-activated kinase 1 (TAK1) signaling complex containing the TAK1 and the TAK-binding proteins TAB1 and TAB2, which are responsible for binding the K63-polyubiquitin chain of RIP1. Ubiquitinated RIP1 also recruits the inhibitor of NF-κB kinase (IκB kinase, IKK) complex, which is composed of the two kinase subunits IKK1/IKKα and IKK2/IKKβ and a regulatory subunit without catalytic activity that is named NF-κB essential modulator (NEMO or IKKγ). TAK1 has been implicated in the activation of NF-κB – and MAPK-signaling, by inducing the phosphorylation of IKKs and upstream kinases of the MAPK cascade.

TNFR1-induced apoptosis is mediated by the assembly of a cytoplasmic signaling complex termed complex ll. Complex ll forms after dissociation of the membrane proximal complex l from TNFR1 [9]. It is thought to consist of TRADD, TRAF2 and deubiquitinated RIP1 that are now free to interact with the adapter molecule Fas-associated DD (FADD). In addition to its DD that endows FADD with the ability to interact with other DD-containing proteins, FADD contains a death effector domain that has been shown to be involved in the recruitment of caspases, in this case particularly the initiator caspase, caspase 8. Caspase 8 is integrated to the signaling complex ll through the death effector domain of the adapter FADD and according to the close proximity model is auto-activated by self-proteolytic cleavage initiating a caspase activation cascade that culminates in the induction of apoptosis.

In order to prevent uncontrolled receptor aggregation and subsequent signaling in the absence of TNF, TNFR1 is kept in an inactive monomeric form by intracellular association with silencer of DDs (SODD). Upon ligation of trimeric TNF molecules to TNFR1, SODD dissociates from TNFR1 and allows the signaling cascade to be initiated.

Nuclear Factor-κB Signaling Pathway

The Rel/NF-κB transcription factors form a family with five members in mammals, namely NF-κB 1 (the precursor p105 and the processed form p50), NF-κB 2 (the precursor p100 and the processed p52), RelA (p65), RelB and c-Rel [10]. NF-κB proteins contain an N-terminal Rel-homology domain, which is required for homo- and heterodimerization and DNA binding. NF-κB dimers bind to specific sites containing a consensus DNA sequence in the promoter regions of NF-κB target genes. Whereas p65, RelB and c-Rel are endowed with a C-terminal transcription activation domain (TAD), p50 and p52 lack a TAD. Therefore, the p50 and p52 proteins need to dimerize with one of the other Rel factors that contain a TAD in order to activate transcription, and p50 or p52 homodimers are believed to suppress instead of activating transcription.

In unstimulated cells, NF-κB molecules are bound to inhibitory proteins belonging to the IκB family containing IκBα, IκBβ and IκBε, BCL3, and IkBζ or IκBNS. IκBs contain several ankyrin repeats in their protein sequence, which allow binding to NF-κB. IκBs largely prevent the translocation of the transcription factor dimers to the nucleus by masking the NF-κB nuclear localization signal. In addition, IκBα exposes a strong nuclear export signal. Therefore, in unstimulated cells NF-κB is found mainly in the cytoplasm, through a permanent shuttling of NF-κB-IκBα complexes between the nucleus and the cytoplasm. NF-κB activation requires the degradation of IκBs, which frees NF-κB dimers allowing them to accumulate in the nucleus where they bind NF-κB sites on the promoters of target genes to regulate transcription.

NF-κB activation is regulated by the IKK complex, which phosphorylates IκB proteins on specific serine residues targeting them for polyubiquitination and subsequent proteasomal degradation [10]. While IκBα, IκBβ and IκBε are fully degraded upon IKK-mediated phosphorylation, p105 and p100, which serve as NF-κB inhibitors due to the ankyrin repeats they harbor in the C-terminal portion of the protein are only partly degraded producing the p50 and p52 proteins, respectively.

Two distinct NF-κB activation pathways have been identified [10]. The classical or canonical NF-κB pathway is induced downstream of proinflammatory cytokines such as TNF and interleukin-1 and innate immune receptors such as Toll-like receptors and NOD-like receptors. The canonical NF-κB pathway involves predominantly the activation of p50, p65 and c-Rel containing NF-κB dimers, by IKK-mediated degradation of IκBα, IκBβ and IκBε. In order to induce NF-κB, the catalytic activity of the IKK complex needs to become activated. The mode of activation of the IKK complex differs in different signaling cascades and has been described in more detail above for the TNFR1 pathway, where the IKK is activated by TAK1 upon recruitment to the ubiquitinated RIP1. Upon activation, the IKK phosphorylates IκBα on 2 serine residues, Ser32 and Ser36. This phosphorylation event targets IκBα for K48-linked polyubiquitination by the βTRCP-SCF complex, followed by the proteasomal degradation of IκBα that allows NF-κB dimers to accumulate in the nucleus and activate gene transcription. Studies in cells lacking different subunits of the IKK complex have shown that NEMO is indispensable for activation of the canonical NF-κB pathway [11]. IKK2 is believed to be the catalytic subunit mainly responsible for the activation of canonical NF-κB signaling; however, cells lacking IKK2 show impaired but not completely inhibited NF-κB activation, as is the case in NEMO-deficient cells, suggesting that in the absence of IKK2 the other IKK, namely IKK1, can compensate to some extent by inducing IκB phosphorylation. Indeed, cells lacking both IKKs showed complete absence of cytokine-induced NF-κB activation, demonstrating that IKK1 and IKK2 show some functional redundancy in activating the canonical NF-κB pathway. Canonical NF-κB signaling regulates the expression of genes controlling immune and inflammatory responses such as cytokines, chemokines and adhesion molecules, and also genes that mediate cell survival in response to stress induced by a variety of factors including cytokines, radiation and reactive oxygen species.

The alternative, or non-canonical, NF-κB pathway is induced by the processing of p100 to p52, which allows p52/RelB NF-κB dimers to translocate to the nucleus and activate the transcription of distinct sets of target genes. p100 processing is induced by phosphorylation of two specific serine residues in the C-terminal part of the protein, Ser866 and Ser870, which targets the molecule for ubiquitination and controlled proteasomal degradation of the inhibitory domain, producing the processed p52 form. Studies in cells deficient in individual IKK subunits showed that p100 processing requires IKK1 but not IKK2 or NEMO [10]. In this pathway, IKK1 is activated by NIK to phosphorylate p100 and induce its processing and production of p52. Cytokine receptors belonging to the TNF receptor superfamily such as the lymphotoxin-β receptor and the BAFF receptor are the main inducers of alternative NF-κB activation. RelB/p52 dimers regulate specific sets of genes including chemokines and cytokines that control the organogenesis and structural organization of lymphoid tissues and also the development and survival of lymphoid cells.

Nuclear Factor κB and TNF Signaling in Mouse Models

Nuclear Factor κB and TNF Signaling in the Fetal Liver

The generation and analysis of knockout mice lacking specific components of the NF-κB pathway demonstrated the importance of this signaling cascade for the maintenance of tissue homeostasis and for the regulation of immune and inflammatory responses. Mice with targeted inactivation of the genes expressing p65, NEMO or IKK2 show early embryonic lethality, while mice lacking p50, p52, c-Rel or RelB are viable. IKK1-deficient mice die at birth displaying skin defects that are not related to NF-κB signaling, but are attributed to an additional function of IKK1 to regulate the production of an as yet unidentified keratinocyte differentiation factor [10].

p65 knockout animals die between embryonic days 14.5 and 16.5, showing internal hemorrhage and signs of liver destruction [12]. Histological examination of p65-deficient embryos revealed massive liver degeneration with extensive apoptosis of liver parenchymal cells but not of hematopoietic precursors. These findings showed that NF-κB-deficient mice die during embryogenesis due to hepatocyte apoptosis. Studies in MEFs isolated from p65 knockout embryos showed that these cells cannot upregulate several NF-κB target genes upon TNF stimulation and are sensitive to TNF-induced apoptosis. As discussed earlier, NF-κB activation protects cells from TNF-induced death by controlling the expression of anti-apoptotic proteins such as cellular FLICE-like inhibitory protein and members of the Bcl2 family but also enzymes scavenging reactive oxygen species. These results suggested that the embryonic lethality of p65 knockout mice could be due to TNF-mediated killing of hepatocytes in the fetal liver. Indeed, genetic experiments provided experimental proof for this hypothesis. When p65-deficient mice were bred into a TNF-knockout genetic background, the embryonic lethality was rescued and the double-knockout mice were born at the expected mendelian ratio [13]. Analysis of embryos at E15–E16 revealed that massive hepatocyte apoptosis and liver degeneration did not occur in p65/TNF double-knockout embryos. Furthermore, in vitro-cultured primary hepatocytes isolated from p65/TNF double-knockout mice showed more pronounced cytotoxicity in response to TNF treatment. These findings demonstrated that p65-deficient mice die during embryogenesis due to TNF-mediated apoptosis of hepatocytes in the fetal liver. These experiments lead to the conclusion that in the fetal liver p65 is absolutely required to assure survival of hepatocytes after TNF challenge. Although also other embryonic tissues were shown to produce TNF and express TNFR1 as well as TNFR2, other organs in the p65-deficient embryos did not show tissue destruction and increased cell apoptosis during embryogenesis. This finding implies that the protective role of p65 in embryonic hepatocytes is characteristic for this particular cell type, while other cells do not require p65 for the prevention of apoptosis during embryonic development. p65/TNFR1 double-knockout mice appeared healthy at birth, but after a few days the mice became sick and most of them died before they reached 3 weeks of age. Histological analysis of dead double-knockout mice revealed signs of bacterial infections in several organs, indicating that the reason of death was impaired immune responses [14]. Bone marrow chimera reconstitution experiments showed that p65/TNFR1 double deficiency in stromal, but not hematopoietic, cells was responsible for impaired leukocyte recruitment and bacterial infections. Later studies reported that when p65/TNFR1 double-deficient mice were kept in a clean animal facility they survived to adulthood without showing signs of disease or inflammation, confirming that opportunistic infections are the cause of death in these animals [15].

Similarly to p65 knockout mice, IKK2-deficient mice were also shown to die during gestation [16]. A detailed analysis of embryos from heterozygous crosses revealed that homozygous IKK2 knockouts showed massive liver degeneration, suggesting that the cause of death was severe liver degeneration due to increased apoptosis of hepatocytes. Deletion of IKK2 confers a slightly more severe phenotype when compared to the phenotype provoked by the knockout of p65, since IKK2-deficient embryos die at around E13.5–E14 of gestation, whereas p65 knockout embryos die around E14.5–E16.5. Genetic experiments showed that the embryonic lethality of IKK2 knockout mice could be rescued by crossing the animals to a TNFR1-deficient background, demonstrating that like in the case of p65 knockout, IKK2-deficient embryos die because of TNF-mediated hepatocyte apoptosis. IKK2/TNFR1 double-knockout mice were born alive but died during the first weeks after birth due to opportunistic infections, similarly to p65/TNFR1 double-deficient animals. These results clearly demonstrated that IKK2 is essential for the protection of embryonic hepatocytes from TNF-induced death. Not surprisingly, targeted inactivation of the X-linked gene encoding NEMO also led to embryonic lethality of NEMO-deficient mice due to massive liver degeneration caused by hepatocyte apoptosis [17, 18]. Death of NEMO-deficient mice occurs at embryonic day 12.5–13.5, which is earlier than the IKK2 or p65 knockout embryos, suggesting that NEMO deficiency causes a more severe defect in NF-κB activation. The similar embryonic phenotypes of mice lacking p65, IKK2 or NEMO demonstrated that canonical NF-κB signaling has an essential function to protect hepatocytes in the fetal liver from TNF-mediated apoptosis. However, the embryonic lethality of these mice also prevented the analysis of the function of canonical NF-κB signaling in different tissues of the adult animal, in particular with relation to the role of TNF in affecting the response of NF-κB-deficient cells and tissues.

Nuclear Factor κB and TNF Signaling in the Adult Liver

The development of site-specific recombinase-mediated conditional gene-targeting methodologies allowing the spatial and temporal control of gene activity in adult mice provided a new method to study the function of genes whose targeted inactivation resulted in embryonic lethal phenotypes. Using Cre/lox recombination-assisted gene targeting, genes can be inactivated in specific cell types of the adult mouse allowing the study of their function. Conditional gene targeting has been used to study the role of NF-κB signaling in the adult liver. Surprisingly, mice lacking IKK2 or p65 specifically in hepatocytes develop normally and reach adulthood without showing any signs of liver degeneration, suggesting that IKK2 and p65 are not required for hepatocyte survival in the adult mouse [19]. Since the timing of p65 or IKK2 ablation by the hepatocyte-specific Cre recombinase transgenes during mouse development has not been studied in detail, it is possible that deletion was not achieved early enough (E13–E15) to sensitize the liver to TNF-induced apoptosis. Therefore, these experiments do not allow direct comparison of the effects of systemic versus tissue-specific knockout of the respective genes in the fetal liver. However, these results clearly showed that adult hepatocytes do not require IKK2 or p65 for survival under steady state conditions.

To study whether IKK2 or p65 are required for hepatocyte survival upon TNF stimulation, mice with hepatocyte-restricted IKK2 or p65 knockout were injected with TNF or lipopolysaccharide (LPS), a well-known inducer of endogenous TNF [19]. These experiments showed that hepatocyte-restricted IKK2 knockouts were resistant to TNF or LPS injection, suggesting that IKK2 is dispensable for the survival of hepatocytes upon TNF stimulation. Mice with hepatocyte-restricted p65 knockout showed increased sensitivity to LPS or TNF injection, revealing that p65 plays an important role for the protection of hepatocytes from TNF-induced apoptosis [19, 20]. Mice with liver parenchymal cell-restricted ablation of NEMO were extremely susceptible to TNF or LPS stimulation in vivo, showing massive liver damage and hepatocyte apoptosis that was much more severe than that observed in the p65 hepatocyte knockouts [21]. These findings demonstrated that canonical NF-κB activity is indispensable for the protection of the liver from TNF-induced toxicity, and suggested that in the absence of IKK2 there is compensatory activation of NF-κB induced by IKK1 that is sufficient to protect the cells from TNF-mediated apoptosis. The generation of mice lacking both IKK1 and IKK2 specifically in liver parenchymal cells provided genetic proof for the existence of functional redundancy between the two IKKs in mediating canonical NF-κB signaling in the liver. Mice with specific ablation of IKK1 in liver parenchymal cells were resistant to LPS/TNF-induced liver toxicity similarly to IKK2 liver knockouts. However, mice lacking both IKK1 and IKK2 in liver parenchymal cells were extremely sensitive to LPS/TNF-mediated liver injury similarly to NEMO hepatocyte knockouts [19]. These results demonstrated that the two IKKs show functional redundancy in mediating canonical NF-κB signaling in hepatocytes and protecting the liver from TNF-mediated apoptosis.

Nuclear Factor κB and TNF Signaling in the Epidermis

NF-κB signaling in epidermal keratinocytes plays a critical role in the maintenance of skin immune homeostasis. Studies in transgenic and conditional knockout mice revealed that NF-κB inhibition specifically in epidermal keratinocytes leads to the development of inflammatory skin lesions. Mice overexpressing a degradation-resistant IκBα inhibitor of NF-κB specifically in keratinocytes under the control of the keratin-5 promoter developed inflammatory skin lesions ultimately resulting in the development of squamous cell carcinomas later in life [22]. Moreover, mice with epidermal keratinocyte-restricted ablation of IKK2 (IKK2 epidermal knock-out, $IKK2^{EKO}$ mice) developed a severe inflammatory skin disease resembling human psoriasis [23]. $IKK2^{EKO}$ mice are born normal and develop skin lesions a few days after birth, resulting in death of the animals before postnatal day 9. The presence of T cells was not required for the lesions to develop in $IKK2^{EKO}$ mice, since $IKK2^{EKO}$ mice that were bred into a T cell receptor-α knockout genetic background, lacking αβ T lymphocytes, also suffered from inflammatory skin disease which progressed with similar kinetics and showed the same severity as in $IKK2^{EKO}$ mice with normal T cell compartment. Due to the very early disease manifestation at P4–P5 and the independence of T cells in disease pathogenesis, skin inflammation in $IKK2^{EKO}$ mice does not seem to be triggered by an antigen-induced immune response. The disease rather appears to be the outcome of a reaction that is driven by the innate immune system.

TNF plays a crucial role in the development of skin inflammation in human psoriasis. The important pathogenic role of TNF in this disease is highlighted by the efficacy of anti-TNF neutralizing antibodies for the treatment of psoriatic lesions. TNF was also shown to have a pivotal pathogenic function for the development of skin inflammation upon NF-κB inhibition in epidermal keratinocytes. When $IKK2^{EKO}$ mice were bred into a TNFR1-deficient genetic background, the double-knockout animals did not develop skin inflammation and survived to adulthood without showing skin lesions [23]. The importance of TNF-signaling in $IKK2^{EKO}$-induced skin disease was further underscored by providing evidence that $IKK2^{EKO}$ mice treated with a TNF-depleting agent showed reduced disease severity. When $IKK2^{EKO}$ mice received daily subcutaneously applied doses of 20 μg/g of a recombinant fusion protein composed of TNFR1 and the Fc part of human IgG (huTNFR:Fc) from P1 and P7, the psoriasis-like skin phenotype was greatly suppressed [24]. These results clearly showed that TNF plays an important disease-promoting role in the pathogenesis of skin inflammation in $IKK2^{EKO}$ mice. It is widely accepted that an impairment of canonical NF-κB signaling due to the lack of NF-κB induced anti-apoptotic factors renders cells more sensitive to TNF-mediated apoptosis [10]. However, cultured IKK2 knockout keratinocytes did not show increased spontaneous apoptosis and responded with only a slight increase in cell death incidence after TNF stimulation, suggesting that TNF-induced apoptosis of keratinocytes is unlikely to be the cause for the induction of skin inflammation in $IKK2^{EKO}$ mice. Moreover, the development of inflammatory skin disease and squamous cell carcinomas was also rescued in mice with keratinocyte-specific expression of degradation-resistant IκBα when they were bred into a TNFR1-deficient background [25].

Deletion of the regulatory subunit NEMO specifically in the epidermis of mice ($NEMO^{EKO}$ mice) was shown to lead to the development of a skin disease that shows a strong similarity to human incontinentia pigmenti (IP) [26]. IP is a rather complex disease affecting humans and was shown to be caused by mutations in the X-linked NEMO gene. IP is characterized by male embryonic lethality whereas heterozygous females, among other pathologies, develop skin disease. Deletion of NEMO completely blocks activation of NF-κB downstream of proinflammatory stimuli and renders keratinocytes more sensitive to TNF-induced cell death. Already at P2, $NEMO^{EKO}$ mice started showing a lack of pigmentation in the skin. At P5 an inflammatory skin phenotype started to develop and the skin defects progressed to severe lesions that resulted in death by P6. Moreover NEMO knockout keratinocytes were undergoing apoptosis more frequently than keratinocytes in wild-type animals at postnatal day 5. The ongoing skin inflammation observed at P5 in $NEMO^{EKO}$ mice and the increased production of TNF observed at this time point, suggested that TNF signaling could be involved in the pathogenesis of the skin lesions upon NEMO deletion in keratinocytes.

Breeding of $NEMO^{EKO}$ mice into a TNFR1-deficient background rescued the development of skin lesions in this model. $NEMO^{EKO}$/TNFR1 knockout animals were protected from the inflammatory skin disease and survived to adulthood [26]. This points towards an essential role of TNFR1-signaling in development of inflammation in the skin of $NEMO^{EKO}$ mice. Though abrogation of TNFR1-signaling prevents disease establishment in $NEMO^{EKO}$ pups, the mice were not completely protected from the development of skin disease. Most of the $NEMO^{EKO}$/TNFR1 knockout animals were shown to suffer from ulcerated skin lesions mostly around the neck area at an age of 4–6 months. Since TNFR1 is absent in these double-knockout mice, the development of the secondary skin lesions in adult mice does not seem to depend on TNF signaling, implying that other inflammatory mediators cause the skin disease in adult mice. Collectively, these results demonstrate that TNFR1 signaling is essential for the pathogenesis of the inflammatory skin disease developing upon epidermis-specific inhibition of NF-κB activation. At present, the molecular mechanisms and the cellular targets by which TNFR1 signaling triggers the pathogenesis of inflammatory skin lesions in mice with epidermal keratinocyte-restricted inhibition of NF-κB remain elusive.

Nuclear Factor κB and TNF Signaling in the Gut

Deletion of NEMO specifically in the intestinal epithelium caused severe spontaneous colitis in mice [27]. Mice lacking NEMO specifically in intestinal epithelial cells

($NEMO^{IEC\text{-}KO}$ mice) were born at the expected mendelian ratio, but already at young age showed severe colonic inflammation which manifests in heavy diarrhea and reduced bodyweight when compared to wild-type littermates. A substantial number of $NEMO^{IEC\text{-}KO}$ mice died at an early age due to wasting caused by pronounced colonic inflammation and severe diarrhea; however, a small number of animals could survive for a year or more. Histological colon cross-sections obtained from $NEMO^{IEC\text{-}KO}$ mice revealed thickening of the mucosal tissue that was accompanied by immune cell infiltration into the mucosa and submucosa.

Increased apoptosis of intestinal epithelial cells (IECs) was detected in the colon of $NEMO^{IEC\text{-}KO}$ mice, suggesting that an impairment of the colonic epithelial barrier could be involved in triggering colitis in this model. In support of this hypothesis, commensal bacteria were found to infiltrate the colonic mucosa of $NEMO^{IEC\text{-}KO}$ mice, presumably due to a breach of epithelial integrity caused by apoptosis of IECs. Lesions in the colonic epithelium of these mice were found to develop in a patchy fashion in young pups. Therefore, at an early time point of disease pathogenesis bacteria seem to invade the mucosa in a rather localized fashion, where they could activate residual innate immune cells. Bacterially induced activation of innate immune cells, like macrophages and dendritic cells, could lead to a transcriptional upregulation of various proinflammatory cytokines, like IL-6, IL-1β and TNF. The secretion of TNF by innate immune cells could further enhance the breakdown of the epithelial barrier by mediating the killing of NEMO-deficient IECs.

By generating $NEMO^{IEC\text{-}KO}$ mice that also lack TNFR1 ($NEMO^{IEC\text{-}KO}$/TNFR1 knockout) or TNF ($NEMO^{IEC\text{-}KO}$/TNF knockout), it could be demonstrated that disease pathogenesis strongly depends on TNF-signaling, since the double-deficient mice were protected from disease development, at least during the first 10 weeks of their life [27]. This finding indicates that functional TNF-signaling plays an important role in colitis development in $NEMO^{IEC\text{-}KO}$ mice. The fact that complete TNFR1 or TNF knockout mice were crossed with $NEMO^{IEC\text{-}KO}$ mice does not allow to identify the cells where TNF signaling acts to induce colon inflammation. TNF could contribute to increase IEC apoptosis and barrier impairment, but it could also act in coordinating the inflammatory response and the recruitment of immune cells to the mucosa causing severe colitis in this model.

Mice lacking p65 specifically in the intestinal epithelium ($p65^{IEC\text{-}KO}$ mice) were mostly healthy, and the appearance of the intestine was grossly normal [28]. Only 10–15% $p65^{IEC\text{-}KO}$ mice developed diarrhea and intestinal bleeding at the early age of P2–P3. Sick $p65^{IEC\text{-}KO}$ animals remained leaner compared to healthy $p65^{IEC\text{-}KO}$ and wild-type littermates, and most of them died before postnatal day 25. Histological examination of the small intestine of these animals revealed an almost complete loss of the normal crypt-villus structure. 85–90% of $p65^{IEC\text{-}KO}$ mice remained disease free and reached adulthood without manifesting signs of intestinal inflammation. Given the prominent role of p65 in NF-κB signaling, it is surprising that $p65^{IEC\text{-}KO}$ mice do not develop colon inflammation similar to $NEMO^{IEC\text{-}KO}$ mice. This finding could be explained by the differential effect of NEMO versus p65 knockout on NF-κB activation. While NEMO deficiency completely blocks NF-κB activation, p65 deficiency in IECs could be compensated by other NF-κB dimers, which could allow normal production of protective NF-κB target genes. Moreover, epithelial-specific knockout of IKK2 did not cause spontaneous intestinal inflammation either [27], suggesting that compensatory signaling of IKK1 could be sufficient to protect IECs and prevent colitis. Indeed, mice lacking both IKK1 and IKK2 developed severe colitis similarly to the $NEMO^{IEC\text{-}KO}$ mice, demonstrating that only complete NF-κB inhibition achieved by either single NEMO deficiency or by double IKK1/IKK2 ablation causes spontaneous colitis in mice [27].

Mice with intestinal epithelial-specific ablation of TAK1 ($TAK1^{IEC\text{-}KO}$ mice) also developed spontaneously intestinal disease with pronounced intestinal bleeding, which already manifested at P0 and resulted in death of neonate animals at P1 [29]. Analysis of $TAK1^{IEC\text{-}KO}$ embryos at E18 revealed a normal intestinal morphology, though increased IEC apoptosis and expression of inflammatory cytokines and chemokines was already detectable. These results imply that TAK1 is essential for the maintenance of cell survival in the intestinal epithelium. In order to address the role of TNF-signaling in disease pathogenesis in $TAK1^{IEC\text{-}KO}$ mice, $TAK1^{IEC\text{-}KO}$/TNFR1 knockout mice were generated. $TAK1^{IEC\text{-}KO}$/TNFR1 double-knockout mice did not develop intestinal disease at perinatal age, correlating with decreased IEC apoptosis compared to $TAK1^{IEC\text{-}KO}$ mice. This finding indicated that TNF is responsible for the induction of IEC apoptosis and inflammation in the small and large intestine of $TAK1^{IEC\text{-}KO}$ mice. Though newborn $TAK1^{IEC\text{-}KO}$ mice were completely protected from intestinal disease by global deletion of TNFR1, up to 50% of $TAK1^{IEC\text{-}KO}$/TNFR1 knockout mice developed intestinal inflammation at 2–3 weeks of age. The affected areas of the bowel showed a heightened expression of inflammatory markers and an increase in apoptosis in IECs. These findings demonstrate that even in the absence of TNF signaling TAK1 is required for the maintenance of the intestinal epithelium and the prevention of intestinal inflammation and hyperplastic epithelial growth.

The more severe intestinal disease in $TAK1^{IEC\text{-}KO}$ mice versus $NEMO^{IEC\text{-}KO}$ mice can be explained by the position the two molecules take in the TNFR1 signaling pathway. Whereas NEMO is required for the activation of canonical NF-κB signaling, TAK1 additionally leads to the activation of both, the NF-κB (through activation of the IKK complex) and activator protein-1 (through activation of MAPKs) signaling cascades. While in the case of NEMO deficiency in IECs, proinflammatory and residual antiapoptotic signaling can still occur via the activation of MAPKs and activator protein-1, deficiency of TAK1 prevents any initiation of a proinflammatory response upon TNFR1 activation. The incomplete rescue of intestinal pathogenesis in $TAK1^{IEC\text{-}KO}$/TNFR1 knockout mice points towards an additional TNF-independent mechanism in disease establishment in $TAK1^{IEC\text{-}KO}$ mice.

References

1 Vassalli P: The pathophysiology of tumor necrosis factors. Annu Rev Immunol 1992;10:411–452.
2 Black RA, Rauch CT, Kozlosky CJ, Peschon JJ, Slack JL, Wolfson MF, Castner BJ, Stocking KL, Reddy P, Srinivasan S, et al: A metalloproteinase disintegrin that releases tumour-necrosis factor-alpha from cells. Nature 1997;385:729–733.
3 Vandenabeele P, Declercq W, Beyaert R, Fiers W: Two tumour necrosis factor receptors: structure and function. Trends Cell Biol 1995;5:392–399.
4 Kontoyiannis D, Pasparakis M, Pizarro TT, Cominelli F, Kollias G: Impaired on/off regulation of TNF biosynthesis in mice lacking TNF AU-rich elements: implications for joint and gut-associated immunopathologies. Immunity 1999;10:387–398.
5 Chen G, Goeddel DV: TNF-R1 signaling: a beautiful pathway. Science 2002;296:1634–1635.
6 Varfolomeev E, Vucic D: (Un)expected roles of c-IAPs in apoptotic and NFkappaB signaling pathways. Cell Cycle 2008;7:1511–1521.
7 Ermolaeva MA, Michallet MC, Papadopoulou N, Utermohlen O, Kranidioti K, Kollias G, Tschopp J, Pasparakis M: Function of TRADD in tumor necrosis factor receptor 1 signaling and in TRIF-dependent inflammatory responses. Nat Immunol 2008;9:1037–1046.
8 Adhikari A, Xu M, Chen ZJ: Ubiquitin-mediated activation of TAK1 and IKK. Oncogene 2007;26:3214–3226.
9 Micheau O, Tschopp J: Induction of TNF receptor I-mediated apoptosis via two sequential signaling complexes. Cell 2003;114:181–190.
10 Hayden MS, Ghosh S: Shared principles in NF-kappaB signaling. Cell 2008;132:344–362.
11 Pasparakis M, Luedde T, Schmidt-Supprian M: Dissection of the NF-kappaB signalling cascade in transgenic and knockout mice. Cell Death Differ 2006;13:861–872.
12 Beg AA, Sha WC, Bronson RT, Ghosh S, Baltimore D: Embryonic lethality and liver degeneration in mice lacking the RelA component of NF-kappa B. Nature 1995;376:167–170.
13 Doi TS, Marino MW, Takahashi T, Yoshida T, Sakakura T, Old LJ, Obata Y: Absence of tumor necrosis factor rescues RelA-deficient mice from embryonic lethality. Proc Natl Acad Sci USA 1999;96:2994–2999.
14 Alcamo E, Mizgerd JP, Horwitz BH, Bronson R, Beg AA, Scott M, Doerschuk CM, Hynes RO, Baltimore D: Targeted mutation of TNF receptor I rescues the RelA-deficient mouse and reveals a critical role for NF-kappa B in leukocyte recruitment. J Immunol 2001;167:1592–1600.
15 Meffert MK, Chang JM, Wiltgen BJ, Fanselow MS, Baltimore D: NF-kappa B functions in synaptic signaling and behavior. Nat Neurosci 2003;6:1072–1078.
16 Li Q, Van Antwerp D, Mercurio F, Lee KF, Verma IM: Severe liver degeneration in mice lacking the IkappaB kinase 2 gene. Science 1999;284:321–325.
17 Schmidt-Supprian M, Bloch W, Courtois G, Addicks K, Israel A, Rajewsky K, Pasparakis M: NEMO/IKK gamma-deficient mice model incontinentia pigmenti. Mol Cell 2000;5:981–992.
18 Rudolph D, Yeh WC, Wakeham A, Rudolph B, Nallainathan D, Potter J, Elia AJ, Mak TW: Severe liver degeneration and lack of NF-kappaB activation in NEMO/IKKgamma-deficient mice. Genes Dev 2000;14:854–862.
19 Luedde T, Heinrichsdorff J, de Lorenzi R, De Vos R, Roskams T, Pasparakis M: IKK1 and IKK2 cooperate to maintain bile duct integrity in the liver. Proc Natl Acad Sci USA 2008;105:9733–9738.
20 Geisler F, Algul H, Paxian S, Schmid RM: Genetic inactivation of RelA/p65 sensitizes adult mouse hepatocytes to TNF-induced apoptosis in vivo and in vitro. Gastroenterology 2007;132:2489–2503.
21 Luedde T, Beraza N, Kotsikoris V, van Loo G, Nenci A, De Vos R, Roskams T, Trautwein C, Pasparakis M: Deletion of NEMO/IKKgamma in liver parenchymal cells causes steatohepatitis and hepatocellular carcinoma. Cancer Cell 2007;11:119–132.
22 van Hogerlinden M, Rozell BL, Ahrlund-Richter L, Toftgard R: Squamous cell carcinomas and increased apoptosis in skin with inhibited Rel/nuclear factor-kappaB signaling. Cancer Res 1999;59:3299–3303.
23 Pasparakis M, Courtois G, Hafner M, Schmidt-Supprian M, Nenci A, Toksoy A, Krampert M, Goebeler M, Gillitzer R, Israel A, et al: TNF-mediated inflammatory skin disease in mice with epidermis-specific deletion of IKK2. Nature 2002;417:861–866.
24 Stratis A, Pasparakis M, Markur D, Knaup R, Pofahl R, Metzger D, Chambon P, Krieg T, Haase I: Localized inflammatory skin disease following inducible ablation of I kappa B kinase 2 in murine epidermis. J Invest Dermatol 2006;126:614–620.
25 Lind MH, Rozell B, Wallin RP, van Hogerlinden M, Ljunggren HG, Toftgard R, Sur I: Tumor necrosis factor receptor 1-mediated signaling is required for skin cancer development induced by NF-kappaB inhibition. Proc Natl Acad Sci USA 2004;101:4972–4977.

26 Nenci A, Huth M, Funteh A, Schmidt-Supprian M, Bloch W, Metzger D, Chambon P, Rajewsky K, Krieg T, Haase I, et al: Skin lesion development in a mouse model of incontinentia pigmenti is triggered by NEMO deficiency in epidermal keratinocytes and requires TNF signaling. Hum Mol Genet 2006;15: 531–542.
27 Nenci A, Becker C, Wullaert A, Gareus R, van Loo G, Danese S, Huth M, Nikolaev A, Neufert C, Madison B, et al: Epithelial NEMO links innate immunity to chronic intestinal inflammation. Nature 2007;446:557–561.
28 Steinbrecher KA, Harmel-Laws E, Sitcheran R, Baldwin AS: Loss of epithelial RelA results in deregulated intestinal proliferative/apoptotic homeostasis and susceptibility to inflammation. J Immunol 2008; 180:2588–2599.
29 Kajino-Sakamoto R, Inagaki M, Lippert E, Akira S, Robine S, Matsumoto K, Jobin C, Ninomiya-Tsuji J: Enterocyte-derived TAK1 signaling prevents epithelium apoptosis and the development of ileitis and colitis. J Immunol 2008;181:1143–1152.

Dr. Manolis Pasparakis
Centre for Molecular Medicine and Cologne Excellence Cluster on Cellular Stress Responses in Aging-Associated Diseases, Institute of Genetics, University of Cologne
Zülpicher Strasse 47, DE–50674 Cologne (Germany)
Tel. +49 221 470 1526, Fax +49 221 470 5163, E-Mail pasparakis@uni-koeln.de

Kollias G, Sfikakis PP (eds): TNF Pathophysiology. Molecular and Cellular Mechanisms.
Curr Dir Autoimmun. Basel, Karger, 2010, vol 11, pp 94–104

Type I Interferon: A New Player in TNF Signaling

Anna Yarilina[a] · Lionel B. Ivashkiv[a,b]

[a]Arthritis and Tissue Degeneration Program, Hospital for Special Surgery, and [b]Graduate Program in Immunology and Microbial Pathogenesis, Weill Graduate School of Medical Sciences of Cornell University, New York, N.Y., USA

Abstract

TNF and type I interferons (IFNs) are induced by microbial stimuli and mediate innate immune responses. They are also involved in the pathogenesis of chronic inflammatory diseases, such as rheumatoid arthritis and systemic lupus erythematosus. Activated macrophages are an important driving force of inflammatory reactions and one of the major producers of TNF in innate immunity and chronic inflammation. Despite the fact that cells at sites of damage are continuously exposed to both cytokines, little is known about mechanisms regulating TNF and type I IFN interactions during inflammation. In this review, we discuss the role of an IFN-β-mediated autocrine loop in the regulation of gene expression program induced by TNF in myeloid cells.

The innate immune system is evolutionarily conserved and represents the first line of defense against pathogens. The mechanisms of recognition by this system are diverse and involve three strategies: recognition of 'microbial non-self', recognition of 'missing self' and recognition of 'induced or altered self'. The first strategy is the most universal in the animal kingdom and is based on the detection of conserved pathogen-associated molecular patterns (PAMPs) that are unique to microbes and are not expressed by the host. PAMPs are recognized by several structurally and functionally distinct classes of pattern recognition receptors (PRRs), including Toll-like receptors (TLRs) [1]. Engagement of PRRs leads to the production of proinflammatory cytokines, such as IL-1, IL-6, IL-12, and TNF, chemokines, and interferons (IFNs). This attracts various cells to the site of damage and creates an inflammatory reaction, resulting in elimination of pathogens and dead tissues, followed by healing. Pathogen clearance and inflammation are tightly controlled and self-limited processes. TNF and type I IFNs are key effectors of innate immune responses. They are also involved in the pathogenesis of chronic inflammatory diseases, such as rheumatoid arthritis and lupus [2, 3].

Virtually all nucleated cells express type I IFN receptors and TNF receptor 1, and thus can sense both cytokines. In this review, we focus on macrophages, one of the main players of the innate immune system. Activated macrophages are known to be major producers of TNF in acute and chronic inflammatory reactions. However, most studies in the TNF field have been focused on activation of cell types other than macrophages, such as fibroblasts and endothelial cells. Macrophages can also secrete low amounts of type I IFN in response to microbial stimuli, and there is evidence in the literature that TNF has some antiviral activities and is able to induce type I IFN in certain cell types [4]. The early events of TNF signaling mediated by nuclear factor κB (NF-κB) and mitogen-activated protein kinases (MAPKs) have been studied extensively, but little is known about the regulation of cellular responses after continuous exposure to this cytokine. Recent works have highlighted the importance of mechanisms and autocrine loops that sustain and regulate activity of signaling pathways and transcription factors after the initial and typically transient response to an extracellular ligand [5–8]. In this review, we will discuss a new autocrine loop induced by TNF, as well as the role of TNF and IFN signaling in the regulation of inflammatory gene expression.

Interferon-β Induction by TNF

IFNs were discovered more than 50 years ago and were named based on their ability to interfere with viral replication in a cell. Three classes of IFNs have been identified and classified according to the structure of their receptors. Type I IFNs are encoded by various genes including IFN-α (composed of a family of 13 genes in human and 14 genes in mice), -β, -ε, -κ, -ω and -δ. Type II IFN corresponds to the single IFN-γ that binds the IFN-γ receptor complex. It is encoded by a single gene structurally unrelated to the type I IFNs, is induced mainly in cells of the immune system such as T cells and natural killer cells, and is important for protection against nonviral pathogens. Type III IFNs include recently identified IFN-λ1 (IL-29), -λ2 (IL-28A), and -λ3 (IL-28B). Type III IFNs signal through receptors containing IFNLR1 (IL-28Rα) and IL-10Rβ, and are induced in virally infected cells by mechanisms similar to those that activate IFN-α and -β genes [9]. In this review, we will mainly focus on the relationship between TNF and type I IFNs.

Type I IFNs are produced mainly in response to microbial stimuli (fig. 1), following the engagement of two types of PRRs: cytosolic, which include retinoic acid-inducible gene I, melanoma differentiation-associated gene 5 and DNA-dependent activator of IFN-regulatory factor (IRF), or transmembrane TLRs [9, 10]. All these pathways culminate in activation of NF-κB, activator protein 1 (AP-1) and IRFs, transcription factors that drive the expression of type I IFN genes and other mediators of innate immune reactions. For example, lipopolysaccharide (TLR4 ligand) and nucleic acids (ligands for TLR3, TLR7, TLR8, and TLR9) strongly induce type I IFN production through activation of IRFs [9]. TLR3 and TLR4 activate IRF3, and TLR7–9 activate IRF7 [6, 9, 11–13].

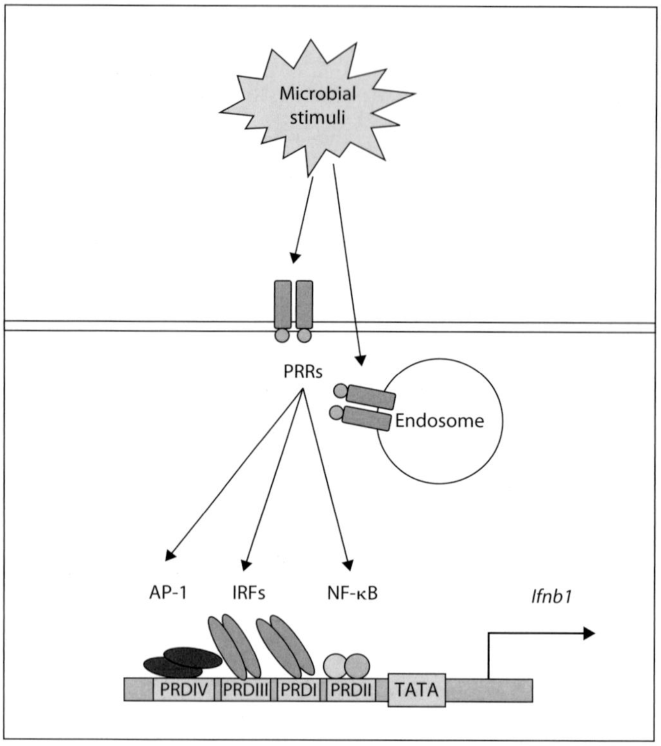

Fig. 1. Type I IFN induction by microbial stimuli. The *ifnb1* promoter region is shown schematically. Microbial stimuli interact with transmembrane or cytosolic PRRs and lead to the activation of NF-κB, AP-1 and IRFs transcription factors which bind to specific PRDs on the promoter region of the IFN-β gene and drive gene expression. The promoter contains four PRDs. PRD I and III are the binding sites for IRFs, PRD II binds NF-κB and PRD IV binds AP-1.

The promoter region of the IFN-β gene contains four regulatory *cis* elements – the positive regulatory domains (PRDs) I, II, III and IV [9]. PRD I and III are the binding sites for IRFs, and PRD II and IV bind NF-κB and AP-1, respectively. After infection of cells with viruses, all of these transcription factors become activated and, together with the high-mobility group protein HMG-I(Y), bind cooperatively to the IFN-β promoter to form an enhanceosome [9]. In the steady state, the transcription start site of the promoter is covered by a nucleosome. Enhanceosome formation recruits histone acetyl transferases to acetylate histones H3 and H4 in the nucleosome, followed by the recruitment of a nucleosome modification complex, which displaces nucleosome and allows IFN-β gene transcription [9]. IRFs are required for the recruitment of chromatin modifiers and nucleosome alteration. TLR-induced IFN production triggers an autocrine loop by binding to its cognate receptor (a heterodimer of IFNAR1 and IFNAR2) and activating the IFN-stimulated gene factor 3 complex (a heterotrimer of STAT1, STAT2 and IRF9), which promotes expression of downstream IFN-induced and STAT-dependent genes [5, 9, 14]. An important function of TLR-induced signaling is the massive systemic production of type I IFNs in order to induce an antiviral state and protect the host, but it can also contribute to endotoxin lethality and autoimmune diseases [15, 16].

Many proinflammatory cytokines are powerful activators of NF-κB and AP-1 [1, 9, 16]. However, we have only limited information on induction of an IFN-mediated autocrine loop by endogenous inflammatory factors. Recently, our group discovered that in myeloid cells, TNF induces low but sustained levels of IFN-β mRNA and protein [17]. This induction occurred early after TNF stimulation and is mediated by IRF1. IRF1 was originally discovered as a transcriptional activator of IFN-β in virus-infected fibroblasts [18, 19]. However, after the discovery of IRF3 and IRF7 as master regulators of type I IFN induction by PRRs, the focus of research on IRF1 shifted to investigation of its role in mediating IFN-γ responses and in T and natural killer cell development and function [19]. In macrophages, IRF1 regulates the induction of inducible nitric oxide synthase and IL-12p35, and has been implicated in the pathogenesis of autoimmune diseases [19, 20]. IRF1 is basally expressed in many cell types and its expression is further increased by various stimuli, including viral infection, IFNs and cytokines [19, 21]. Resting macrophages express IRF1 that is partially localized in the nucleus [Yarilina, unpubl. obs.], and TNF treatment increases IRF1 gene expression in a direct fashion [7, 17]. Recent work revealed that mechanisms of IRF1-mediated *ifnb1* expression depend on cell type and ligand [21, 22]. This work demonstrated a 'licensing' process whereby interaction of IRF1 with TLR9-activated MyD88 increases the rate of IRF1 nuclear translocation and transcriptional activation of target genes relative to that observed with IFN-γ-induced IRF1 that has not been licensed by MyD88. The mechanisms and molecular basis for licensing and/or activation of IRF1 function remain unknown, although posttranslational modification and/or activation of IRF1 was originally suggested in 1991, and recent work suggests potential phosphorylation [21, 22]. TNF does not utilize MyD88 for signaling or enhance IRF1 nuclear translocation [Yarilina, unpubl. obs.]. Also, production of IFN-β after TNF stimulation was several orders of magnitude lower than that after TLR stimulation. Thus, similar to IFN-γ, TNF increases the expression of 'unlicensed' IRF1. As unlicensed IRF1 is a much weaker activator of gene expression [21], the gene-activating function of IRF1 induced by endogenous factors and cytokines such as IFN-γ and TNF will be much more dependent on synergistic interactions with additional signals and transcription factors such as NF-κB.

TNF Induces Sustained IFN Response Gene Expression

Similar to previous work categorizing TLR-induced genes into different response categories [5, 6], we were able to separate TNF-induced genes into three groups based on the kinetics of expression and requirement for new protein synthesis [17]. Recently, a similar three-group pattern was described for TNF-activated mouse fibroblasts and bone marrow-derived macrophages, but classification was based on differences in mRNA stability between the genes in each group [7].

The first group includes well-known TNF-induced 'primary response' genes including NF-κB targets *Tnf* and *Il1b*, whose expression increased rapidly and transiently after incubation with TNF and did not require new protein synthesis (fig. 2, upper section).

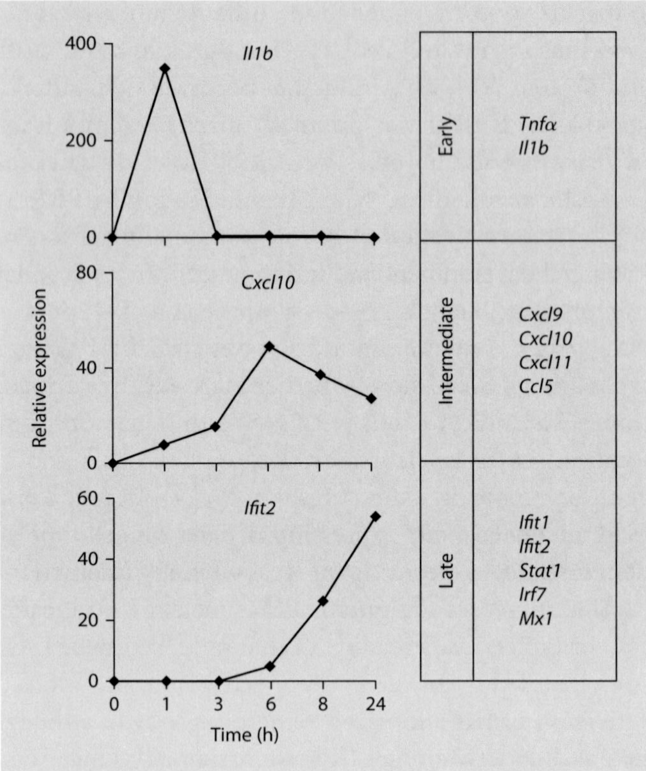

Fig. 2. Kinetics of TNF-induced gene expression. Real-time PCR analysis of expression of indicated genes in mouse bone marrow-derived macrophage treated with mTNF for indicated periods of time, relative to untreated cells at the same time point (set at 1). Genes were divided into three groups according to the kinetics of their response to TNF.

A second group of 'intermediate response' genes that included the T helper type 1 chemokines *Cxcl9, Cxcl10, Cxcl11* and *Ccl5* increased in expression with slower kinetics, reaching maximal expression 3–8 h after TNF stimulation (fig. 2, middle section). Expression of these genes is known to be regulated by NF-κB-, STAT1- and IRF1-dependent pathways. At earlier time points, gene expression was independent of new protein synthesis, thus fitting into the delayed primary response category. However, sustained expression of these genes at later time points was dependent on autocrine IFN-β production, demonstrating a transition to the secondary response category.

The third group of genes ('late response' genes; fig. 2, lower section) was induced with delayed kinetics (induction first detected at about 6 h after TNF stimulation) in a manner entirely dependent on IFN-β and IRF-1, and included classical IFN response genes important for antiviral responses *(Mx1, Isg54, Isg56)*, and for augmented subsequent responses to microbial products or inflammatory cytokines *(Irf7, Ikke, Stat1)* [5, 9, 11, 14, 23]. Consistent with a process involving autocrine type I IFN action, TNF induced activation of STAT proteins: in TNF-stimulated macrophages, both STAT1 and STAT2 were phosphorylated on the activating tyrosine residues (701 and 689, respectively). TNF also increased phosphorylation on STAT1 serine 727 residue that enhances its transcriptional activity. Thus, TNF induces a cascade of gene expression with an increasing requirement for IFN-β synthesis to sustain 'intermediate' gene expression after initial TNF stimulation and to induce the set of antiviral genes. The TNF-activated IRF1-IFN-Jak-STAT signaling pathway contributes to the proinflammatory functions of TNF, and provides evidence that induction of IFN-mediated autocrine loops is not limited to PRRs.

Synergy between TNF-Induced Nuclear Factor-κB and Interferon-β

Direct early events of cytokine signal transduction are relatively well studied. However, the mechanisms of signal propagation over time are not clear, and the importance of autocrine loops in regulating cellular responses has gained increasing attention [5, 6]. As was noted before, the modest increase in chemokine gene expression observed early after TNF stimulation was not dependent on autocrine IFN-β, but demonstrated an increased requirement for IFN-β signaling over time. Notably, both IRF1 and IFN-β expression fit the profile of 'intermediate response' genes that are basally expressed, and then are further induced by TNF in a manner dependent on IFNAR signaling, supporting the existence of a positive feedback loop [22]. Disruption of NF-κB signaling prevented induction of 'intermediate response' genes by TNF, and, in contrast to 'late response' genes, low amounts of IFN-β were not sufficient to rescue gene activation. Furthermore, simultaneous addition of TNF and low concentrations of IFN-β (1 U/ml) at early time points (prior to production of endogenous IFN-β) induced higher expression of 'intermediate response' genes. Thus, the temporal expression of 'intermediate' genes required a direct TNF-induced signal (activated NF-κB) as well as indirect signaling through an IFN-β-mediated autocrine loop.

Synergistic actions of IFNs and members of TNF family have been studied in the context of viral infection [4, 24]. It has been shown in several animal models that anti-HBV effects depend on the synergy between TNF and both type I and II IFNs, and another TNF family member – lymphotoxin cooperates with IFN-β to arrest the replication of HCMV in fibroblasts [25]. Adenoviral (Ad) vectors are commonly used as a model to study immune response to viruses. Ad gene transfer is one of the most efficient techniques available for in vivo gene transduction. However, the successful use of Ad vectors in gene therapy is limited by their rapid elimination from the circulation. The host response to Ad vectors is dependent on both type I IFNs and TNF [26]. Neutralization of autocrine TNF attenuated adenovirus-induced expression of IFN-β and IFN-dependent genes in human macrophages, providing additional insights into the mechanism of TNF-related clearance of Ad vectors via the synergy with type I IFN. Another important biological role of low concentrations of type I IFNs is 'priming' cells to produce more IFN upon subsequent stimulation. Low amounts of IFN-β induced by TNF enhanced type I IFN expression in human macrophages stimulated with TLR7/8 and TLR9 ligands, which are known as weak IFN inducers in macrophages. This mechanism may be involved in antiviral responses mediated by TNF.

Table 1. Expression of IFN response genes in rheumatoid arthritis (RA) synovial macrophages

Description	Gene symbol	Fold increase RA vs. control	Mean signal intensity in RA
IFN-α-inducible protein 27	*IFI27*	12.88	1,092
Guanylate-binding protein 1, IFN-inducible, 67 kDa	*GBP1*	7.2	527
IFN-induced transmembrane protein 1 (9–27)	*IFITM1*	5.9	906.6
Signal transducer and activator of transcription 1, 91 kDa	*STAT1*	3.3	943
Eukaryotic translation initiation factor 2-α kinase 2	*EIF2AK2*	2.5	1,500
IFN-induced protein with tetratricopeptide repeats 3	*IFIT3*	2.2	1,001
Chemokine (C-X-C motif) ligand 9	*CXCL9*	80.5	33
Indoleamine-pyrrole 2,3 dioxygenase	*INDO*	6.8	499
Chemokine (C-X-C motif) ligand 10	*CXCL10*	6.2	282
Chemokine (C-X-C motif) ligand 11	*CXCL11*	5.7	81
IRF1	*IRF1*	3.2	983
TNF (TNF superfamily, member 2)	*TNF*	3.95	702
INF-α_4[1]	*IFNA4*	2.1	4.75
INF-β_1, fibroblast[1]	*IFNB1*	2.8	11.75

Listed genes were elevated 2-fold or more in RA synovial macrophages (n = 5) relative to control macrophages isolated from peripheral blood of healthy donors (samples were pooled into 3 sets, 5 donors each), $p < 0.05$ (Welch's t test or Welch's analysis of variance with the Benjamini-Hochberg correction for false discovery rate multiple testing).

[1] Expression levels were low. Difference was not statistically significant.

Another significant outcome of the synergy between TNF and IFN is strong induction of proinflammatory chemokines. Despite the modest induction of type I IFN by TNF compared to TLR ligands, the production of chemokines eventually reached similar amounts, and remained elevated for at least 2 days after TNF stimulation, supporting the functional importance of late TNF-induced gene induction in macrophages.

TNF induces a similar pattern of chemokine and IFN-mediated gene expression in vivo, when injected into mice intraperitoneally [17]. We also observed increased expression of type I IFN and IFN-induced genes in the joints of mice expressing transgenic TNF [Yarilina, unpubl. obs.]. These mice develop spontaneous arthritis. Furthermore, these findings can explain the elevated expression of type I IFNs and type I IFN response genes in synovial macrophages from patients with rheumatoid arthritis, a sterile inflammatory condition that is driven by TNF (table 1). This phenomenon is known as the 'IFN signature' and has been described by several groups [3, 23, 27]. We recently found that blocking of endogenous TNF by the soluble TNF receptor etanercept prevented the increase in type I IFN-dependent genes and chemokines in macrophages isolated from synovial fluids of patients with rheumatoid arthritis [Kalliolias, unpubl. obs.].

Regulation of Autocrine Type I IFN Production by Toll-Like Receptor Ligands and TNF

It is interesting to compare activation of type I IFN-mediated autocrine loops downstream PRRs (such as TLR3 and TLR4) that respond to microbial pathogens with that activated by TNF. Activation of TRIF-dependent signaling pathways by TLR3 and TLR4 immediately results in TBK1-IKKε-mediated phosphorylation and activation of IRF3 (or IRF7 in plasmacytoid DCs or primed cells) with a rapid and massive production of IFNs and expression of IFN response genes [11–14]. This type of robust response has the advantage of efficiently mobilizing host defense against infection, and large amounts of IFNs act systemically to broadly induce an antiviral state [5, 9]. However, production of large amounts of systemic IFN can be toxic and contributes to endotoxin lethality [15], and in a chronic setting can predispose to autoimmunity [16]. In contrast, TNF utilizes unlicensed IRF1 to induce the production of small amounts of IFN-β and only minimal amounts of IFN-α mRNA and protein that act locally and require synergistic interactions with additional TNF-induced signals for effective induction of inflammatory genes. Thus, the toxicity associated with systemic IFN production is avoided, while sustained and synergistic induction of gene expression primes cells for strong responses to infectious pathogens if needed.

Conclusions

Our results suggest a model whereby TNF-induced gene expression is sustained and amplified by sequential induction of IRF1, IFN-β and STAT1 (fig. 3). Acute stimulation with TNF activates NF-κB and MAPK pathways, which lead to rapid expression of inflammatory genes, and to increased expression of IRF1. IRF1 works together with NF-κB and AP-1 to drive production of low amounts of IFN-β. IFN-β, in turn, activates Jak-STAT signaling that synergizes with other TNF-induced signals to sustain inflammatory chemokine and IFN-β expression, induce slow and delayed accumulation of mRNAs encoding canonical IFN response genes, increase expression of signaling components such as IKKε, IRF7 and STAT1 that are known to further amplify activation of genes by low concentrations of IFNs and shift the balance of macrophage responses in an inflammatory direction [5, 14, 23, 28]. Regulation of sequential gene expression by autocrine loops induced by initial stimuli is well known in the IFN field [5, 6]. Our study focused only on a limited array of specific genes, related to type I IFN and induced by TNF. TNF induces many more target genes and pathways, and this activation is regulated on multiple levels and by different mechanisms. For example, microarray analysis of genes increased by TNF in mouse cells revealed hundreds of genes induced by TNF with similar kinetic patterns, but this study demonstrated that the kinetics of gene expression was dependent on mRNA stability [7].

Work from several laboratories have demonstrated that low type I IFN signaling observed under physiological conditions in the absence of infection [29] is important

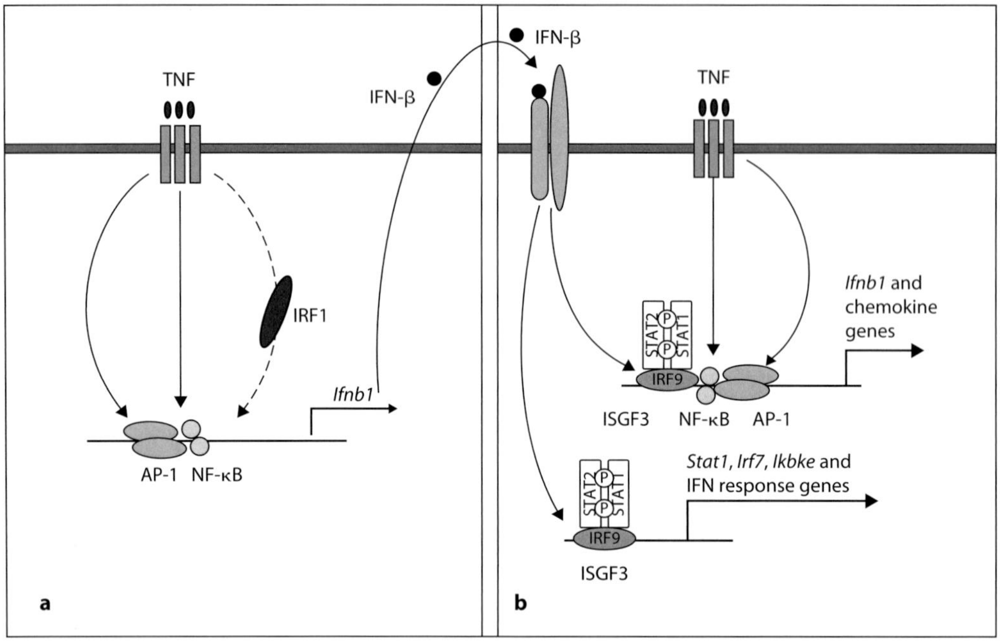

Fig. 3. Model for induction of a gene activation program by TNF. Stimulation with TNF (**a**) activates NF-κB and MAPK pathways leading to rapid expression of inflammatory genes and to increasing expression of IRF1. IRF1 works together with NF-κB and AP-1 to drive production of low amounts of IFN-β. IFN-β, in turn, activates STAT signaling (**b**) that synergizes with other TNF-induced signals to sustain inflammatory chemokine expression and *Ifnb1* expression, and to induce canonical IFN response genes and increase expression of signaling components such as IKKε, IRF7 and STAT1. ISGF3 = IFN-stimulated gene factor 3.

for a robust response to microbial pathogens. The factors that regulate basal IFN production and signaling in the absence of pathogens are not well understood, but include ITAM-coupled immunoreceptors including Fc receptors [28, 30]. Our findings implicate TNF as a new endogenous inducer of type I IFN production and signaling. It is unlikely that the low amounts of IFN induced by TNF contribute significantly to overall IFN production in the setting of an infection where multiple PRRs are engaged. However, TNF-induced local production of IFN-β may regulate inflammation of noninfectious origin such as after tissue damage, and in chronic sterile TNF-dominated inflammation, such as in the joint lining in rheumatoid arthritis [23, 27].

Acknowledgements

This work was supported by NIH grants AR050401, AR46713 and AI46712 to L.B.I.

References

1 Medzhitov R, Janeway CA Jr: Decoding the patterns of self and nonself by the innate immune system. Science 2002;296:298–300.

2 Banchereau J, Pascual V: Type I interferon in systemic lupus erythematosus and other autoimmune diseases. Immunity 2006;25:383–392.

3 McInnes IB, Schett G: Cytokines in the pathogenesis of rheumatoid arthritis. Nat Rev Immunol 2007;7: 429–442.

4 Benedict CA: Viruses and the TNF-related cytokines, an evolving battle. Cytokine Growth Factor Rev 2003;14:349–357.

5 Marie I, Durbin JE, Levy DE: Differential viral induction of distinct interferon-alpha genes by positive feedback through interferon regulatory factor-7. EMBO J 1998;17:6660–6669.

6 Doyle S, Vaidya S, O'Connell R, Dadgostar H, Dempsey P, Wu T, Rao G, Sun R, Haberland M, Modlin R, et al: IRF3 mediates a TLR3/TLR4-specific antiviral gene program. Immunity 2002;17: 251–263.

7 Hao S, Baltimore D: The stability of mRNA influences the temporal order of the induction of genes encoding inflammatory molecules. Nat Immunol 2009;10:281–288.

8 Ramirez-Carrozzi VR, Nazarian AA, Li CC, Gore SL, Sridharan R, Imbalzano AN, Smale ST: Selective and antagonistic functions of SWI/SNF and Mi-2beta nucleosome remodeling complexes during an inflammatory response. Genes Dev 2006;20: 282–296.

9 Honda K, Takaoka A, Taniguchi T: Type I interferon (corrected) gene induction by the interferon regulatory factor family of transcription factors. Immunity 2006;25:349–360.

10 Wang Z, Choi MK, Ban T, Yanai H, Negishi H, Lu Y, Tamura T, Takaoka A, Nishikura K, Taniguchi T: Regulation of innate immune responses by DAI (DLM-1/ZBP1) and other DNA-sensing molecules. Proc Natl Acad Sci USA 2008;105:5477–5482.

11 Sharma S, tenOever BR, Grandvaux N, Zhou GP, Lin R, Hiscott J: Triggering the interferon antiviral response through an IKK-related pathway. Science 2003;300:1148–1151.

12 Fitzgerald KA, McWhirter SM, Faia KL, Rowe DC, Latz E, Golenbock DT, Coyle AJ, Liao SM, Maniatis T: IKKepsilon and TBK1 are essential components of the IRF3 signaling pathway. Nat Immunol 2003; 4:491–496.

13 Sakaguchi S, Negishi H, Asagiri M, Nakajima C, Mizutani T, Takaoka A, Honda K, Taniguchi T: Essential role of IRF-3 in lipopolysaccharide-induced interferon-beta gene expression and endotoxin shock. Biochem Biophys Res Commun 2003; 306:860–866.

14 Honda K, Yanai H, Negishi H, Asagiri M, Sato M, Mizutani T, Shimada N, Ohba Y, Takaoka A, Yoshida N, et al: IRF-7 is the master regulator of type-I interferon-dependent immune responses. Nature 2005; 434:772–777.

15 Karaghiosoff M, Steinborn R, Kovarik P, Kriegshauser G, Baccarini M, Donabauer B, Reichart U, Kolbe T, Bogdan C, Leanderson T, et al: Central role for type I interferons and Tyk2 in lipopolysaccharide-induced endotoxin shock. Nat Immunol 2003;4:471–477.

16 Baccala R, Hoebe K, Kono DH, Beutler B, Theofilopoulos AN: TLR-dependent and TLR-independent pathways of type I interferon induction in systemic autoimmunity. Nat Med 2007: 543–551.

17 Yarilina A, Park-Min KH, Antoniv T, Hu X, Ivashkiv LB: TNF activates an IRF1-dependent autocrine loop leading to sustained expression of chemokines and STAT1-dependent type I interferon-response genes. Nat Immunol 2008;9:378–387.

18 Miyamoto M, Fujita T, Kimura Y, Maruyama M, Harada H, Sudo Y, Miyata T, Taniguchi T: Regulated expression of a gene encoding a nuclear factor, IRF-1, that specifically binds to IFN-beta gene regulatory elements. Cell 1988;54:903–913.

19 Taniguchi T, Ogasawara K, Takaoka A, Tanaka N: IRF family of transcription factors as regulators of host defense. Annu Rev Immunol 2001;19:623–655.

20 Liu J, Guan X, Tamura T, Ozato K, Ma X: Synergistic activation of interleukin-12 p35 gene transcription by interferon regulatory factor-1 and interferon consensus sequence-binding protein. J Biol Chem 2004;279:55609–55617.

21 Negishi H, Fujita Y, Yanai H, Sakaguchi S, Ouyang X, Shinohara M, Takayanagi H, Ohba Y, Taniguchi T, Honda K: Evidence for licensing of IFN-gamma-induced IFN regulatory factor 1 transcription factor by MyD88 in Toll-like receptor-dependent gene induction program. Proc Natl Acad Sci USA 2006; 103:15136–15141.

22 Schmitz F, Heit A, Guggemoos S, Krug A, Mages J, Schiemann M, Adler H, Drexler I, Haas T, Lang R, et al: Interferon-regulatory-factor 1 controls Toll-like receptor 9-mediated IFN-beta production in myeloid dendritic cells. Eur J Immunol 2007;37:315–327.

23 Hu X, Herrero C, Li WP, Antoniv TT, Falck-Pedersen E, Koch AE, Woods JM, Haines GK, Ivashkiv LB: Sensitization of IFN-gamma Jak-STAT signaling during macrophage activation. Nat Immunol 2002;3:859–866.
24 Mestan J, Digel W, Mittnacht S, Hillen H, Blohm D, Moller A, Jacobsen H, Kirchner H: Antiviral effects of recombinant tumour necrosis factor in vitro. Nature 1986;323:816–819.
25 Benedict CA, Banks TA, Senderowicz L, Ko M, Britt WJ, Angulo A, Ghazal P, Ware CF: Lymphotoxins and cytomegalovirus cooperatively induce interferon-beta, establishing host-virus detente. Immunity 2001;15:617–626.
26 Elkon KB, Liu CC, Gall JG, Trevejo J, Marino MW, Abrahamsen KA, Song X, Zhou JL, Old LJ, Crystal RG, et al: Tumor necrosis factor alpha plays a central role in immune-mediated clearance of adenoviral vectors. Proc Natl Acad Sci USA 1997;94:9814–9819.
27 Sweeney SE, Mo L, Firestein GS: Antiviral gene expression in rheumatoid arthritis: role of IKKepsilon and interferon regulatory factor 3. Arthritis Rheum 2007;56:743–752.
28 Tassiulas I, Hu X, Ho H, Kashyap Y, Paik P, Hu Y, Lowell CA, Ivashkiv LB: Amplification of IFN-alpha-induced STAT1 activation and inflammatory function by Syk and ITAM-containing adaptors. Nat Immunol 2004;5:1181–1189.
29 Taniguchi T, Takaoka A: A weak signal for strong responses: interferon-alpha/beta revisited. Nat Rev Mol Cell Biol 2001;2:378–386.
30 Dhodapkar KM, Banerjee D, Connolly J, Kukreja A, Matayeva E, Veri MC, Ravetch JV, Steinman RM, Dhodapkar MV: Selective blockade of the inhibitory Fcgamma receptor (FcgammaRIIB) in human dendritic cells and monocytes induces a type I interferon response program. J Exp Med 2007;204:1359–1369.

Dr. Lionel B. Ivashkiv
Arthritis and Tissue Degeneration Program, Hospital for Special Surgery
535 East 70th Street
New York, NY 10021 (USA)
Tel. +1 212 606 1653, Fax +1 212 774 2337, E-Mail ivashkivl@hss.edu

Kollias G, Sfikakis PP (eds): TNF Pathophysiology. Molecular and Cellular Mechanisms.
Curr Dir Autoimmun. Basel, Karger, 2010, vol 11, pp 105–118

T Cells as Sources and Targets of TNF: Implications for Immunity and Autoimmunity

Ioannis Chatzidakis[a,b] · Clio Mamalaki[a]

[a]Institute of Molecular Biology and Biotechnology, Foundation for Research and Technology, [b]Department of Biology, University of Crete, Heraklion, Greece

Abstract

TNF is a pleiotropic cytokine produced by many cell types upon different stimuli and in various physiological and pathological conditions. In this review, we focus on the role of TNF in T cell responses as demonstrated by in vitro and in vivo observations in mice and humans. TNF has an impact on all aspects of T cell biology such as development in the thymus, peripheral homeostasis, primary antigenic responses, apoptosis, effector functions, memory cell formation and tolerance induction and maintenance. In most cases, TNF has an immunostimulatory role in T cell responses; however, under certain conditions, TNF can exert immunomodulatory effects on T cells. We also review how T cell-derived TNF is an important component of T cell immunity as exemplified by many studies involving intracellular pathogens and tumors. Finally, we summarize how TNF T cells interplay contributes to pathology in autoimmune disorders and what is known about the effect of widely used TNF blockers on T cell differentiation/function.

Thymus and TNF

There is plenty of evidence that TNF is expressed in the developing thymus of mice [1, 2] as well as in humans [3]. TNF is synthesized also in thymocytes in adult mice [4] and its mRNA is localized in the subcortical regions of medulla, a region related to thymocyte selection processes [5].

But what can be the role of TNF in the fetal or adult thymus? Most studies point towards a role for TNF in regulating lineage commitment and early thymocyte development by promoting both differentiation/proliferation (most likely by membrane-bound TNF, mTNF) and apoptosis of immature thymocytes [5, 6]. TNF is also shown to augment human T cell lymphopoiesis in irradiated NOD/SCID mice [7]. Altogether, it is proposed that TNF in the thymus plays an important role in thymocyte production by delivering both positive and negative signals at early stages of differentiation.

Overproduction of TNF after lipopolysaccharide (LPS) injection results in apoptosis of double-positive thymocytes, which is completely abrogated when a-TNF Ab is coadministered [8], suggestive for a strong effect of TNF in thymocyte homeostasis under stressful conditions. The role of TNF in negative selection is less clear. Although a possible role in promoting negative selection was suggested by experiments with TNFRKO mice [9] or with fetal thymic organ cultures [10], other studies using class I and class II restricted TCR-transgenic (TCR-tg) mice failed to demonstrate a direct role of TNF in mediating thymocyte negative selection [11, 12]. It is probable that systems used do not really reflect physiological negative selection events and double-positive cell death observed can be due to cytokines or steroids resulting from activated T cells in the periphery. It is also possible that multiple coreceptors are cooperatively involved in negative selection [10, 13] and blocking one of them could lead to subtle defects only. TNF function in thymus can affect T cell tolerance not only by modulating negative selection but also by regulating Treg production [14], a topic reviewed elsewhere in this volume.

Role of TNF in T Cell Responses

Initial evidence that TNF plays a role in T cell responses came from studies on normal human T lymphocytes. Recombinant human TNF was demonstrated to enhance T cell proliferation in response to a variety of stimuli such as IL-2 [15], a-CD3 [16], alloantigen [17], or phorbol esters [18]. Moreover, rTNF promoted upregulation of MHC molecules [15], IFN-γ production [15, 19], expression of high-affinity IL-2R [18] and TNFR2 [19]. These effects can be well attributed to optimal NF-κB activation by TNF [18, 20].

Studies in human T cells highlighted a role for TNFR2, but not TNFR1, in delivering costimulatory signals distinct of CD28 [19]. The role of TNFR2 as T cell costimulatory molecule was confirmed in studies using TNFR2KO mice. TNFR2 ablation was shown to decrease proliferative capacity of CD4 and CD8 T cells, and reduces IFN-γ, TNF and IL-2 expression in response to a-CD3 crosslinking or antigenic stimulation [21, 22]. TNFR2 signaling was found to modulate AKT activity, as well as NF-κB activation. Remarkably, CD28 coligation was unable to rescue either defect [23]. In these studies, it was also shown that TNFR2 is important for the survival of T cells during the proliferative response and this was associated with upregulation of Bcl-x_L, Bcl-2 and survivin expression [21, 22].

Studies from our lab extended the role of TNF in enhancing TCR-generated signals. TNFKO, polyclonal or TCR-tg, CD8 T cells showed an abnormally high TCR signal threshold for T cell responses, and this defect was restored by TNF provided from activated wild-type (WT) cells as shown in coculture experiments [24]. Apart from the role of TNF as a costimulatory ligand in responses of naïve T cells to antigen, we showed that TNF has a serious impact on T cell tolerance, since $TNF^{-/-}$ T cells exhibited defective anergy induction and clonal deletion, as a result of altered signaling thresholds in response to self-antigen recognition. Additionally, we revealed a previously unrecognized role of endogenous TNF in the homeostasis of naïve CD8 T cells since TCR-tg $TNF^{-/-}$ naïve CD8 T cells had a survival defect when compared to their $TNF^{+/+}$ counterparts. Moreover, naïve TCR-tg or polyclonal CD8 $TNF^{-/-}$ T cells, when transferred to lymphopenic recipients, undergo impaired homeostatic expansion, a process that induces TNFR2 expression on T cells [24]. In line with the role of endogenous TNF to activate T cells is the observation that B6.gld/gld$TNF^{-/-}$ mice exhibited an attenuated course of the generalized lymphoproliferative disorder and that was correlated with decreased peripheral T cell activation and lower concentration of IFN-γ in the serum [25].

Although relatively brief exposure to TNF has a costimulatory effect on T cells, its role after prolonged exposure appears quite different. Chronic exposure to rTNF led to impaired production of cytokines such as IL-2, IFN-γ, IL-4, IL-10, TNF and LT from both human T cells and T cells from TCR-tg mice [26, 27]. Attenuation of T cell activation was associated with defective Ca^{2+} responses and could be partly attributed to decreased surface CD3ζ expression and attenuated LAT and PLCγ tyrosine phosphorylation [28] or downregulation of CD28 expression [29], indicating that intact proximal TCR signaling can be disrupted by chronic TNFR signaling. This inhibitory effect of chronic TNF on human T cell activation was shown to be TNFR2-dependent [30], whereas in mouse and Jurkat T cells TNFR1 was found to be responsible for impairment of proximal TCR signaling [31].

The role of TNF/TNFR pathway in T cell death was first demonstrated by induction of apoptotic death of human and mouse T cell blasts in a Fas-independent manner [32]. In line with this finding, rTNF could enhance activation-induced death in memory and naïve T cells from aged humans [33, 34].

However, in two studies using TCR-tg TNFKO mice infected with LCMV, no apparent role of TNF in mediating CD8 T cell death was observed [35, 36]. Accordingly, in vivo neutralization of TNF had no impact on CD4 T cell apoptosis unless Fas pathway was defective, too [37]. Nevertheless, in vivo neutralization of TNF in another TCR-tg experimental model showed that TNF is responsible for controlling intrahepatic apoptosis of activated T cells and thus regulating peripheral T cell numbers [38].

Both receptors can mediate the cytotoxic effect of TNF on activated T cells as shown by early studies using agonistic a-TNFR Abs [33]. The role of TNFRs was directly assessed by studies using TNFRKO mice. Experiments with TNFR1KO- and TNFR2KO-activated T cells revealed that TNF mediates activation-induced cell death of CD8 T cells, but not CD4, through TNFR2 signaling [39, 40]. On the other hand, in vitro experiments with TNFR2KO CD8 T cells [23] showed that TNFR2 actually delivers survival signals during TCR stimulation probably through upregulation of Bcl-x_L and activation of NF-κB [23], a factor that has been shown to protect from TNF-induced cell death [41]. These data were also confirmed in vivo after engineered-LM infection of TCR-tg TNFR2KO mice, where increased CD8 T cell apoptosis was correlated with diminished expression of Bcl-2 and survivin [21].

Two rather similar studies involving infection of TCR-tg TNFR1KO mice with LCMV, revealed no role of TNFR1 in activated CD8 T cell death [35, 36]; however, it was shown that Fas and TNFR1 synergize to promote peptide-induced T cell deletion under limited conditions [36]. Other in vivo studies with TCR-tg TNFR1KO mice revealed a role of TNFR1 in activation-induced cell death of CD8 T cells only after challenge with low antigen concentrations, whereas in high concentrations both TNFR1KO and WT CD8 T cells declined with similar kinetics [42].

Studies in humans revealed that under certain pathological conditions T cells can be rather susceptible to TNF-induced cell death. In one study, T cells only from HIV-infected patients were sensitive to either a-TNFR1- or a-TNFR2-mediated apoptosis, but this was not correlated with differences in TNFR expression. Notably, susceptibility to TNFR-mediated death was associated with disease progress and was significantly decreased in patients treated with antiretroviral agents. What made T cells from HIV-infected donors poised for death was the lack of protection due to decreased levels of Bcl-2 and at the same time expression of active caspases 3 and 8 [43]. In another study, TNF- or TNFR2-agonistic antibodies could selectively kill activated CD8 T lymphocytes from type I diabetes (T1D) patients. It is of exceptional interest that only antigen-specific autoreactive CD8 T cells, but not other activated T cells, were susceptible to TNFR2-mediated apoptosis [44].

Taken together, the TNF/TNFR pathway(s) has definitely a role in apoptotic death of T lymphocytes (mostly CD8). However, the role of TNF is manifested mainly in conditions of low antigenic stimulation, whereas in T cells activated by high antigen doses and/or fully mature dendritic cells (DCs) more pathways are involved and may compensate for the lack of intact TNFR pathway(s).

Not only are there conditions where TNF is dispensable for T cell activation-induced cell death, but TNF can actually inhibit it. It is reported that a membrane form of TNF found on exosomes produced by synovial fibroblasts from rheumatoid arthritis (RA) patients can delay T cell death and sustain primary CD4 T cell proliferation after a-CD3 stimulation [45].

TNF and T Cells in Immunity

Early experiments in TNFKO or TNFRKO mice demonstrated the role of TNF in immune responses as manifested by reduced LPS-induced lethality, or by defective responses against a variety of pathogens such as *Corynebacterium parvum*, increased susceptibility to *Candida albicans* and *Listeria monocytogenes* [46–49] and inability to control infections with *Mycobacterium tuberculosis, M. avium and M. bovis* [50–52]. Moreover, with the development and widespread use of TNF blockers for treatment of RA, Crohn's disease, psoriasis and ankylosing spondylitis, it became evident that neutralization of TNF in humans can lead to reactivation of latent tuberculosis and development of lymphomas (both rare but life-threatening side effects) [53]. Since TNF is produced by many cell types including activated macrophages, activated CD4 and CD8 T cells, NK cells, DCs and other immune and nonimmune cells [54], it is conceivable that its production by a specific cell type has distinct biological consequences. It has been shown that naïve T cells from both humans and mice transcribe TNF mRNA early after initial TCR engagement, and especially in the case of CD8 T cells TNF protein can be detected (either membrane-bound or soluble) as early as 5 h after stimulation [55 and pers. obs.]. Proinflammatory and cytotoxic properties of T cell-derived TNF can have a dramatic impact on different cell types of the host but also can affect differentiation and/or promote apoptosis of T cells themselves. The role of T cell-derived TNF in protection against *M. tuberculosis* infection was assessed by Saunders et al. [56] by transferring T cells from WT or TNFKO mice to infected Rag1KO hosts. These experiments showed that only transfer of T cells competent for producing TNF could increase survival of infected mice probably through granuloma formation. However, T cell-derived TNF was not enough to control bacterial growth in most tissues, and mice finally succumbed, albeit at later time points. A later study from the same group [57] using mice expressing only mTNF showed that transmembrane TNF contributes to T cell migration and subsequent initial granuloma formation; transfer of mTNF-expressing T cells to Rag1KO or TNFKO mice was sufficient to control the acute phase of *M. tuberculosis* infection and prolonged survival of host mice. Accordingly, another study demonstrated that mTNF on memory CD8 T cells enhanced greatly (and to a lesser extent for CD4 T cells) their ability to control *Francisella tularensis* live vaccine strain intramacrophage growth in vitro [58]. Experiments with Armstrong LCMV-infected WT or TNFKO mice hosting P14/WT or P14/TNFKO transferred CD8 T cells indicated that TNF is not critical for the primary LCMV immune response [59]. However, at later time points, LCMV-specific T cell numbers were significantly higher in TNFKO hosts. This is consistent with observations in experiments using LCMV-infected TNFKO, TNFRDKO, or WT mice and measuring LCMV-specific CD4 T cells at relatively late time points [60]. However, the authors showed that this was due to an indirect role of TNF on T cells, since T cell-derived TNF can induce apoptosis of nonplasmacytoid DCs and in that way it can compromise T cell activation and proliferation. A consequence in LCMV-infected TNFRDKO mice was increased number of memory CD4 T cells, implying that TNF has a suppressive role in the protective responses of CD4 T cells against viruses [60]. Similar role for TNF in CD8 T cell memory cells was reported by the same group, but in this case TNF was directly regulating effector CD8 T cell apoptosis [61]. In another study using adenovirus infected TNFKO, TNFR1KO and TNFR2KO mice, it was demonstrated that TNF-TNFR2 interactions are required for optimal generation of CD8 T cell effector functions which result in optimal clearance of virally infected hepatocytes [62].

Another study which highlighted the role of TNF in adaptive recall responses to intracellular bacteria involved secondary infection of mice with *Ixodes ovatus* ehrlichia. This resulted in fatal outcome which was CD8 T cell-mediated, and TNF neutralization in vivo could protect mice from immunopathology. Although infected macrophages could elicit TNF production by CD8 T cells in vitro, it was not clear whether this happened indeed in vivo in the sites of inflammation and tissue damage. Also the authors could not discriminate between the possibility of a TNF-dependent inflammatory pathology or TNF-mediated immunosuppression since fatal outcome was accompanied by uncontrolled bacterial infection [63].

A more clear insight into the role of T cell-derived TNF was provided by using T cell-specific TNFKO mice. T cell-specific ablation of TNF did not protect mice against LPS-induced septic shock, but reduced lethality from shock induced by SEB plus D-Gal [64]. Results from infection with *Listeria* suggested that T cell-derived TNF provides protection against high bacterial loads where in lower doses of infection it becomes dispensable. A different study using the same mouse model in an acute LPS-induced respiratory dysfunction model demonstrated a protective role of T cell-derived TNF by downregulating airway resistance and pulmonary neutrophil recruitment after exposure to endotoxin [65].

It is evident from the above that T cell-derived TNF holds a nonredundant role in regulating immune response against pathogens and especially viruses and intracellular bacteria. Its function can be immune-promoting or immunomodulatory, beneficial or detrimental to the host. More infectious disease animal models should be studied in order to fully evaluate the contribution of T cell-derived TNF to immunity or pathogenesis.

Although TNF was originally identified because of its antitumor properties in mice, its role in tumor progression is still obscure. While TNF has an immunosurveillance role in tumor growth, there are many studies implicating TNF in tumor development, due to its proinflammatory properties.

CD8 T cells capable of lysing tumor cells produce high amounts of TNF, suggesting that T cell-derived TNF is important for antitumor immunity in mice [66–68]. Prevost-Blondel et al. [69] used a lung carcinoma cell transfer model, where CD8 T cells induce a protective tumor-specific response and found that tumor was not rejected when carcinoma cells were inoculated to $TNF^{-/-}$ hosts. Poehlein et al. [70] demonstrated a role of TNF in T cell responses against tumor cells, evident only if T cells were unable to produce IFN-γ and perforin. Specifically, IFN-γ/perforin DKO effector T cells were able to produce TNF and mediate tumor regression of established pulmonary metastases of melanoma cells in WT recipients. However, regression was inhibited when soluble TNFR2-Fc fusion protein was administered to tumor-bearing mice. Nevertheless, blocking TNF in tumor-bearing mice receiving WT effector T cells did not have an effect on tumor regression. More direct evidence about the role of CD8 T cell-derived TNF came from studies in which transferred WT CD8 T cells could circumvent immune privilege in the eye and mediate intraocular tumor rejection in severe combined immunodeficiency (SCID) mice, whereas TNFKO CD8 T cells could not protect recipient mice from progressive tumor growth [71]. In an experimental system utilizing modeled neo-antigen-expressing pancreatic tumors and transfers of $TNF^{-/-}$ TCR-tg CD8 T cells to $TNF^{+/+}$ hosts and vice versa, it was shown that optimal immune responses against tumor antigen required TNF production on both T cell and non-T cell populations [59]. In the same study, it was demonstrated that TNFR2, but not TNFR1, mediates the costimulatory effect of TNF on T cells in response to tumor antigen.

It is of particular notice that in most cases, mTNF on T cells, rather than soluble, was proposed to mediate tumor killing/rejection [66, 67, 70, 71].

Most data point towards an important role of CD8 T cell-derived TNF in tumor surveillance and rejection; however, its role in killing tumor cells is not evident in all mouse models, since sensitivity to TNF-induced cytotoxicity is not a global property of tumors.

TNF and T Cells in Graft-versus-Host Disease

Much knowledge has been accumulated about the role of TNF in T cell alloresponses or in graft-versus-host disease (GVHD). In particular, TNFR1KO T cells, but not TNFR2KO, exhibited a proliferative defect and produced lower levels of Th1 cytokines in a mixed lymphocyte culture. In bone marrow (BM) transplantation experiments, resulting in GVHD, recipients of TNFR1KO cells had significantly reduced mortality and morbidity as compared with recipients of WT or TNFR2KO T cells. Moreover, it was shown that the absence of TNFR1 from T cells rather than BM cells was responsible for the decreased graft responses [72]. Similar results were obtained when CD4 T cells were transferred to class II MHC-mismatched hosts, where absence of TNFR2 from donor T cells, or expression of a TNF inhibitor in the host, resulted in reduced weight loss and intestinal GVHD by IL-12-independent mechanisms [73, 74]. Studies with a class I MHC disparate GVHD model, demonstrated a role of TNF in CD8 alloproliferative response and concomitant hepatic histopathology [75].

Importantly, and in agreement with the role of mTNF in antitumor T cell activity, it was recently demonstrated that transferred T cells expressing only mTNF resulted in less severe liver and intestinal GVHD, whereas the graft-versus-tumor activity remained intact, indicating that selective blockade of sTNF or inhibition of TNF-converting enzyme could be a candidate therapy for BM recipients [76].

TNF and T Cells in Autoimmunity

It is now clear that TNF plays a key role in the pathogenesis of several inflammatory and autoimmune disorders in humans, such as RA, Crohn's disease, psoriasis, and in mouse models of autoimmunity like T1D, collagen-induced arthritis (CIA) or experimental autoimmune encephalomyelitis (EAE). However, the role of T cell-derived TNF, or the role of TNF on pathogenic T cells, in promoting or ameliorating disease is only partially understood.

There is a plethora of studies that highlight the role of TNF in T1D and especially in the NOD mouse model. T cells are key components of autoimmune diabetes; both CD8 and CD4 T cells can mediate islet destruction, and several CD4 and CD8 diabetogenic clones have been isolated [77]. By transferring diabetogenic Th1 clones in NOD/SCID mice, it was demonstrated that these cells were major sources of TNF (along with IFN-γ), mostly soluble, in the pancreas during the effector phase of autoimmune attack [78]. Interestingly, microarray analysis of diabetogenic clones chronically exposed to TNF in vitro revealed modulation of several genes related to TCR signaling and to decreased T cell responses [79]. Injection of TNF to NOD mice early in life accelerated the incidence of diabetes [80], whereas TNF administration later in life in fact prevented disease [81]. Similar temporal correlation of disease progression with onset of TNF expression was found in NOD mice that expressed TNF strictly in islets early or later in life [82, 83]. Notably, treatment of T cells from diabetic patients with TNF- or TNFR2-agonist induced selective death of insulin-specific autoreactive T cells while sparing normal T cells [44]. Constitutive expression of costimulatory molecules along with TNF expression by islets is sufficient to drive diabetes in C57BL/6 mice by breaking peripheral tolerance of autoreactive T cells [84]. CD8 T cells have a prominent role in the pathogenesis in this model, since elimination of CD8 T cells fully protects mice from diabetes. Interestingly, using KO mice it was shown that β-cell destruction is totally independent of perforin and Fas, but when islets lacked TNFR2, diabetes –and even insulitis to some extent – was prevented, suggesting that T cells utilize mTNF for killing islet cells [85]. Elegant experiments where TNF was inducibly expressed/repressed in islets showed that there is a minimum duration of TNF expression, independent of the maturity of immune system, required for overcoming peripheral T cell tolerance, subsequent recruitment of effector autoaggressive CD8 T cells to the islets and autoimmunity [86]. Collectively, TNF and T cell interplay has a central role in development of autoimmune diabetes and probably may take the scene in pharmaceutical approaches, but the time frame of treatment and the target receptor(s) and cells for potential intervention remain critical issues.

It is supported by a vast amount of evidence that TNF is produced in inflamed joints in most types of inflammatory arthritis in humans and experimental animals and has a dominant role in the pathogenesis of these syndromes. TNF in joints can be produced mostly by macrophages, synovial fibroblasts and T cells [87]. Although T cells are one of the most abundant cell types in RA synovium and accumulating evidence suggests a pivotal role of T cells in RA [88], there is little knowledge about how T cells are affected by TNF overproduction or about the contribution of T cell-derived TNF in the pathophysiology of disease.

Initial studies with TNF-transgenic mice showed that overexpression of hTNF in T cells exclusively resulted in arthritis, wasting syndrome and organ necrosis [89].

Experiments by the same group using knock-in mice with deleted mRNA-destabilizing ARE sequence from TNF gene resulted in development of inflammatory arthritis and inflammatory bowel disease. However, by crossing these mice with RagKO it was shown that T cell-derived TNF was fully dispensable for the development of arthritis, but not for bowel inflammation [90]. Data about the role of TNF in RA T cells emerged from patients treated with TNF antagonists. Much evidence exists about a role of TNF in compromising regulatory T cell function (covered extensively elsewhere in this volume), a defect that can be reversed by anti-TNF therapy [91]. Anti-TNF treatment also affects T cell migration, thus preventing T cell accumulation in joints of RA patients and collagen-induced arthritic mice [92, 93]. Noteworthy, TNFKO mice still develop CIA, albeit at reduced severity [94], whereas T cell-specific TNFKO mice exhibit a fully manifested arthritic phenotype [95]. Surprisingly, although TNF blockade in CIA reduced arthritis severity, it expanded populations of pathogenic Th1 and Th17, probably by upregulating expression of IL-12/IL-23 common subunit, p40 [92].

Multiple sclerosis (MS) and EAE, a mouse model of MS, are demyelinating diseases of the central nervous system (CNS). Mainly CD4, but also CD8, T cells, are found in infiltrates and play an important role in the pathogenesis of disease [96, 97]. TNF is produced in the CNS, primarily by microglia and astrocytes, in response to inflammation and other pathological processes [54]. In EAE, T cell apoptosis precedes clinical remission, and TNF may mediate the apoptosis of activated T cells as implied by markedly diminished apoptosis in EAE lesions of TNF/LTa DKO mice. On the other hand, TNF induces production of T cell chemokines in the CNS (such as IP-10 and MCP-1) leading to invasion by T cells [98]. Although the onset of EAE symptoms was delayed in TNFKO mice, they developed a chronic, progressive form of EAE in which myelin-specific T cell reactivity and expansion of activated/memory T cells failed to regress [97]. Moreover, treatment with TNF decreases EAE severity both in TNFKO and WT mice [99]. Interestingly, TNFR1KO mice did not manifest late EAE symptoms, suggesting that defective TNFR1-mediated apoptosis can not fully account for prolonged symptomatology in TNFKO mice [97]. Preclinical studies revealed that treatment of MS patients with TNF antagonists can exacerbate the disease, probably by decreasing apoptosis of pathogenic T cells [95]. Furthermore, anti-TNF treatment in RA patients can aggravate known MS, or even may be associated with onset of MS-like demyelinating disease [100].

Crohn's disease is a chronic inflammatory bowel disease, and excessive T cell-mediated immune responses contribute to its pathogenesis. Treatment with TNF-blocking agent infliximab leads to even complete clinical remission, with a yet unknown mechanism. It has been demonstrated that infliximab, but not etanercept which does not bind mTNF, apart from its general anti-inflammatory effect, leads to apoptosis of activated lamina propria T cells. The precise mechanism has not been identified, but it is possible that it involves reverse signaling of mTNF on activated T cells [101]. Recent work has shown that infliximab mediates restoration of perturbed mucosal homeostasis by affecting activation and possibly expansion of mucosal regulatory T cells [102].

Psoriasis is an autoimmune, chronic inflammatory disease, targeting primarily skin and nails but can also affect joints in a small fraction of patients. The role of T cells in pathogenesis of psoriasis is still obscure but it is suggested that Th1 and Th17 cells are implicated in disease development [103]. There is evidence that TNF antagonist treatment in psoriatic patients modulates DC activation and maturation and subsequent T cell activation, function and migration [104]. More recently, it was proposed that TNF blockade negatively regulates production of IL-23 by DCs resulting in reduction in numbers and activity of Th17 cells [105].

References

1 Ohsawa T, Natori S: Expression of tumor necrosis factor at a specific developmental stage of mouse embryos. Dev Biol 1989;135:459–461.

2 Deman J, Martin MT, Delvenne P, Humblet C, Boniver J, Defresne MP: Analysis by in situ hybridization of cells expressing mRNA for tumor-necrosis factor in the developing thymus of mice. Dev Immunol 1992;2:103–109.

3 Krangel MS, Yssel H, Brocklehurst C, Spits H: A distinct wave of human T cell receptor gamma/delta lymphocytes in the early fetal thymus: evidence for controlled gene rearrangement and cytokine production. J Exp Med 1990;172:847–859.

4 Giroir BP, Brown T, Beutler B: Constitutive synthesis of tumor necrosis factor in the thymus. Proc Natl Acad Sci USA 1992;89:4864–4868.

5 Baseta JG, Stutman O: TNF regulates thymocyte production by apoptosis and proliferation of the triple negative (CD3–CD4–CD8–) subset. J Immunol 2000;165:5621–5630.

6 Zuniga-Pflucker JC, Di J, Lenardo MJ: Requirement for TNF-alpha and IL-1 alpha in fetal thymocyte commitment and differentiation. Science 1995;268:1906–1909.

7 Samira S, Ferrand C, Peled A, Nagler A, Tovbin Y, Ben-Hur H, Taylor N, Globerson A, Lapidot T: Tumor necrosis factor promotes human T-cell development in nonobese diabetic/severe combined immunodeficient mice. Stem Cells 2004;22:1085–1100.

8 Kato Y, Morikawa A, Sugiyama T, Koide N, Jiang GZ, Takahashi K, Yokochi T: Role of tumor necrosis factor-alpha and glucocorticoid on lipopolysaccharide (LPS)-induced apoptosis of thymocytes. FEMS Immunol Med Microbiol 1995;12:195–204.

9 Page DM, Roberts EM, Peschon JJ, Hedrick SM: TNF receptor-deficient mice reveal striking differences between several models of thymocyte negative selection. J Immunol 1998;160:120–133.

10 Page DM: Cutting edge: thymic selection and autoreactivity are regulated by multiple coreceptors involved in T cell activation. J Immunol 1999;163:3577–3581.

11 Jondal M, Okret S, McConkey D: Killing of immature CD4+ CD8+ thymocytes in vivo by anti-CD3 or 5′-(N-ethyl)-carboxamide adenosine is blocked by glucocorticoid receptor antagonist RU-486. Eur J Immunol 1993;23:1246–1250.

12 Martin S, Bevan MJ: Antigen-specific and nonspecific deletion of immature cortical thymocytes caused by antigen injection. Eur J Immunol 1997;27:2726–2736.

13 Kishimoto H, Sprent J: Several different cell surface molecules control negative selection of medullary thymocytes. J Exp Med 1999;190:65–73.

14 Wu AJ, Hua H, Munson SH, McDevitt HO: Tumor necrosis factor-alpha regulation of CD4+CD25+ T cell levels in NOD mice. Proc Natl Acad Sci USA 2002;99:12287–12292.

15 Scheurich P, Thoma B, Ucer U, Pfizenmaier K: Immunoregulatory activity of recombinant human tumor necrosis factor (TNF)-alpha: induction of TNF receptors on human T cells and TNF-alpha-mediated enhancement of T cell responses. J Immunol 1987;138:1786–1790.

16 Yokota S, Geppert TD, Lipsky PE: Enhancement of antigen- and mitogen-induced human T lymphocyte proliferation by tumor necrosis factor-alpha. J Immunol 1988;140:531–536.

17 Shalaby MR, Espevik T, Rice GC, Ammann AJ, Figari IS, Ranges GE, Palladino MA Jr: The involvement of human tumor necrosis factors-alpha and -beta in the mixed lymphocyte reaction. J Immunol 1988;141:499–503.

18 Pimentel-Muinos FX, Munoz-Fernandez MA, Fresno M: Control of T lymphocyte activation and IL-2 receptor expression by endogenously secreted lymphokines. J Immunol 1994;152:5714–5722.

19 Aspalter RM, Eibl MM, Wolf HM: Regulation of TCR-mediated T cell activation by TNF-RII. J Leukoc Biol 2003;74:572–582.
20 Banerjee D, Liou HC, Sen R: c-Rel-dependent priming of naive T cells by inflammatory cytokines. Immunity 2005;23:445–458.
21 Kim EY, Priatel JJ, Teh SJ, Teh HS: TNF receptor type 2 (p75) functions as a costimulator for antigen-driven T cell responses in vivo. J Immunol 2006; 176:1026–1035.
22 Kim EY, Teh HS: TNF type 2 receptor (p75) lowers the threshold of T cell activation. J Immunol 2001;167:6812–6820.
23 Kim EY, Teh HS: Critical role of TNF receptor type-2 (p75) as a costimulator for IL-2 induction and T cell survival: a functional link to CD28. J Immunol 2004;173:4500–4509.
24 Chatzidakis I, Fousteri G, Tsoukatou D, Kollias G, Mamalaki C: An essential role for TNF in modulating thresholds for survival, activation, and tolerance of CD8+ T cells. J Immunol 2007;178:6735–6745.
25 Wiede F, Roomberg A, Cretney E, Lechner A, Fromm P, Wren L, Smyth MJ, Korner H: Age-dependent, polyclonal hyperactivation of T cells is reduced in TNF-negative gld/gld mice. J Leukoc Biol 2009;85:108–116.
26 Cope AP, Liblau RS, Yang XD, Congia M, Laudanna C, Schreiber RD, Probert L, Kollias G, McDevitt HO: Chronic tumor necrosis factor alters T cell responses by attenuating T cell receptor signaling. J Exp Med 1997;185:1573–1584.
27 Cope AP, Londei M, Chu NR, Cohen SB, Elliott MJ, Brennan FM, Maini RN, Feldmann M: Chronic exposure to tumor necrosis factor (TNF) in vitro impairs the activation of T cells through the T cell receptor/CD3 complex; reversal in vivo by anti-TNF antibodies in patients with rheumatoid arthritis. J Clin Invest 1994;94:749–760.
28 Isomaki P, Panesar M, Annenkov A, Clark JM, Foxwell BM, Chernajovsky Y, Cope AP: Prolonged exposure of T cells to TNF down-regulates TCR zeta and expression of the TCR/CD3 complex at the cell surface. J Immunol 2001;166:5495–5507.
29 Bryl E, Vallejo AN, Weyand CM, Goronzy JJ: Down-regulation of CD28 expression by TNF-alpha. J Immunol 2001;167:3231–3238.
30 Aspalter RM, Wolf HM, Eibl MM: Chronic TNF-alpha exposure impairs TCR-signaling via TNF-RII but not TNF-RI. Cell Immunol 2005;237:55–67.
31 Church LD, Hessler G, Goodall JE, Rider DA, Workman CJ, Vignali DAA, Bacon PA, Gulbins E, Young SP: TNFR1-induced sphingomyelinase activation modulates TCR signaling by impairing store-operated Ca^{2+} influx. J Leukoc Biol 2005;78: 266–278.
32 Sarin A, Conan-Cibotti M, Henkart PA: Cytotoxic effect of TNF and lymphotoxin on T lymphoblasts. J Immunol 1995;155:3716–3718.
33 Aggarwal S, Gollapudi S, Gupta S: Increased TNF-alpha-induced apoptosis in lymphocytes from aged humans: changes in TNF-alpha receptor expression and activation of caspases. J Immunol 1999;162: 2154–2161.
34 Gupta S, Gollapudi S: TNF-alpha-induced apoptosis in human naive and memory CD8+ T cells in aged humans. Exp Gerontol 2006;41:69–77.
35 Reich A, Korner H, Sedgwick JD, Pircher H: Immune down-regulation and peripheral deletion of CD8 T cells does not require TNF receptor-ligand interactions nor CD95 (Fas, APO-1). Eur J Immunol 2000;30:678–682.
36 Nguyen LT, McKall-Faienza K, Zakarian A, Speiser DE, Mak TW, Ohashi PS: TNF receptor 1 (TNFR1) and CD95 are not required for T cell deletion after virus infection but contribute to peptide-induced deletion under limited conditions. Eur J Immunol 2000;30:683–688.
37 Sytwu HK, Liblau RS, McDevitt HO: The roles of Fas/APO-1 (CD95) and TNF in antigen-induced programmed cell death in T cell receptor transgenic mice. Immunity 1996;5:17–30.
38 Murray DA, Crispe IN. TNF-alpha controls intrahepatic T cell apoptosis and peripheral T cell numbers: J Immunol 2004;173:2402–2409.
39 Zheng L, Fisher G, Miller RE, Peschon J, Lynch DH, Lenardo MJ: Induction of apoptosis in mature T cells by tumour necrosis factor. Nature 1995;377: 348–351.
40 Teh H-S, Seebaran A, Teh S-J: TNF receptor 2-deficient CD8 T cells are resistant to Fas/Fas ligand-induced cell death. J Immunol 2000;165:4814–4821.
41 Van Antwerp DJ, Martin SJ, Kafri T, Green DR, Verma IM: Suppression of TNF-alpha-induced apoptosis by NF-kappa B. Science 1996;274:787–789.
42 Speiser DE, Sebzda E, Ohteki T, Bachmann MF, Pfeffer K, Mak TW, Ohashi PS: Tumor necrosis factor receptor p55 mediates deletion of peripheral cytotoxic T lymphocytes in vivo. Eur J Immunol 1996;26:3055–3060.
43 de Oliveira Pinto LM, Garcia S, Lecoeur H, Rapp C, Gougeon M-L: Increased sensitivity of T lymphocytes to tumor necrosis factor receptor 1 (TNFR1)- and TNFR2-mediated apoptosis in HIV infection: relation to expression of Bcl-2 and active caspase-8 and caspase-3. Blood 2002;99:1666–1675.
44 Ban L, Zhang J, Wang L, Kuhtreiber W, Burger D, Faustman DL: Selective death of autoreactive T cells in human diabetes by TNF or TNF receptor 2 agonism. Proc Natl Acad Sci 2008;105:13644–13649.

45 Zhang H-G, Liu C, Su K, Yu S, Zhang L, Zhang S, Wang J, Cao X, Grizzle W, Kimberly RP: A membrane form of TNF-alpha presented by exosomes delays T cell activation-induced cell death. J Immunol 2006;176:7385–7393.
46 Marino MW, Dunn A, Grail D, Inglese M, Noguchi Y, Richards E, Jungbluth A, Wada H, Moore M, Williamson B, Basu S, Old LJ: Characterization of tumor necrosis factor-deficient mice. Proc Natl Acad Sci USA 1997;94:8093–8098.
47 Pasparakis M, Alexopoulou L, Episkopou V, Kollias G: Immune and inflammatory responses in TNF alpha-deficient mice: a critical requirement for TNF alpha in the formation of primary B cell follicles, follicular dendritic cell networks and germinal centers, and in the maturation of the humoral immune response. J Exp Med 1996;184:1397–1411.
48 Pfeffer K, Matsuyama T, Kundig TM, Wakeham A, Kishihara K, Shahinian A, Wiegmann K, Ohashi PS, Kronke M, Mak TW: Mice deficient for the 55 kD tumor necrosis factor receptor are resistant to endotoxic shock, yet succumb to *L. monocytogenes* infection. Cell 1993;73:457–467.
49 Rothe J, Lesslauer W, Lotscher H, Lang Y, Koebel P, Kontgen F, Althage A, Zinkernagel R, Steinmetz M, Bluethmann H: Mice lacking the tumour necrosis factor receptor 1 are resistant to TNF-mediated toxicity but highly susceptible to infection by *Listeria monocytogenes*. Nature 1993;364:798–802.
50 Ehlers S, Benini J, Kutsch S, Endres R, Rietschel ET, Pfeffer K: Fatal granuloma necrosis without exacerbated mycobacterial growth in tumor necrosis factor receptor p55 gene-deficient mice intravenously infected with *Mycobacterium avium*. Infect Immun 1999;67:3571–3579.
51 Flynn JL, Goldstein MM, Chan J, Triebold KJ, Pfeffer K, Lowenstein CJ, Schreiber R, Mak TW, Bloom BR: Tumor necrosis factor-alpha is required in the protective immune response against *Mycobacterium tuberculosis* in mice. Immunity 1995;2: 561–572.
52 Mohan VP, Scanga CA, Yu K, Scott HM, Tanaka KE, Tsang E, Tsai MM, Flynn JL, Chan J: Effects of tumor necrosis factor alpha on host immune response in chronic persistent tuberculosis: possible role for limiting pathology. Infect Immun 2001;69: 1847–1855.
53 Bongartz T, Sutton AJ, Sweeting MJ, Buchan I, Matteson EL, Montori V: Anti-TNF antibody therapy in rheumatoid arthritis and the risk of serious infections and malignancies: systematic review and meta-analysis of rare harmful effects in randomized controlled trials. JAMA 2006;295:2275–2285.
54 Bradley JR: TNF-mediated inflammatory disease. J Pathol 2008;214:149–160.
55 Brehm MA, Daniels KA, Welsh RM: Rapid production of TNF-alpha following TCR engagement of naive CD8 T cells. J Immunol 2005;175:5043–5049.
56 Saunders BM, Briscoe H, Britton WJ: T cell-derived tumour necrosis factor is essential, but not sufficient, for protection against *Mycobacterium tuberculosis* infection. Clin Exp Immunol 2004;137: 279–287.
57 Saunders BM, Tran S, Ruuls S, Sedgwick JD, Briscoe H, Britton WJ: Transmembrane TNF is sufficient to initiate cell migration and granuloma formation and provide acute, but not long-term, control of *Mycobacterium tuberculosis* infection. J Immunol 2005;174:4852–4859.
58 Cowley SC, Sedgwick JD, Elkins KL: Differential requirements by CD4+ and CD8+ T cells for soluble and membrane TNF in control of *Francisella tularensis* live vaccine strain intramacrophage growth. J Immunol 2007;179:7709–7719.
59 Calzascia T, Pellegrini M, Hall H, Sabbagh L, Ono N, Elford AR, Mak TW, Ohashi PS: TNF-alpha is critical for antitumor but not antiviral T cell immunity in mice. J Clin Invest 2007;117:3833–3845.
60 Singh A, Wuthrich M, Klein B, Suresh M: Indirect regulation of CD4 T-cell responses by tumor necrosis factor receptors in an acute viral infection. J Virol 2007;81:6502–6512.
61 Singh A, Suresh M: A role for TNF in limiting the duration of CTL effector phase and magnitude of CD8 T cell memory. J Leukoc Biol 2007;82:1201–1211.
62 Kafrouni MI, Brown GR, Thiele DL: The role of TNF-TNFR2 interactions in generation of CTL responses and clearance of hepatic adenovirus infection. J Leukoc Biol 2003;74:564–571.
63 Bitsaktsis C, Winslow G: Fatal recall responses mediated by CD8 T cells during intracellular bacterial challenge infection. J Immunol 2006;177:4644–4651.
64 Grivennikov SI, Tumanov AV, Liepinsh DJ, Kruglov AA, Marakusha BI, Shakhov AN, Murakami T, Drutskaya LN, Forster I, Clausen BE, Tessarollo L, Ryffel B, Kuprash DV, Nedospasov SA: Distinct and nonredundant in vivo functions of TNF produced by T cells and macrophages/neutrophils: protective and deleterious effects. Immunity 2005;22:93–104.
65 Togbe D, Grivennikov SI, Noulin N, Couillin I, Maillet I, Jacobs M, Maret M, Fick L, Nedospasov SA, Quesniaux VF, Schnyder B, Schnyder-Candrian S: T cell-derived TNF down-regulates acute airway response to endotoxin. Eur J Immunol 2007;37:768–779.
66 Ratner A, Clark WR: Role of TNF-alpha in CD8+ cytotoxic T lymphocyte-mediated lysis. J Immunol 1993;150:4303–4314.

67 Gonthier M, Llobera R, Arnaud J, Rubin B: Self-reactive T cell receptor-reactive CD8+ T cells inhibit T cell lymphoma growth in vivo. J Immunol 2004;173:7062–7069.

68 Perret R, Ronchese F: Effector CD8+ T cells activated in vitro confer immediate and long-term tumor protection in vivo. Eur J Immunol 2008;38: 2886–2895.

69 Prevost-Blondel A, Roth E, Rosenthal FM, Pircher H: Crucial role of TNF-alpha in CD8 T cell-mediated elimination of 3LL-A9 Lewis lung carcinoma cells in vivo. J Immunol 2000;164:3645–3651.

70 Poehlein CH, Hu HM, Yamada J, Assmann I, Alvord WG, Urba WJ, Fox BA: TNF plays an essential role in tumor regression after adoptive transfer of perforin/IFN-gamma double knockout effector T cells. J Immunol 2003;170:2004–2013.

71 Dace DS, Chen PW, Niederkorn JY: CD8+ T cells circumvent immune privilege in the eye and mediate intraocular tumor rejection by a TNF-alpha-dependent mechanism. J Immunol 2007;178: 6115–6122.

72 Hill GR, Teshima T, Rebel VI, Krijanovski OI, Cooke KR, Brinson YS, Ferrara JL: The p55 TNF-alpha receptor plays a critical role in T cell alloreactivity. J Immunol 2000;164:656–663.

73 Brown GR, Lee E, Thiele DL: TNF-TNFR2 interactions are critical for the development of intestinal graft-versus-host disease in MHC class II-disparate (C57BL/6J->C57BL/6J x bm12)F1 mice. J Immunol 2002;168:3065–3071.

74 Brown GR, Lee EL, Thiele DL: TNF enhances CD4+ T cell alloproliferation, IFN-gamma responses, and intestinal graft-versus-host disease by IL-12-independent mechanisms. J Immunol 2003;170: 5082–5088.

75 El-Hayek JM, Rogers TE, Brown GR: The role of TNF in hepatic histopathological manifestations and hepatic CD8+ T cell alloresponses in murine MHC class I disparate GVHD. J Leukoc Biol 2005; 78:1001–1007.

76 Borsotti C, Franklin AR, Lu SX, Kim TD, Smith OM, Suh D, King CG, Chow A, Liu C, Alpdogan O, van den Brink MR: Absence of donor T-cell-derived soluble TNF decreases graft-versus-host disease without impairing graft-versus-tumor activity. Blood 2007;110:783–786.

77 Anderson MS, Bluestone JA: The NOD mouse: a model of immune dysregulation. Annu Rev Immunol 2005;23:447–485.

78 Cantor J, Haskins K: Effector function of diabetogenic CD4 Th1 T cell clones: a central role for TNF-alpha. J Immunol 2005;175:7738–7745.

79 Lee LF, Lih CJ, Huang CJ, Cao T, Cohen SN, McDevitt HO: Genomic expression profiling of TNF-alpha-treated BDC2.5 diabetogenic CD4+ T cells. Proc Natl Acad Sci USA 2008;105:10107–10112.

80 Jacob CO, Aiso S, Michie SA, McDevitt HO, Acha-Orbea H: Prevention of diabetes in nonobese diabetic mice by tumor necrosis factor (TNF): similarities between TNF-alpha and interleukin 1. Proc Natl Acad Sci USA 1990;87:968–972.

81 Yang XD, Tisch R, Singer SM, Cao ZA, Liblau RS, Schreiber RD, McDevitt HO: Effect of tumor necrosis factor alpha on insulin-dependent diabetes mellitus in NOD mice. I. The early development of autoimmunity and the diabetogenic process. J Exp Med 1994;180:995–1004.

82 Green EA, Eynon EE, Flavell RA: Local expression of TNFalpha in neonatal NOD mice promotes diabetes by enhancing presentation of islet antigens. Immunity 1998;9:733–743.

83 Grewal IS, Grewal KD, Wong FS, Picarella DE, Janeway CA Jr, Flavell RA: Local expression of transgene encoded TNF alpha in islets prevents autoimmune diabetes in nonobese diabetic (NOD) mice by preventing the development of auto-reactive islet-specific T cells. J Exp Med 1996;184:1963–1974.

84 Guerder S, Picarella DE, Linsley PS, Flavell RA: Costimulator B7–1 confers antigen-presenting-cell function to parenchymal tissue and in conjunction with tumor necrosis factor alpha leads to autoimmunity in transgenic mice. Proc Natl Acad Sci USA 1994;91:5138–5142.

85 Herrera PL, Harlan DM, Vassalli P: A mouse CD8 T cell-mediated acute autoimmune diabetes independent of the perforin and Fas cytotoxic pathways: possible role of membrane TNF. Proc Natl Acad Sci USA 2000;97:279–284.

86 Green EA, Flavell RA: The temporal importance of TNFalpha expression in the development of diabetes. Immunity 2000;12:459–469.

87 Ostor AJ: Abatacept: a T-cell co-stimulation modulator for the treatment of rheumatoid arthritis. Clin Rheumatol 2008;27:1343–1353.

88 Isaacs JD: Therapeutic T-cell manipulation in rheumatoid arthritis: past, present and future. Rheumatology (Oxford) 2008;47:1461–1468.

89 Probert L, Keffer J, Corbella P, Cazlaris H, Patsavoudi E, Stephens S, Kaslaris E, Kioussis D, Kollias G: Wasting, ischemia, and lymphoid abnormalities in mice expressing T cell-targeted human tumor necrosis factor transgenes. J Immunol 1993;151: 1894–1906.

90 Kontoyiannis D, Pasparakis M, Pizarro TT, Cominelli F, Kollias G: Impaired on/off regulation of TNF biosynthesis in mice lacking TNF AU-rich elements: implications for joint and gut-associated immunopathologies. Immunity 1999;10:387–398.

91 Ehrenstein MR, Evans JG, Singh A, Moore S, Warnes G, Isenberg DA, Mauri C: Compromised function of regulatory T cells in rheumatoid arthritis and reversal by anti-TNFalpha therapy. J Exp Med 2004;200:277–285.

92 Notley CA, Inglis JJ, Alzabin S, McCann FE, McNamee KE, Williams RO: Blockade of tumor necrosis factor in collagen-induced arthritis reveals a novel immunoregulatory pathway for Th1 and Th17 cells. J Exp Med 2008;205:2491–2497.

93 Taylor PC, Peters AM, Paleolog E, Chapman PT, Elliott MJ, McCloskey R, Feldmann M, Maini RN: Reduction of chemokine levels and leukocyte traffic to joints by tumor necrosis factor alpha blockade in patients with rheumatoid arthritis. Arthritis Rheum 2000;43:38–47.

94 Campbell IK, O'Donnell K, Lawlor KE, Wicks IP: Severe inflammatory arthritis and lymphadenopathy in the absence of TNF. J Clin Invest 2001;107: 1519–1527.

95 Kruglov AA, Kuchmiy A, Grivennikov SI, Tumanov AV, Kuprash DV, Nedospasov SA: Physiological functions of tumor necrosis factor and the consequences of its pathologic overexpression or blockade: mouse models. Cytokine Growth Factor Rev 2008;19:231–244.

96 Huseby ES, Liggitt D, Brabb T, Schnabel B, Ohlen C, Goverman J: A pathogenic role for myelin-specific CD8+ T cells in a model for multiple sclerosis. J Exp Med 2001;194:669–676.

97 Kassiotis G, Kollias G: Uncoupling the proinflammatory from the immunosuppressive properties of tumor necrosis factor (TNF) at the p55 TNF receptor level: implications for pathogenesis and therapy of autoimmune demyelination. J Exp Med 2001; 193:427–434.

98 Probert L, Eugster HP, Akassoglou K, Bauer J, Frei K, Lassmann H, Fontana A: TNFR1 signalling is critical for the development of demyelination and the limitation of T-cell responses during immune-mediated CNS disease. Brain 2000;123:2005–2019.

99 Liu J, Marino MW, Wong G, Grail D, Dunn A, Bettadapura J, Slavin AJ, Old L, Bernard CC: TNF is a potent anti-inflammatory cytokine in autoimmune-mediated demyelination. Nat Med 1998;4:78–83.

100 Fromont A, De Seze J, Fleury MC, Maillefert JF, Moreau T: Inflammatory demyelinating events following treatment with anti-tumor necrosis factor. Cytokine 2009;45:55–57.

101 Van den Brande JM, Braat H, van den Brink GR, Versteeg HH, Bauer CA, Hoedemaeker I, van Montfrans C, Hommes DW, Peppelenbosch MP, van Deventer SJ: Infliximab but not etanercept induces apoptosis in lamina propria T-lymphocytes from patients with Crohn's disease. Gastroenterology 2003;124:1774–1785.

102 Ricciardelli I, Lindley KJ, Londei M, Quaratino S: Anti tumour necrosis-alpha therapy increases the number of FOXP3 regulatory T cells in children affected by Crohn's disease. Immunology 2008;125: 178–183.

103 Hueber AJ, McInnes IB: Immune regulation in psoriasis and psoriatic arthritis – recent developments. Immunol Lett 2007;114:59–65.

104 Gottlieb AB, Chamian F, Masud S, Cardinale I, Abello MV, Lowes MA, Chen F, Magliocco M, Krueger JG: TNF inhibition rapidly down-regulates multiple proinflammatory pathways in psoriasis plaques. J Immunol 2005;175:2721–2729.

105 Zaba LC, Cardinale I, Gilleaudeau P, Sullivan-Whalen M, Suarez-Farinas M, Fuentes-Duculan J, Novitskaya I, Khatcherian A, Bluth MJ, Lowes MA, Krueger JG: Amelioration of epidermal hyperplasia by TNF inhibition is associated with reduced Th17 responses. J Exp Med 2007;204:3183–3194.

Ioannis Chatzidakis, PhD
Institute of Molecular Biology and Biotechnology, Foundation for Research and Technology, Hellas
Nikolaou Plastira 100
GR–70013 Heraklion, Crete (Greece)
Tel. +30 2810391164, Fax +30 2810 391101, E-Mail chatzida@imbb.forth.gr

Kollias G, Sfikakis PP (eds): TNF Pathophysiology. Molecular and Cellular Mechanisms.
Curr Dir Autoimmun. Basel, Karger, 2010, vol 11, pp 119–134

TNF-α: An Activator of CD4+FoxP3+TNFR2+ Regulatory T Cells

Xin Chen[a] · Joost J. Oppenheim[b]

[a]Basic Science Program, SAIC-Frederick, Inc., NCI-Frederick, and [b]Laboratory of Molecular Immunoregulation, CIP, Center for Cancer Research, NCI-Frederick, Frederick, Md., USA

Abstract

TNF-α (TNF) is a pleiotropic cytokine which can have proinflammatory or immunosuppressive effects, depending on the context, duration of exposure and disease state. The basis for the opposing actions of TNF remains elusive. The growing appreciation of CD4+FoxP3+ regulatory T cells (Tregs), which comprise ~10% of peripheral CD4 cells, as pivotal regulators of immune responses has provided a new framework to define the cellular and molecular basis underlying the contrasting action of TNF. TNF by itself can overcome the profound anergic state of T cell receptor-stimulated Tregs. Furthermore, in concert with IL-2, TNF selectively activates Tregs, resulting in proliferation, upregulation of FoxP3 expression and increases in their suppressive activity. Both human and mouse Tregs predominantly express TNFR2, making it possible for TNF to enhance Treg activity, which helps limit the collateral damage caused by excessive immune responses and eventually terminates immune response. TNFR2-expressing CD4+FoxP3+ Tregs comprise ~40% of peripheral Tregs in normal mice and present the maximally suppressive subset of Tregs. In this review, studies describing the action of TNF on Treg function will be discussed. The role of Tregs in the autoimmune disorders and cancer as well as the effect of anti-TNF therapy on Tregs, especially in rheumatoid arthritis, will also be considered.

Introduction

CD4+FoxP3+ Regulatory T Cells Are Pivotal Regulators of Immune Responses

The evidence that suppressive T cells downregulate antigen-specific response of effector T cells and maintain immune tolerance was reported as early as 1970s [1]. In the

The content of this chapter does not necessarily reflect the views or policies of the Department of Health and Human Services, nor does mention of trade names, commercial products, or organization imply endorsement by the U.S. Government.

mid-1990s, Sakaguchi et al. [2] identified CD4 cells which constitutively coexpressed CD25, the IL-2 receptor α-chain, in normal rodents as potent suppressive regulatory T cells (Tregs), and showed that elimination of this population of cells elicited autoimmune responses. Subsequent studies extending over more than a decade have provided compelling evidence that CD4+FoxP3+ Tregs, comprising ~10% of peripheral CD4 cells, play an indispensable role in maintaining immune homeostasis and in suppressing deleterious excessive immune responses [3].

Two major sources of Tregs, namely naturally occurring Tregs (nTregs) and induced Tregs (iTregs), are engaged in normal tolerogenic surveillance of self-antigens and prevent potential autoimmune responses. nTregs develop in the thymus and are exported to the periphery [3], and iTregs are converted from naïve CD4 cells by TGF-β in conjunction with T cell receptor (TCR) stimulation in the periphery [4]. Tregs are preferentially self-reactive since their TCRs have higher affinity for self-antigens, and are similar to the TCR used by self-reactive pathogenic effector T cells (Teffs) [5]. Foxp3, a member of the forkhead/winged-helix family of transcription factors, is a master regulator of Treg development and function [6], as shown by deficiency of Tregs and lethal autoimmunity caused by the mutation of FoxP3 in human patients with IPEX (immune dysregulation, polyendocrinopathy, enteropathy, X-linked) and its murine counterpart scurfy [reviewed in 7]. The characteristic phenotype of Tregs such as high expression of CD25, CTLA-4, GITR and low expression of CD127 was shown to be regulated by FoxP3 [reviewed in 8].

The activation, proliferation and effector functions of a large spectrum of immunocompetent cells, such as CD4 cells [9], CD8 cells [10], NK cells [11], NKT cells [12], dendritic cells [13], macrophages [14] and B cells [15] are susceptible to Treg-mediated suppression. The induction of Treg-suppressive activity is specific and requires antigenic stimulation through the TCR; however, the suppression exerted by Tregs is not antigen specific [16]. Therefore, a wide range of immune responses can be inhibited by Tregs through 'bystander' suppression [17]. In addition to suppressing immune responses to auto-antigens, Tregs also attenuate host defense responses against pathogens [reviewed in 18] and tumor antigens [reviewed in 19].

The exact mechanism(s) of Treg-mediated suppression remain incompletely understood. The in vitro suppressive activity of Tregs depends on cell-to-cell contact as over a short distance [9]. In addition, several molecules, such as IL-10, TGF-β, CTLA-4, indoleamine 2,3-dioxygenase and granzyme/perforin are reported to contribute to the suppressive activity of Tregs [reviewed in 20]. IL-35 is reportedly expressed by mouse FoxP3+ Tregs and contributes to Treg function [21]; however, this immunosuppressive cytokine is not expressed by human Tregs [22, 23]. Tregs express CD39/ENTPD1 and CD73/ecto-5′-nucleotidase, ectoenzymes which have the capacity to generate pericellular adenosine from extracellular nucleotides. The coordinated expression of CD39/CD73 on Tregs and the adenosine A2A receptor on activated Teffs therefore may generate an immunosuppressive loop [24]. Although Tregs are likely to use multiple mechanisms to suppress immune responses, CTLA-4 may have a dominant role. It has been recently shown that CTLA-4 was critically required for the function of Tregs in vivo by inhibiting activities of antigen-presenting cells (APCs) [25, 26].

Besides FoxP3+ Tregs, there are other types of Tregs that can be induced from naïve CD4 cells in the periphery, such as IL-10- and TGF-β-producing Tr1 cells and TGF-β-producing Th3 cells [3]. Thus, various Tregs by using distinct mechanism are likely to operate collaboratively to regulate the duration and magnitude of an immune response.

Contrasting Roles of TNF-α in Autoimmune Diseases

TNF-α (TNF) is a pleiotropic cytokine that is a major participant in the initiation and orchestration of complex events in inflammation and immunity [27]. TNF has well-documented proinflammatory effects. Nevertheless, increasing evidence reveals that TNF also has unexpected anti-inflammatory and immunosuppressive effects, especially after prolonged exposure [reviewed in 28–30]. For example, as expected, several transgenic mouse strains overproducing TNF consistently develop autoimmune disorders [reviewed in 31]. However, transgenic NOD mice overexpressing TNF in their pancreatic islets failed to develop autoimmune diabetes [32] and repeated injection of TNF suppressed both type I diabetes in NOD mice and lupus nephritis in susceptible mouse strains [33, 34]. Furthermore, NZB mice deficient in TNF exhibited acceleration of autoimmunity and lupus nephritis [35]. C67BL/6.129 mice deficient in TNF developed mild autoimmunity resembling the initial stages of lupus nephritis [36]. TNF knockout (KO) mice developed prolonged and exacerbated experimental autoimmune encephalomyelitis (EAE), although with a delayed onset, after EAE induction [37]. In multiple sclerosis patients, treatment with anti-TNF agents resulted almost uniformly in immune activation and exacerbation of disease [36]. Perhaps reflecting the strikingly contrasting activities of TNF, anti-TNF therapy in rheumatoid arthritis (RA) and inflammatory bowel disease, although impressively beneficial to the majority of patients, led to the development of lupus and neuroinflammatory diseases in some patients [36].

Regulatory T Cell Levels Are Increased at Autoimmune Inflamed Sites

It has been established for more than a decade that the breakdown of immune tolerance maintained by Tregs can cause organ-specific autoimmune responses in animal models [2]. The essential role of intact FoxP3 for Treg function in prevention of autoimmune responses has been convincingly confirmed by the development of autoimmunity in human patients with IPEX and in the homologous *scurfy* mouse [7]. It has therefore been proposed that the abnormality autoimmune patients generally have in common is either low Treg numbers with normal function or normal Treg numbers with compromised suppressive function [38]. However, the clinical and laboratory data lend very little support to this simplified notion and, instead, provide a much more complicated and often counterintuitive scenario. For example, the frequency and function of Tregs in the peripheral blood (PB) of RA patients is still controversial [reviewed in 38, 39]. The mounting evidence clearly indicates that the frequency of Tregs in RA patients with an activated phenotype and enhanced suppressive potential in the synovial fluid of inflamed joints was increased, as compared with those in the periphery [40–43]. Similarly, activated Tregs also accumulated in the inflamed joint of other arthropathies such as juvenile idiopathic arthritis and spondyloarthropathies [44–46]. Markedly elevated levels of Tregs were found in the synovial fluid in K/BxN mouse model of spontaneous inflammatory arthritis [47]. By crossing them with FoxP3^{gfp} mice, the population of Tregs, which could be unequivocally identified by GFP expression, was also increased during development of arthritis in K/BxN mice, especially at sites of inflammation [48]. Consequently, functional Tregs are often increased at the site of autoimmune inflammation, presumably resulting from active recruitment and in situ proliferative expansion.

Elucidation of the effect of TNF on Treg activity is critically important in an era where the use of biological agents to block TNF in the autoimmune patients is becoming a routine therapy. This review focuses on the new developments that provide some insight into how TNF stimulates Treg activity and how this may explain the puzzling immunosuppressive property of TNF in chronic inflammation.

Inflammation in General Activates Regulatory T Cells

Inflammation is the hallmark of a wide variety of diseases, in addition to autoimmunity, including infection and cancer. Studies of both human patients and animal models show that the frequency and suppressive function of Tregs were increased in sepsis, which contributes to the postseptic immunosuppressive phase and its fatal consequence [reviewed in 49], in a TNFR2-dependent manner [50]. In chronic infections, Tregs are activated and accumulate at the site of infection, which limits the magnitude of effector responses to control infection, but also reduces collateral tissue damage caused by excessively vigorous antimicrobial immune responses [reviewed in 18]. Inflammation in the tumor microenvironment promotes tumor progression [reviewed in 51]. Tregs accumulate in the tumor microenvironment, which can dampen natural or induced immune responses against tumor antigens and is predictive of a poor prognosis [reviewed in 19]. Thus, activation of Tregs has been reported in various types of inflammatory responses, which may represent a negative feedback mechanism to curtail excessive inflammation and prevent self-tissue destruction.

TNF Activates Mouse Tregs

Administration of TNF Expands Tregs in Young Adult Autoimmune-Prone Mice

A study examining the effect of intraperitoneal administration of TNF on the number of Tregs in NOD mice shed some light on the in vivo effect of TNF on Treg activity. Injection of TNF into newborn NOD mice led to an accelerated development of diabetes, while TNF treatment of adult NOD mice inhibited the development of diabetes [34]. Wu et al. [52] found that administration of TNF into neonatal NOD as well as neonatal B6.NOD mice resulted in a reduction of CD4+CD25+ cells in the spleen. In contrast, administration of TNF into young adult NOD mice increased the number of splenic CD4+CD25+ cells. It is now known that self-reactive T cells are exported from the thymus prior to Tregs, and neonatal mice are virtually deficient in Treg in the periphery [3]. Thus, it is likely that the stimulatory action of TNF may preferentially activate Teffs in neonatal NOD mice. In adult NOD mice, when the peripheral Treg pool has reached its full size, TNF action is likely to expand Tregs and tip the immune balance maintained by Tregs and Teffs toward an immune-tolerant direction.

TNF Mediated the Capacity of Pertussis Toxin to Activate Tregs

Our studies showed that pertussis toxin in the immunizing cocktail to induce EAE was solely responsible for the resulting reduction in Treg activity [53, 54]. Furthermore, pertussis toxin inhibited TGF-β-induced FoxP3 expression by wild-type (WT) naïve CD4 cells cocultured with WT bone marrow-derived dendritic cells. However, although pertussis toxin markedly reduced Tregs number in WT mice, administration of pertussis toxin paradoxically expanded Tregs in IL-6 KO mice. Pertussis toxin also promoted FoxP3 induction when marrow-derived dendritic cells were from IL-6 KO mice [55]. Therefore, this led us to hypothesize that mediator(s) induced by pertussis toxin in the absence of IL-6 should be able to stimulate Tregs. We confirmed previous reports [56, 57] that proinflammatory cytokines (IL-1β, IL-6, TNF) and Th1 cytokine (IFN-γ) were produced by pertussis toxin-treated splenocytes. This led us to identify TNF as the sole mediator with the capacity to expand Tregs in this study [58].

TNF Stimulates Proliferative Responses of Both Regulatory T Cells and Effector T Cells

Our subsequent studies revealed that TNF actually stimulated proliferation of both CD25– as well as CD25+ subsets of mouse CD4 cells, and therefore enhanced proliferation of cocultures containing Tregs and Teffs. Thus, we showed for the first time that TNF had the ability to overcome the profound anergy of Tregs to TCR stimulation in vitro. TNF has the capacity to directly stimulate purified CD4 Treg cells free of APCs. Incubation of cocultures containing Tregs and Teffs over a shorter time of 48 h with TNF (0.5–10 ng/ml) partially reversed Treg-suppressive activity. However, after a more prolonged incubation of 72 h with TNF, Treg suppression prevailed and the degree of inhibition in co-cultures was restored to a normal level [50]. These data suggest that over the short-term, TNF, as seen in the early phases of inflammatory response, may enable Teffs to proliferate despite the presence of Tregs, whereas more prolonged exposure to TNF favors the expansion and activation of functional Tregs.

In Concert with Interleukin-2, TNF Selectively Activates Mouse Regulatory T Cells

TNF by itself was not sufficient to support the survival of Tregs in vitro [50]. To maintain the in vitro survival of Tregs, Treg cultures were supplemented with IL-2. In the presence of IL-2, TNF markedly increased level of the FoxP3 expression (MFI) by Tregs in a dose- dependent (0.1~10 ng/ml) and time-dependent (24~72 h) manner. IL-1β and IL-6 lacked this activity. Unlike TGF-β, TNF did not induce FoxP3 expression by anti-CD3-stimulated naïve CD4+CD25– T cells. In conjunction with IL-2, TNF was able to selectively expand the subset of CD4+FoxP3+ T cells and selectively increased the phosphorylation of Stat5 in Tregs. Importantly, Tregs pretreated with TNF and IL-2 were markedly more suppressive than Tregs pretreated with IL-2 alone [50]. Thus, activation of Tregs by TNF has proliferative and functional consequences.

IL-2 has been appreciated for its nonredundant role in maintaining Treg survival and function [reviewed in 59]. Our study showed that TNF selectively upregulated CD25 expression on Tregs [50], while IL-2 preferentially upregulated TNFR2 expression on Tregs (our unpubl. data). This suggests that TNF and IL-2 form a reciprocating receptor amplification circuit and synergistically upregulate Treg suppressive activity. Collectively, our data clearly demonstrate that TNF, in conjunction with IL-2, has the capacity to selectively activate Tregs, resulting in proliferation and upregulation of FoxP3 expression and Stat5 phosphorylation, and consequently to enhance the suppressive potential of Tregs.

TNFR2 Is Predominantly Expressed on Human and Mouse Regulatory T Cells

TNF mediates its biological functions through two structurally distinct receptors: TNFR1 (p55) and TNFR2 (p75); the latter is largely confined to cells of the immune system [60]. Unlike TNFR1, which contains a death domain in its cytoplasmic tail, the primary function of TNFR2 is to promote lymphocyte proliferation and survival [27]. Agonist antibodies against p75, like TNF, have the capacity to enhance T cell proliferative response, whereas the specific activation of p55 had no such effect [61]. Kim et al. [62–64] have convincingly demonstrated that TNFR2 has important costimulatory function and markedly enhanced the responses of lymphocyte to TCR-mediated signaling.

nTregs develop in the thymus as a distinct lineage of functionally mature suppressor T cell subset [3]. All human thymic CD4+CD25+ Tregs constitutively express TNFR2, while thymic CD4+CD25– cells do not express this receptor [65]. Interestingly, immunosuppressive human thymic CD4-CD8+CD25+ T cells also express TNFR2 mRNA and protein [66]. We found that the majority (~80%) of mouse thymic Tregs (CD8–CD4+CD25+) are also TNFR2-expressing cells [67]. The normal thymus is the only organ that constitutively expresses TNF [68]. Thus, TNF may regulate Treg development and differentiation in the thymus.

In human PB, all CD4+CD25^{hi} T cells express the highest level of TNFR2, while 20~30% of CD4+CD25– T cells express a low level of this receptor [69 and our unpubl. data]. We confirmed that TNFR2 expression was highest on, but not exclusively confined to, human peripheral CD4+FoxP3+ T cells [69]. We demonstrated that 30–40% of CD4+CD25+ cells which comprised >90% FoxP3+ cells in peripheral lymphoid tissues of normal Balb/c mice and C57BL/6 (B6) mice expressed TNFR2, while only 8% of CD4+CD25– cells (which contain 25–40% of FoxP3+ cells) were TNFR2+ cells. After TCR stimulation, surface expression of TNFR2 on both CD4+CD25+ and CD4+CD25– T cells was upregulated. However, the activated CD4+CD25+ T cells still expressed considerably higher levels of TNFR2 on 47% of cells as compared with 32% of activated CD4+CD25– T cells. TNFR1 was not detectable by FACS on either CD4+CD25+ or CD4+CD25– T cells [50, 67]. In contrast to human circulating CD4+CD25^{hi} cells, virtually all of which express high levels of TNFR2 [69], less than 10% of mouse PB CD4+CD25+ cells are TNFR2+ cells [67], presumably because the laboratory mice live in a pathogen-free environment.

TNFR2-Expressing Mouse Tregs Are Maximally Suppressive Cells

The diverse biological activities of TNF may be caused as a distinct signaling consequence of a particular receptor [37, 70]. Several lines of evidence suggest that immunosuppressive action of TNF is mediated by TNFR2 [37, 71, 72]. For example, in the EAE mouse model, TNFR1-deficient mice were completely resistant to induction of disease, while TNFR2-deficient mice exhibited more severe EAE [37, 71]. The transmembrane form of TNF preferentially activates TNFR2 [73], and thus its role in selective stimulation of Tregs warrants future research. Although activation of Tregs can be a result of antigen-specific responses, proinflammatory cytokines such as TNF are also likely to enhance the activation of Tregs. Thus, TNFR2 is likely to costimulate Tregs in conjunction with TCR signaling and yields more activated functional Tregs.

To test this hypothesis, we have compared the phenotype and suppressive function of TNFR2+ Tregs with TNFR2– Tregs from normal C57BL/6 mice. The expression of TNFR2 actually defined a subset of activated/effector Tregs which were $CD45RB^{lo}CD62L^{lo}CD44^{hi}$. Although TNFR2+ and TNFR2– Tregs expressed comparable levels of FoxP3 (91.0 and 92.8%, respectively), TNFR2+ Tregs expressed a much higher level of CTLA4 (83.6%, MFI: 101.1) than TNFR2– Tregs (41.0%, MFI: 50.9). More importantly, TNFR2+ Tregs were highly suppressive in vitro. In sharp contrast, CD4+CD25+TNFR2– T cells exhibited a naïve phenotype and only have minimal suppressive activity. CD103-expressing Tregs are generally known to be the most suppressive subset of mouse Tregs [74]. The suppressive activity of TNFR2+ Tregs was even superior to that of CD103-expressing Tregs. Further, the number of TNFR2+ Tregs was 5–7 times greater than that of CD103+ Tregs. Although not identified as Tregs previously, even CD4+CD25–TNFR2+ T cells had moderate suppressive activity [67], which correlated with expression of FoxP3 (our unpubl. data). Thus, TNFR2 expression identified maximally suppressive FoxP3+ Tregs.

Our studies showed that tumor-associated Tregs were characterized by highly suppressive activity and by high level of TNFR2 expression which may increase their capacity to respond to the elevated TNF production by tumors [67]. We characterized Tregs in several mouse tumor models. In both 4T1 breast tumor in Balb/c mice and Lewis lung carcinoma tumors in C57BL/6 mice, the majority of tumor-infiltrating Tregs were TNFR2+ Tregs with highly suppressive activity [67]. It has been reported that TGF-β, a cytokine crucial for de novo generation of Tregs [4], was also able to induce TNFR2 expression on CD4 cells [75]. Therefore, tumor-derived TGF-β may contribute to the increase in TNFR2+ Tregs in the tumor. Our study suggests that TNFR2+ Tregs may play a crucial role in immune evasion of the tumor by potently dampening host immune responses to tumor antigens. Thus, TNFR2+ Tregs may provide a therapeutic target. In support of this hypothesis, a recent study proposed that cyclophosphamide eradicated tumor by selectively eliminating TNFR2+ Tregs [76].

van Mierlo et al. [69] reported that both human and mouse Tregs not only strongly expressed TNFR2, but they also shed large amounts of soluble TNFR2 upon stimulation with anti-CD3/CD28 in conjunction with IL-2. This was paralleled by their ability to inhibit the action of TNF. In vivo, Tregs suppressed IL-6 production in response to LPS injection in mice. In contrast, Treg cells from TNFR2 KO mice were unable to do so despite their unhampered capacity to suppress T cell proliferation in a conventional in vitro suppression assay. This study suggests that shed TNFR2 represents a novel mechanism by which Tregs can inhibit the inflammatory action of TNF [69].

Taken together, TNFR2 expression on Tregs has either phenotypic or functional implications. Although we favor the idea the high expression of CTLA-4 account for the potent suppressive activity of TNFR2-expressing Tregs [67], other mechanism such as shedding TNFR2 molecules [69] may also contribute.

Effects of TNF on Human Treg Activity

Our understanding of the action of TNF on human Treg activity has been gained mainly from studies examining the effect of anti-TNF therapy in autoimmune patients. Due to the limitation in the study of human Tregs in inflammatory conditions, in vitro experiments have been used to clarify the direct action of TNF on human Treg activity. Using conventional in vitro Treg function assays, Ehrenstein et al. [77] first examined the effect of exogenous TNF (0.1–5 ng/ml) on cocultures containing CD4+CD25^{hi} cells and CD4+CD25– cells from normal human healthy donors or from RA patients responsive to anti-TNF therapy. Their results showed that exogenous TNF neither increased nor decreased Treg activity, as evidenced by no alteration of percentage inhibition of proliferation in the cocultures. This was confirmed by another study [44]. Valencia et al. [78] examined the effect of higher concentrations of TNF (50 ng/ml) on in vitro normal human Treg activity. They reported that exogenous TNF downregulated FoxP3 expression and blocked Treg suppressive activity by signaling through TNFR2 and that an agonist monoclonal antibody to TNFR2 also reversed the suppressive activity of healthy donor PB CD4+CD25^{hi} cells. In contrast, our own preliminary data suggest that TNF at the same concentration ranges has stimulatory activity on normal human Tregs. Thus, the current data regarding direct action of TNF on human Tregs is divergent, which is partially due to the lack of uniform criteria for the identification and isolation of functional human Tregs.

Anti-TNF Therapy Promotes Tr1 and Th3 Cells

TNF is a major inflammatory cytokine contributing to the pathogenesis of RA, which provides rationale for the development of anti-TNF biological agents in the treatment of RA [79]. The impressive clinical benefit of anti-TNF therapy in the majority of patients prompted investigators to examine the effect of anti-TNF treatment on the number and function of Tregs. Our findings that TNF promotes the proliferation and function of Tregs would predict that anti-TNF therapy should decrease the Treg levels in treated patients.

It was reported that 83.2% of PB CD4+CD25^{hi} T cells from healthy donors expressed FoxP3 and exhibited suppressive activity, while only 37.1% of CD4+CD25^{hi} T cells isolated from active RA patients were FoxP3+ and failed to exhibit suppressive function [78]. In contrast to the prediction, anti-TNF therapy reportedly increased FoxP3 expression as well as the suppressive activity of RA CD4+CD25^{hi} cells [78]. Apparently, CD4+CD25^{hi} population in active RA patients consist of a mixture of Tregs and activated effector cells. Anti-TNF therapy can attenuate inflammatory response and consequently reduce the number of 'contaminating' effector cells in CD4+CD25^{hi} population, resulting in a relative increase in Treg activity. However, identification of Tregs in inflammatory states based only on CD25 may lead to incorrect enumeration and consequently inappropriate evaluation of Treg activity, simply because activated CD4 effector cells can also express high levels of CD25.

It was also reported that Tregs in RA patients exhibited a tendency to undergo apoptosis, which could be reversed by anti-TNF therapy [80]. However, as convincingly demonstrated in a study using the K/BxN-FoxP_{gfp} mouse model of spontaneous arthritis, which is characterized by elevated level of TNF [79], Tregs exhibited both increased replication and more apoptosis, thereby maintaining equilibrium with Teff cells [48]. Thus, anti-TNF treatment may reduce Treg replication as well as Treg apoptosis.

Ehrenstein et al. [77] reported that PB CD4+CD25+ T cells of active RA patients were fully competent in inhibiting the proliferation of responder CD4+CD25– T cells, but they were unable to suppress the production of proinflammatory cytokines by activated T cells and monocytes. This failure of Tregs to inhibit cytokine production was reportedly restored by anti-TNF therapy. Furthermore, the number of 'CD4+CD25^{hi} Tregs' was increased after anti-TNF therapy in responding patients. A subsequent study from the same group concluded that anti-TNF therapy did not expand or restore the function of preexisting nTregs in RA patients, but instead, resulted in the generation of a distinct population of suppressor cells lacking CD62L expression derived from naïve CD4 cells isolated from RA patients, but not from healthy donors. These induced suppressor cells are presumably Tr1 and/or Th3 cells since their suppressive activity was IL-10 and TGF-β dependent [81]. Similarly, Deepe and Gibbons [82] reported that neutralization of TNF gave rise to a population of antigen-specific CD4+CD25+ suppressor cells that inhibit protective immunity in murine histoplasmosis. These induced CD4+CD25+ suppressor cells did not express FoxP3, and their suppressive activity was IL-10 dependent. Thus, they were likely to be Tr1 cells rather than naturally occurring FoxP3+ Tregs. These studies clearly show that in RA patients anti-TNF therapy induces Tr1 and/or Th3 suppressor cells, but does not promote activity of nTregs which exert their suppressive function in a direct cell-to-cell contact manner. Taken together, current published evidence does not lend support to the notion that anti-TNF therapy can simply restore or promote Treg activity in RA patients.

Directly contradictory to its well-known anti-inflammatory action, anti-TNF therapy also has at times been shown to promote inflammation. A recent study showed that anti-TNF therapy in collagen-induced arthritis model expanded pathogenic Th1 and Th17 T cells in the LNs, mediated by the upregulation of IL-12/IL-23 p40 expression. However, anti-TNF therapy blocked the accumulation of Th1/Th17 cells in the synovium, providing an explanation for the paradox that anti-TNF therapy ameliorates experimental arthritis despite increasing numbers of pathogenic T cells in LNs [83]. This observation is consistent with previous studies showing that TNF selectively inhibits p40 expression in human and mouse myeloid cells [84, 85]. Furthermore, another study demonstrated that neutralization of TNF in the heart of healthy baboon resulted in myocarditis [86], which was hypothetically caused by eliminating Treg activity [87]. Together with studies showing that neutralization of TNF could reverse the attenuated TCR signaling resulting from chronic TNF exposure [reviewed in 29], these recent publications provide mechanistic insight into the puzzling autoimmune disorders caused by anti-TNF therapy in human patients, including induction of anti-dsDNA production and lupus, as well as neuroinflammatory diseases [reviewed in 30]. The basis for the proinflammatory effect of anti-TNF remains to be more precisely determined.

Nevertheless, anti-TNF therapy ameliorates clinical symptoms and laboratory parameters of inflammation in the majority of RA patients [79], which is probably achieved mainly by eliminating the TNF-TNFR costimulation pathway on activated pathogenic Teffs because they most likely expressed elevated level of TNFR2. In addition, immune complexes formed by cytokine and the antibody against it can markedly enhance the biological activity of the respective cytokine [recently reviewed in 88]. One such example is that IL-2:anti-IL2 Ab complexes expand Tregs more efficiently and consequently inhibit allergic inflammatory responses [89]. Whether complexes of endogenous TNF and therapeutic biological antagonists of TNF are able to stimulate Tregs through TNFR2 in autoimmune patients needs to be further investigated.

Conclusions

In normal Balb/c or C57BL/6 mice, TNFR2 is constitutively expressed by ~40% of Tregs, but TNFR2 is expressed by less than 10% of resting CD4 Teffs, suggesting that TNF may participate in the maintenance of Treg activity and immune homeostasis. Expression of TNFR2 by Teffs is also upregulated upon TCR stimulation [50], which may allow the TNF signal mediated by TNFR2 to effectively stimulate Teffs and render Teffs more resistant to Treg-mediated inhibition. The report showing that TNF downregulated Treg activity [38] is likely to reflect the stimulating effect of TNF on Teffs. Thus, liberating Teffs from Treg-mediated inhibition may account for proinflammatory effects of TNF. However, prolonged exposure to TNF favors the activation of Tregs [50]. Furthermore, in concert with IL-2 and perhaps with other common γ-chain cytokines, TNF selectively activates Tregs. Higher levels of TNFR2 expression by Tregs mediated by chronic inflammatory responses may enable them to outcompete Teffs for TNF. TNF-TNFR2-costimulated activation of Tregs may therefore account for the accumulation of activated Tregs found in inflammatory responses such as autoimmunity, sepsis, infection and tumors, and may also account for immunosuppression seen in the chronic exposure to TNF [28, 90]. The stimulatory effect of TNF on Tregs thus represents an important negative feedback mechanism that results in the attenuation and termination of prolonged or excessive immune responses, which otherwise may cause severe collateral damage (fig. 1). The polarizing action of TNF on Tregs and Teffs may be determined by timing, location, cytokine milieu, receptor usage, form (free vs. membrane bound) and cellular source of TNF. Further clarification of the pathways or cofactor(s) which polarize the stimulating activity of TNF will not only improve our understanding of cellular and molecular basis underlying the

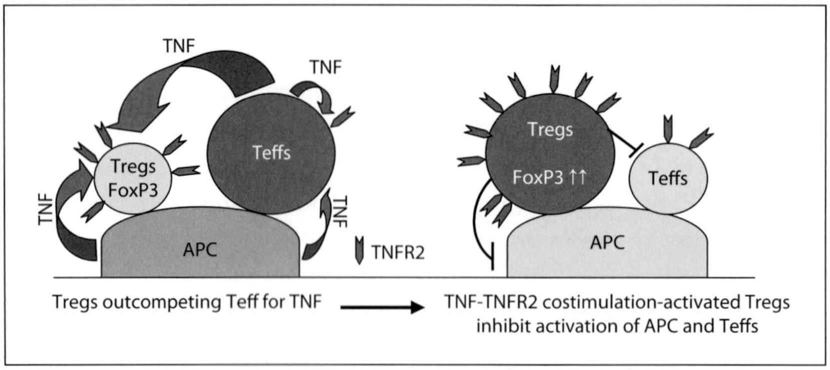

Fig. 1. Tregs predominately express TNFR2 and outcompete Teffs for TNF. TNFR2 is predominately expressed by both human and mouse Tregs. In the inflammatory responses, pathogenic Teffs and APCs are able to produce TNF. The chronic TNF exposure may favor the activation of Tregs by TNFR2 costimulation pathways. Accumulation of activated Tregs at the inflammatory site suppresses the activation of both innate immune cells as well as adaptive immune cells, and therefore may present an important negative feedback mechanism to limit the magnitude of the immune responses and avoid collateral damage.

contrasting action of TNF, but may also allow us to identify better means of manipulating this powerful cytokine to the benefit of our patients.

Acknowledgements

This project has been funded in whole or in part with federal funds from the National Cancer Institute, National Institutes of Health, under contract N01-CO-12400. This research was supported (in part) by the Intramural Research Program of the NIH, National Cancer Institute, Center for Cancer Research.

The authors thank Drs. Arthur A. Hurwitz, O.M. Zack Howard and Jonathan M. Weiss and Ryoko Hamano for discussion and critical review of the manuscript, and are grateful to the members of the Laboratory of Molecular Immunoregulation, CIP, CCR, NCI-Frederick and to the collaborators for their contributions to the research discussed here. The authors apologize to those researchers whose work could not be cited due to space limitations.

References

1 Gershon RK, Kondo K: Cell interactions in the induction of tolerance: the role of thymic lymphocytes. Immunology 1970;18:723–737.
2 Sakaguchi S, Sakaguchi N, Asano M, Itoh M, Toda M: Immunologic self-tolerance maintained by activated T cells expressing IL-2 receptor alpha-chains (CD25). Breakdown of a single mechanism of self-tolerance causes various autoimmune diseases. J Immunol 1995;155:1151–1164.
3 Sakaguchi S, Yamaguchi T, Nomura T, Ono M: Regulatory T cells and immune tolerance. Cell 2008; 133:775–787.
4 Chen W, Jin W, Hardegen N, Lei KJ, Li L, Marinos N, McGrady G, Wahl SM: Conversion of peripheral CD4+CD25- naive T cells to CD4+CD25+ regulatory T cells by TGF-beta induction of transcription factor Foxp3. J Exp Med 2003;198:1875–1886.

5 Hsieh CS, Zheng Y, Liang Y, Fontenot JD, Rudensky AY: An intersection between the self-reactive regulatory and nonregulatory T cell receptor repertoires. Nat Immunol 2006;7:401–410.
6 Zheng Y, Rudensky AY: Foxp3 in control of the regulatory T cell lineage. Nat Immunol 2007;8:457–462.
7 Torgerson TR, Ochs HD: Immune dysregulation, polyendocrinopathy, enteropathy, X-linked: forkhead box protein 3 mutations and lack of regulatory T cells. J Allergy Clin Immunol 2007;120:744–750; quiz 751–742.
8 Campbell DJ, Ziegler SF: FOXP3 modifies the phenotypic and functional properties of regulatory T cells. Nat Rev Immunol 2007;7:305–310.
9 Thornton AM, Shevach EM: CD4+CD25+ immunoregulatory T cells suppress polyclonal T cell activation in vitro by inhibiting interleukin 2 production. J Exp Med 1998;188:287–296.
10 Piccirillo CA, Shevach EM: Cutting edge: control of CD8+ T cell activation by CD4+CD25+ immunoregulatory cells. J Immunol 2001;167:1137–1140.
11 Trzonkowski P, Szmit E, Mysliwska J, Dobyszuk A, Mysliwski A: CD4+CD25+ T regulatory cells inhibit cytotoxic activity of T CD8+ and NK lymphocytes in the direct cell-to-cell interaction. Clin Immunol 2004;112:258–267.
12 Azuma T, Takahashi T, Kunisato A, Kitamura T, Hirai H: Human CD4+ CD25+ regulatory T cells suppress NKT cell functions. Cancer Res 2003;63: 4516–4520.
13 Cederbom L, Hall H, Ivars F: CD4+CD25+ regulatory T cells down-regulate co-stimulatory molecules on antigen-presenting cells. Eur J Immunol 2000;30: 1538–1543.
14 Fu Y, Yi S, Wu J, Jimenez E, Simond D, Hawthorne WJ, O'Connell PJ: In vitro suppression of xenoimmune-mediated macrophage activation by human CD4+CD25+ regulatory T cells. Transplantation 2008;86:865–874.
15 Seo SJ, Fields ML, Buckler JL, Reed AJ, Mandik-Nayak L, Nish SA, Noelle RJ, Turka LA, Finkelman FD, Caton AJ, Erikson J: The impact of T helper and T regulatory cells on the regulation of anti-double-stranded DNA B cells. Immunity 2002;16:535–546.
16 Thornton AM, Shevach EM: Suppressor effector function of CD4+CD25+ immunoregulatory T cells is antigen nonspecific. J Immunol 2000;164:183–190.
17 Karim M, Feng G, Wood KJ, Bushell AR: CD25+CD4+ regulatory T cells generated by exposure to a model protein antigen prevent allograft rejection: antigen-specific reactivation in vivo is critical for bystander regulation. Blood 2005;105: 4871–4877.
18 Belkaid Y, Rouse BT: Natural regulatory T cells in infectious disease. Nat Immunol 2005;6:353–360.
19 Zou W: Regulatory T cells, tumour immunity and immunotherapy. Nat Rev Immunol 2006;6:295–307.
20 Miyara M, Sakaguchi S: Natural regulatory T cells: mechanisms of suppression. Trends Mol Med 2007; 13:108–116.
21 Collison LW, Workman CJ, Kuo TT, Boyd K, Wang Y, Vignali KM, Cross R, Sehy D, Blumberg RS, Vignali DA: The inhibitory cytokine IL-35 contributes to regulatory T-cell function. Nature 2007;450: 566–569.
22 Allan SE, Song-Zhao GX, Abraham T, McMurchy AN, Levings MK: Inducible reprogramming of human T cells into Treg cells by a conditionally active form of FOXP3. Eur J Immunol 2008;38:3282–3289.
23 Bardel E, Larousserie F, Charlot-Rabiega P, Coulomb-L'Hermine A, Devergne O: Human CD4+ CD25+ Foxp3+ regulatory T cells do not constitutively express IL-35. J Immunol 2008;181:6898–6905.
24 Deaglio S, Dwyer KM, Gao W, Friedman D, Usheva A, Erat A, Chen JF, Enjyoji K, Linden J, Oukka M, Kuchroo VK, Strom TB, Robson SC: Adenosine generation catalyzed by CD39 and CD73 expressed on regulatory T cells mediates immune suppression. J Exp Med 2007;204:1257–1265.
25 Wing K, Onishi Y, Prieto-Martin P, Yamaguchi T, Miyara M, Fehervari Z, Nomura T, Sakaguchi S: CTLA-4 control over Foxp3+ regulatory T cell function. Science 2008;322:271–275.
26 Friedline RH, Brown DS, Nguyen H, Kornfeld H, Lee J, Zhang Y, Appleby M, Der SD, Kang J, Chambers CA: CD4+ regulatory T cells require CTLA-4 for the maintenance of systemic tolerance. J Exp Med 2009;206:421–434.
27 Aggarwal BB: Signalling pathways of the TNF superfamily: a double-edged sword. Nat Rev Immunol 2003;3:745–756.
28 Cope AP: Regulation of autoimmunity by proinflammatory cytokines. Curr Opin Immunol 1998;10: 669–676.
29 Clark J, Vagenas P, Panesar M, Cope AP: What does tumour necrosis factor excess do to the immune system long term? Ann Rheum Dis 2005; 64(suppl 4):iv70–iv76.
30 Kollias G, Kontoyiannis D: Role of TNF/TNFR in autoimmunity: specific TNF receptor blockade may be advantageous to anti-TNF treatments. Cytokine Growth Factor Rev 2002;13:315–321.

31 Kruglov AA, Kuchmiy A, Grivennikov SI, Tumanov AV, Kuprash DV, Nedospasov SA: Physiological functions of tumor necrosis factor and the consequences of its pathologic overexpression or blockade: mouse models. Cytokine Growth Factor Rev 2008;19:231–244.

32 Grewal IS, Grewal KD, Wong FS, Picarella DE, Janeway CA Jr, Flavell RA: Local expression of transgene encoded TNF alpha in islets prevents autoimmune diabetes in nonobese diabetic (NOD) mice by preventing the development of auto-reactive islet-specific T cells. J Exp Med 1996;184:1963–1974.

33 Jacob CO, McDevitt HO: Tumour necrosis factor-alpha in murine autoimmune 'lupus' nephritis. Nature 1988;331:356–358.

34 Jacob CO, Aiso S, Michie SA, McDevitt HO, Acha-Orbea H: Prevention of diabetes in nonobese diabetic mice by tumor necrosis factor (TNF): similarities between TNF-alpha and interleukin 1. Proc Natl Acad Sci USA 1990;87:968–972.

35 Kontoyiannis D, Kollias G: Accelerated autoimmunity and lupus nephritis in NZB mice with an engineered heterozygous deficiency in tumor necrosis factor. Eur J Immunol 2000;30:2038–2047.

36 Kollias G, Kontoyiannis D, Douni E, Kassiotis G: The role of TNF/TNFR in organ-specific and systemic autoimmunity: implications for the design of optimized 'anti-TNF' therapies. Curr Dir Autoimmun 2002;5:30–50.

37 Kassiotis G, Kollias G: Uncoupling the proinflammatory from the immunosuppressive properties of tumor necrosis factor (TNF) at the p55 TNF receptor level: implications for pathogenesis and therapy of autoimmune demyelination. J Exp Med 2001; 193:427–434.

38 Valencia X, Lipsky PE: CD4+CD25+FoxP3+ regulatory T cells in autoimmune diseases. Nat Clin Pract Rheumatol 2007;3:619–626.

39 Leipe J, Skapenko A, Lipsky PE, Schulze-Koops H: Regulatory T cells in rheumatoid arthritis. Arthritis Res Ther 2005;7:93.

40 Cao D, Malmstrom V, Baecher-Allan C, Hafler D, Klareskog L, Trollmo C: Isolation and functional characterization of regulatory CD25brightCD4+ T cells from the target organ of patients with rheumatoid arthritis. Eur J Immunol 2003;33:215–223.

41 van Amelsfort JM, Jacobs KM, Bijlsma JW, Lafeber FP, Taams LS: CD4(+)CD25(+) regulatory T cells in rheumatoid arthritis: differences in the presence, phenotype, and function between peripheral blood and synovial fluid. Arthritis Rheum 2004;50:2775–2785.

42 Mottonen M, Heikkinen J, Mustonen L, Isomaki P, Luukkainen R, Lassila O: CD4+ CD25+ T cells with the phenotypic and functional characteristics of regulatory T cells are enriched in the synovial fluid of patients with rheumatoid arthritis. Clin Exp Immunol 2005;140:360–367.

43 Jiao Z, Wang W, Jia R, Li J, You H, Chen L, Wang Y: Accumulation of FoxP3-expressing CD4+CD25+ T cells with distinct chemokine receptors in synovial fluid of patients with active rheumatoid arthritis. Scand J Rheumatol 2007;36:428–433.

44 Ruprecht CR, Gattorno M, Ferlito F, Gregorio A, Martini A, Lanzavecchia A, Sallusto F: Coexpression of CD25 and CD27 identifies FoxP3+ regulatory T cells in inflamed synovia. J Exp Med 2005;201:1793–1803.

45 de Kleer IM, Wedderburn LR, Taams LS, Patel A, Varsani H, Klein M, de Jager W, Pugayung G, Giannoni F, Rijkers G, Albani S, Kuis W, Prakken B: CD4+CD25bright regulatory T cells actively regulate inflammation in the joints of patients with the remitting form of juvenile idiopathic arthritis. J Immunol 2004;172:6435–6443.

46 Cao D, van Vollenhoven R, Klareskog L, Trollmo C, Malmstrom V: CD25brightCD4+ regulatory T cells are enriched in inflamed joints of patients with chronic rheumatic disease. Arthritis Res Ther 2004; 6:R335–R346.

47 Nguyen LT, Jacobs J, Mathis D, Benoist C: Where FoxP3-dependent regulatory T cells impinge on the development of inflammatory arthritis. Arthritis Rheum 2007;56:509–520.

48 Monte K, Wilson C, Shih FF: Increased number and function of FoxP3 regulatory T cells during experimental arthritis. Arthritis Rheum 2008;58:3730–3741.

49 Venet F, Chung CS, Monneret G, Huang X, Horner B, Garber M, Ayala A: Regulatory T cell populations in sepsis and trauma. J Leukoc Biol 2008;83:523–535.

50 Chen X, Baumel M, Mannel DN, Howard OM, Oppenheim JJ: Interaction of TNF with TNF receptor type 2 promotes expansion and function of mouse CD4+CD25+ T regulatory cells. J Immunol 2007;179:154–161.

51 Coussens LM, Werb Z: Inflammation and cancer. Nature 2002;420:860–867.

52 Wu AJ, Hua H, Munson SH, McDevitt HO: Tumor necrosis factor-alpha regulation of CD4+CD25+ T cell levels in NOD mice. Proc Natl Acad Sci USA 2002;99:12287–12292.

53 Chen X, Oppenheim JJ, Winkler-Pickett RT, Ortaldo JR, Howard OM: Glucocorticoid amplifies IL-2-dependent expansion of functional FoxP3+CD4+ CD25+ T regulatory cells in vivo and enhances their capacity to suppress EAE. Eur J Immunol 2006; 36:2139–2149.
54 Chen X, Winkler-Pickett RT, Carbonetti NH, Ortaldo JR, Oppenheim JJ, Howard OM: Pertussis toxin as an adjuvant suppresses the number and function of CD4+CD25+ T regulatory cells. Eur J Immunol 2006;36:671–680.
55 Cassan C, Piaggio E, Zappulla JP, Mars LT, Couturier N, Bucciarelli F, Desbois S, Bauer J, Gonzalez-Dunia D, Liblau RS: Pertussis toxin reduces the number of splenic Foxp3+ regulatory T cells. J Immunol 2006; 177:1552–1560.
56 Tonon S, Goriely S, Aksoy E, Pradier O, Del Giudice G, Trannoy E, Willems F, Goldman M, De Wit D: Bordetella pertussis toxin induces the release of inflammatory cytokines and dendritic cell activation in whole blood: impaired responses in human newborns. Eur J Immunol 2002;32:3118–3125.
57 Ausiello CM, Fedele G, Urbani F, Lande R, Di Carlo B, Cassone A: Native and genetically inactivated pertussis toxins induce human dendritic cell maturation and synergize with lipopolysaccharide in promoting T helper type 1 responses. J Infect Dis 2002;186:351–360.
58 Chen X, Howard OM, Oppenheim JJ: Pertussis toxin by inducing IL-6 promotes the generation of IL-17-producing CD4 cells. J Immunol 2007;178: 6123–6129.
59 Malek TR, Bayer AL: Tolerance, not immunity, crucially depends on IL-2. Nat Rev Immunol 2004;4: 665–674.
60 Carpentier I, Coornaert B, Beyaert R: Function and regulation of tumor necrosis factor type 2. Curr Med Chem 2004;11:2205–2212.
61 Tartaglia LA, Goeddel DV, Reynolds C, Figari IS, Weber RF, Fendly BM, Palladino MA Jr: Stimulation of human T-cell proliferation by specific activation of the 75-kDa tumor necrosis factor receptor. J Immunol 1993;151:4637–4641.
62 Kim EY, Priatel JJ, Teh SJ, Teh HS: TNF receptor type 2 (p75) functions as a costimulator for antigen-driven t cell responses in vivo. J Immunol 2006;176: 1026–1035.
63 Kim EY, Teh HS: TNF type 2 receptor (p75) lowers the threshold of T cell activation. J Immunol 2001; 167:6812–6820.
64 Kim EY, Teh HS: Critical role of TNF receptor type-2 (p75) as a costimulator for IL-2 induction and T cell survival: a functional link to CD28. J Immunol 2004;173:4500–4509.
65 Annunziato F, Cosmi L, Liotta F, Lazzeri E, Manetti R, Vanini V, Romagnani P, Maggi E, Romagnani S: Phenotype, localization, and mechanism of suppression of CD4+CD25+ human thymocytes. J Exp Med 2002;196:379–387.
66 Cosmi L, Liotta F, Lazzeri E, Francalanci M, Angeli R, Mazzinghi B, Santarlasci V, Manetti R, Vanini V, Romagnani P, Maggi E, Romagnani S, Annunziato F: Human CD8+CD25+ thymocytes share phenotypic and functional features with CD4+CD25+ regulatory thymocytes. Blood 2003;102:4107–4114.
67 Chen X, Subleski JJ, Kopf H, Howard OM, Mannel DN, Oppenheim JJ: Cutting edge: expression of TNFR2 defines a maximally suppressive subset of mouse CD4+CD25+FoxP3+ T regulatory cells: applicability to tumor-infiltrating T regulatory cells. J Immunol 2008;180:6467–6471.
68 Baseta JG, Stutman O: TNF regulates thymocyte production by apoptosis and proliferation of the triple negative (CD3-CD4-CD8-) subset. J Immunol 2000;165:5621–5630.
69 van Mierlo GJ, Scherer HU, Hameetman M, Morgan ME, Flierman R, Huizinga TW, Toes RE: Cutting edge: TNFR-shedding by CD4+CD25+ regulatory T cells inhibits the induction of inflammatory mediators. J Immunol 2008;180:2747–2751.
70 Kollias G: TNF pathophysiology in murine models of chronic inflammation and autoimmunity. Semin Arthritis Rheum 2005;34:3–6.
71 Suvannavejh GC, Lee HO, Padilla J, Dal Canto MC, Barrett TA, Miller SD: Divergent roles for p55 and p75 tumor necrosis factor receptors in the pathogenesis of MOG(35–55)-induced experimental autoimmune encephalomyelitis. Cell Immunol 2000;205:24–33.
72 Ebach DR, Riehl TE, Stenson WF: Opposing effects of tumor necrosis factor receptor 1 and 2 in sepsis due to cecal ligation and puncture. Shock 2005;23: 311–318.
73 Grell M, Douni E, Wajant H, Lohden M, Clauss M, Maxeiner B, Georgopoulos S, Lesslauer W, Kollias G, Pfizenmaier K, Scheurich P: The transmembrane form of tumor necrosis factor is the prime activating ligand of the 80 kDa tumor necrosis factor receptor. Cell 1995;83:793–802.
74 Lehmann J, Huehn J, de la Rosa M, Maszyna F, Kretschmer U, Krenn V, Brunner M, Scheffold A, Hamann A: Expression of the integrin alpha Ebeta 7 identifies unique subsets of CD25+ as well as CD25- regulatory T cells. Proc Natl Acad Sci USA 2002; 99:13031–13036.
75 Gray JD, Liu T, Huynh N, Horwitz DA: Transforming growth factor beta enhances the expression of CD154 (CD40L) and production of tumor necrosis factor alpha by human T lymphocytes. Immunol Lett 2001;78:83–88.

76 van der Most RG, Currie AJ, Mahendran S, Prosser A, Darabi A, Robinson BW, Nowak AK, Lake RA: Tumor eradication after cyclophosphamide depends on concurrent depletion of regulatory T cells: a role for cycling TNFR2-expressing effector-suppressor T cells in limiting effective chemotherapy. Cancer Immunol Immunother 2009;58:1219–1228.
77 Ehrenstein MR, Evans JG, Singh A, Moore S, Warnes G, Isenberg DA, Mauri C: Compromised function of regulatory T cells in rheumatoid arthritis and reversal by anti-TNFalpha therapy. J Exp Med 2004;200:277–285.
78 Valencia X, Stephens G, Goldbach-Mansky R, Wilson M, Shevach EM, Lipsky PE: TNF downmodulates the function of human CD4+CD25hi T regulatory cells. Blood 2006;108:253–261.
79 Feldmann M: Development of anti-TNF therapy for rheumatoid arthritis. Nat Rev Immunol 2002;2:364–371.
80 Toubi E, Kessel A, Mahmudov Z, Hallas K, Rozenbaum M, Rosner I: Increased spontaneous apoptosis of CD4+CD25+ T cells in patients with active rheumatoid arthritis is reduced by infliximab. Ann N Y Acad Sci 2005;1051:506–514.
81 Nadkarni S, Mauri C, Ehrenstein MR: Anti-TNF-alpha therapy induces a distinct regulatory T cell population in patients with rheumatoid arthritis via TGF-beta. J Exp Med 2007;204:33–39.
82 Deepe GS Jr, Gibbons RS: TNF-alpha antagonism generates a population of antigen-specific CD4+ CD25+ T cells that inhibit protective immunity in murine histoplasmosis. J Immunol 2008;180:1088–1097.
83 Notley CA, Inglis JJ, Alzabin S, McCann FE, McNamee KE, Williams RO: Blockade of tumor necrosis factor in collagen-induced arthritis reveals a novel immunoregulatory pathway for Th1 and Th17 cells. J Exp Med 2008;205:2491–2497.
84 Zakharova M, Ziegler HK: Paradoxical anti-inflammatory actions of TNF-alpha: inhibition of IL-12 and IL-23 via TNF receptor 1 in macrophages and dendritic cells. J Immunol 2005;175:5024–5033.
85 Ma X, Sun J, Papasavvas E, Riemann H, Robertson S, Marshall J, Bailer RT, Moore A, Donnelly RP, Trinchieri G, Montaner LJ: Inhibition of IL-12 production in human monocyte-derived macrophages by TNF. J Immunol 2000;164:1722–1729.
86 McTiernan CF, Mathier MA, Zhu X, Xiao X, Klein E, Swan CH, Mehdi H, Gibson G, Trichel AM, Glorioso JC, Feldman AM, McCurry KR, London B: Myocarditis following adeno-associated viral gene expression of human soluble TNF receptor (TNFRII-Fc) in baboon hearts. Gene Ther 2007;14:1613–1622.
87 Hajjar RJ, Zsebo K: AAV vectors and cardiovascular disease: targeting TNF receptor in the heart: clue to way forward with AAV? Gene Ther 2007;14:1611–1612.
88 Mostbock S: Cytokine/antibody complexes: an emerging class of immunostimulants. Curr Pharm Des 2009;15:809–825.
89 Wilson MS, Pesce JT, Ramalingam TR, Thompson RW, Cheever A, Wynn TA: Suppression of murine allergic airway disease by IL-2:anti-IL-2 monoclonal antibody-induced regulatory T cells. J Immunol 2008;181:6942–6954.
90 Cope AP, Liblau RS, Yang XD, Congia M, Laudanna C, Schreiber RD, Probert L, Kollias G, McDevitt HO: Chronic tumor necrosis factor alters T cell responses by attenuating T cell receptor signaling. J Exp Med 1997;185:1573–1584.

Dr. Xin Chen
Basic Science Program, SAIC-Frederick, Inc., NCI-Frederick
PO Box B, Bldg 560, Rm 31-19
Frederick, MD 21702-1201 (USA)
Tel. +1 301 846 1347, Fax +1 301 846 6752, E-Mail chenxin@mail.nih.gov

Kollias G, Sfikakis PP (eds): TNF Pathophysiology. Molecular and Cellular Mechanisms.
Curr Dir Autoimmun. Basel, Karger, 2010, vol 11, pp 135–144

TNF and Bone

Jean-Pierre David · Georg Schett

Rheumatology and Immunology, Department of Internal Medicine 3, University of Erlangen-Nuremberg, Erlangen, Germany

Abstract

Bone is subject to permanent remodeling during development and through life. This activity is essential for (a) proper shaping and growth of each bone during development; (b) maintenance of bone mass as well as structural integrity of the microarchitecture of bone through adult life, and (c) tissue repair needed for healing of fracture as well as of micro-damage. In addition to genetically linked rare developmental diseases, disturbances in bone remodeling are causing common bone pathologies, which severely impair the quality of life of patients. Among them are postmenopausal osteoporosis and local as well as systemic bone loss observed in chronic inflammatory diseases such as rheumatoid arthritis. The role of TNF-α in mediating bone remodeling will be presented and discussed in this chapter.

Bone Remodeling

Bone remodeling results from coordinated activity of three cell types: the bone-forming cells or osteoblasts, their final stage of differentiation termed osteocytes, and the bone-resorbing cells or osteoclasts. Osteoblasts differentiate from mesenchymal stromal/stem cells from which chondrocytes, adipocytes and myocytes are also derived. They are the bone-lining cells that will secrete and mineralize bone matrix. Osteocytes are the most abundant cells of bone. They are embedded into mineralized matrix and are connected between each other by cellular extension going through a network of canalicules. This network is believed to serve as physical sensor of changes affecting the bone. Osteoclasts are the only cells of the organism able to resorb bone. They are multinucleated cells formed by the fusion of monocytes, and thus are of hematopoietic origin [1, 2].

Bone formation is a dynamic process temporally and locally regulated inside a functional unit called 'bone remodeling unit'. Apoptosis-mediated cell death of the osteocytes is believed to serve as mechanoreceptor sensing the need to form new bone by most likely reacting to changes in the local pressure (mechanical loading)

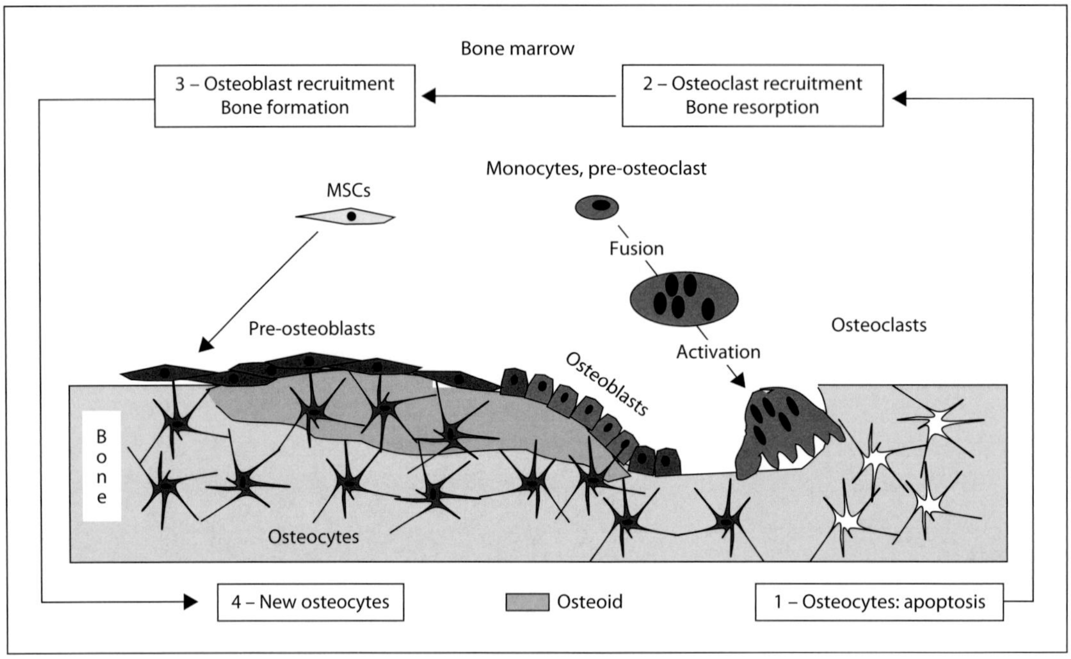

Fig. 1. Organization of the bone remodeling unit. MSCs = Mesenchymal stromal/stem cells.

and thus initiate activation of the bone remodeling unit. Inside the bone remodeling unit, ancient bone matrix is first demineralized and resorbed by locally recruited osteoclasts and progressively replaced by new matrix or osteoid secreted and mineralized by osteoblasts. Embedded osteoblasts will become new osteocytes that will negatively regulate osteoblast activity and therefore deactivate the bone remodeling unit (fig. 1).

TNF-α and Its Receptors

TNF-α is the prototype of a big family of cytokines that binds as trimers to transmembrane receptors activating various cell-signaling pathways [3]. In addition, secreted decoy receptors have been characterized for some but not all of the TNF members, which titrate the ligand. The various TNF members are involved in cell-specific control of differentiation and apoptosis. Most notably, TNF-α has been shown to play an essential role in the development of lymphoid organs. TNF-α is also a major regulator of acute or chronic inflammation as well as host defense against pathogens. Thus, TNF-α is a key player of both adaptive and innate immune responses. TNF-α can be synthesized into two different forms: a membrane bound and a circulating one, suggesting its effect to be both local and systemic. In agreement, the different physiological functions of TNF-α seem to depend on the cellular source of TNF-α. Finally, TNF-α can bind to two different receptors TNFR1 or p55 and TNFR2 or p75. TNFR1 but not TNFR2 contains a death domain that can directly couple TNF-α to apoptosis. However, so far TNFR1 seemed to play a predominant role in most of the biological effect of TNF-α. Thus, the resulting biological effect of TNF-α depends on the combination of all these parameters [4].

In bone, osteoclast precursors, osteoclasts and osteoblasts are all expressing TNFR1 and TNFR2, TNFR1 being predominant [5–7]. In addition, although not being a major source of TNF-α, both cell types can produce TNF-α in response to external stimulation. All these suggested a potential direct role for TNF-α in bone remodeling.

TNF-α, Osteoclast Differentiation and Function

Osteoclasts are multinucleated bone-specialized cells that differentiate from hematopoietic precursors along the monocytic lineage [2]. Two cytokines were shown to be playing an essential role in osteoclast differentiation: the macrophage-colony stimulating factor (M-CSF) and the receptor activator of NF-κB ligand (RANKL) [8]. M-CSF binds to a tyrosine kinase receptor encoded by *cfms* to promote proliferation and survival of the monocyte progenitors at the expense of the other myelomonocytic lineages such as dendritic cells and granulocytes) [9]. RANKL is the necessary osteoclastogenic differentiation factor that drives the monocytes toward premature and mature osteoclasts [10, 11]. RANKL binds to the receptor RANK [12] and via intracellular coupling that involves the TNFR-activating factor 6 (TRAF6) [13, 14], induces synthesis and activation of NF-κB, AP-1 and NFATc1, three groups of transcription factors required for osteoclast differentiation [1].

RANKL (TRANCE or Tnfsf11) is a member of the TNF family of cytokines, and RANK (or Tnfrsf11a) a member of the non-death domain-containing TNFRs that shares most but not all of its signaling properties with TNFR1 [15]. These suggested that TNF-α could also act as an osteoclast-differentiating factor. Indeed, TNF-α can induce osteoclast differentiation of mouse and human bone marrow cells in vitro [5, 16–18]. Today, it is still debated whether TNF-α can independently induce osteoclastogenesis. Based on the lack of effect of a RANK-neutralizing antibody and of the addition of RANKL decoy receptor OPG, Kobayashi et al. [17] concluded that TNF-α could directly drive osteoclastogenesis. In contrast, Teitelbaum's group concluded that priming of the progenitors with a permissive dose of RANKL is required for the pro-osteoclastogenic effect of TNF-α [19]. In agreement, TNF-α was shown to stimulate RANKL-induced osteoclastogenesis by increasing the level of expression of RANK [20]. Similarly, while all the studies agreed on the essential role of TNFR1 in mediating the pro-osteoclastogenic effect of TNF-α, divergences about the role of TNFR2 appeared. Again, a positive role for both TNFR1 and TNFR2 was proposed using neutralizing antibody [17], while genetic experiments using TNFR1- or TNFR2-deficient bone marrow cells demonstrated that TNFR1 is necessary for TNF-α-induced osteoclastogenesis and that TNFR2 exerted a negative effect on TNF-α-induced osteoclast differentiation [16, 20]. Interestingly, these late studies also reported a similar effect on RANKL-induced osteoclasts that might be explained by the increased TNF-α production observed in response to RANKL stimulation of osteoclast precursors [21]. Importantly, osteoclast precursors lacking RANKL or RANK could be differentiated into osteoclasts in response to TNF-α and TGF-β, clearly indicating that TNF-α can bypass the RANKL/RANK pathway to promote osteoclastogenesis [22]. Interestingly, presence of rare osteoclast-like cells was also reported at the site of injection of TNF-α in the bone of RANK-deficient mice [23].

A common result of all the in vitro data is that TNF-α, in contrast to RANKL, is a poor activator of osteoclast activity. Indeed, if TNF-α can regulate differentiation of osteoclasts, it cannot efficiently induce their resorptive activity that is only observed in combination with IL-1 [5, 17]. This lack of activation of osteoclasts by TNF-α seemed to be independent of apoptosis since both RANKL and TNF-α can similarly promote the survival of mature osteoclasts [24, 25]. It is rather due to the incapacity of TNF-α to activate the downstream signaling pathway necessary for the organization of the resorbing cellular structure that can be restored by IL-1 [26]. The TRAFs are key components in the signaling transduction of the various members of TNFR superfamily. Not all receptors can bind to all TRAFs, and coupling of a TRAF to a TNFR determines the specificity of the downstream signaling [27, 28]. The recruitment of a specific TRAF may explain the differences between TNF-α and RANKL as well as the incapacity of TNF-α to directly regulate the bone resorptive activity of the osteoclasts. Indeed, although the pro-osteoclastogenic activity of RANKL and TNF-α requires coupling with TRAF6 and participation of TRAF5 [29, 30], TRAF2 that is not mediating RANK signaling in osteoclasts is required for TNFR1-differentiating activity [31]. In addition, the strength of TRAF6-mediated signaling was found to be crucial for osteoclastogenic activity of the TNFR members [32]. IL-1 is well known to act via coupling with TRAF6 and indeed, IL-1 is stimulating the formation of the actin ring necessary for bone resorption via a signaling pathway involving TRAF6 and Src [26].

Thus, whether TNF-α can directly regulate osteoclastogenesis and whether TNF-α can stimulate osteoclast differentiation in physiological conditions is still unclear. However, independent or dependent on the presence of RANKL, TNF-α, via its binding to TNFR1, can promote the different stages of osteoclastogenesis in vitro (fig. 2).

TNF-α, Osteoblast Differentiation and Function

While numerous studies reported a proapoptotic effect of TNF-α on osteoblasts [7], apoptosis might only be relevant in case of pathological decrease in bone formation

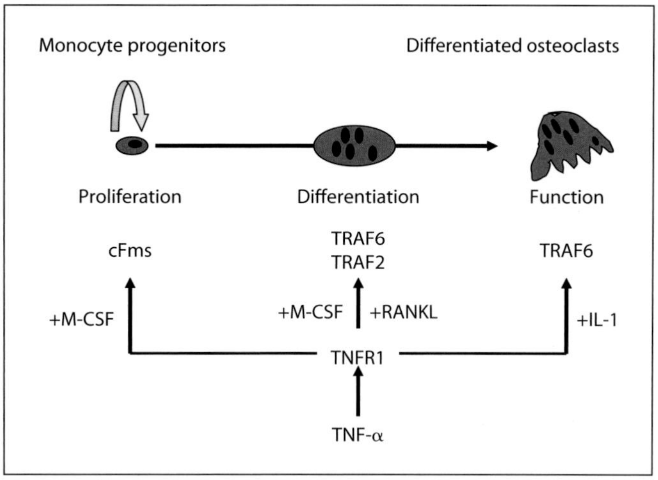

Fig. 2. Effects of TNF-α on osteoclast differentiation.

[33, 34]. The clear effect of TNF-α on osteoblastogenesis in vitro is blockage of osteoblast differentiation [35] by inhibiting the expression of two transcription factors essential for bone formation: Runx2 and its downstream target Osterix (Osx). It results in a decreased level of expression of osteoblast-specific differentiation markers [7, 36]. The inhibition of osteoblastogenesis is clearly mediated via TNFR1 engagement and only moderately regulated by TNFR2 [33, 37]. At the molecular level, it appears that TNF-α is blocking Runx2 expression at multiple levels. It inhibits *runx2* transcription and destabilizes its messenger [38, 39], and it promotes degradation of Runx2 protein via the upregulation of the E3 ubiquitin ligases Smurf1 and Smurf2 [39]. In addition, TNF-α stimulates NF-κB activation that inhibits TGF-β/BMP-induced Smad signaling in osteoblasts and decreases *Runx2* expression [37]. The importance of NF-κB activation in negative regulation of bone formation was recently demonstrated by Chang et al. [40]. They proposed that inhibition of NF-kB cell autonomously promotes bone formation by osteoblasts by increasing the expression of Fra-1, a member of the transcription factor AP-1 that regulates bone matrix synthesis [40–42]. Another mechanism by which TNF-α might be repressing bone formation is by interfering with Wnt signaling. Wnt was recently identified as the major pathway regulating bone formation [43]. Wnt is a family of ligands that bind to complex transmembrane receptors formed by association of Frizzled (Fz) and LRP5 and 6 to activate the translocation of β-catenin into the nucleus, thereby regulating the transcription of its target genes. Numerous negative regulators play an essential role in the termination of the Wnt signaling. Among them are the secreted Fz-related proteins (SFRPs), members of the DKK family and sclerostin (Sost) [44, 45]. They can either act as decoy receptors limiting the availability of Wnt ligands in the case of the SFRPs, or, as DKK and Sost, by interfering with the recruitment of coreceptor LRP5/6 to Fz receptor. Interestingly, most of these inhibitors are highly expressed in osteocytes in agreement

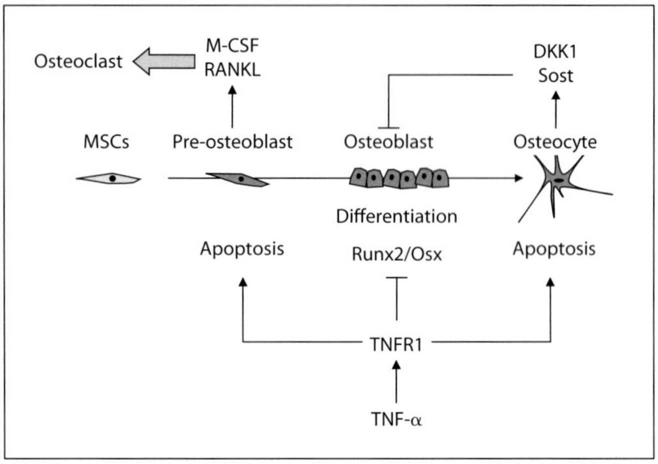

Fig. 3. Effects of TNF-α on osteoblast differentiation.

with the suggested role of these cells, which is to terminate bone formation. TNF-α-induced DKK1 was recently shown to be mediated by TNFR1 and activation of MKK3-p38MAPK pathway and to regulate the decreased bone formation observed in TNF-α-induced inflammatory arthritis [46] and TNF-α-increased Sost expression to be involved in TNF-α-induced apoptosis of human osteoblasts [47].

Osteoblasts, or at least their progenitors, are considered to be supportive to osteoclastogenesis. Indeed, they secrete all essential osteoclastogenic cytokines (M-CSF and RANKL) as well as numerous factors known to positively or negatively modulate osteoclastogenesis such as the decoy RANKL receptor OPG and TNF-α itself. TNF-α stimulation of osteoblasts was shown to stimulate the resorbing activity of osteoclasts [48] and to transcriptionally increase the production of M-CSF by the osteoblasts [49] as well as of RANKL [50]. Thus, in vitro experiments revealed the ability of TNF-α to regulate all steps of osteoblast activity (fig. 3).

TNF-α and Bone in vivo

In vitro, TNF-α can decrease osteoblastogenesis and increase osteoclastogenesis. Is the in vivo situation reflecting the in vitro observations, and in which case? Until recently, no clear bone phenotypes were reported in mice lacking TNF-α or its receptors. Bone histomorphometry was performed that indicated an increased bone mass in the TNF-α-deficient and in TNFR1-deficient mice but not in the absence of TNFR2. While osteoclast parameters were unaffected, increased bone mass was due to an increased bone formation, as suggested by the increased numbers of osteoblasts that correlated with increased levels of circulating osteocalcin [37]. Thus, surprisingly, although not essential for bone remodeling, the TNF-α/TNFR1 axis was found to inhibit physiological bone formation rather than to increase bone resorption.

What is the role of TNF-α in the case of pathological bone loss? Postmenopausal osteoporosis and inflammation-induced bone loss, as observed in rheumatoid arthritis or in chronic bacterial infection, are among the most common pathologies of bone remodeling in humans. Injection of lipopolysaccharide (LPS) is a model for local inflammation of bone. LPS induces local bone destruction due to increased numbers of osteoclasts. Mice lacking TNFR1 but not TNFR2 are fully protected against LPS-induced osteoclastogenesis [51]. Transgenic mice overexpressing a stabilized form of human TNF-α (*hTNF-α*-tg) develop severe inflammatory arthritis with many hallmarks of rheumatoid arthritis including bone and cartilage destruction [52] and general osteopenia [53]. Administration of OPG can protect these mice against bone destruction [54, 55] as well as against the general bone loss [53], suggesting that increased bone resorption by osteoclasts is crucial for local and systemic bone loss induced by TNF-α. This was confirmed by showing that mice lacking osteoclasts, although highly inflamed, were fully protected against TNF-α-induced bone destruction [55]. Interestingly, it was recently reported that *hTNF-α*-tg mice injected with IL-1 receptor antagonist or *hTNF-α*-tg mice lacking IL-1 were still developing local inflammation and local bone loss but were protected from the TNF-α-induced systemic bone loss [56, 57]. These data argued in favor of an indirect stimulatory effect of TNF-α on osteoclast differentiation via IL-1 in the case of inflammatory arthritis. However, one direct effect of TNF-α on osteoclastogenesis is the induction of the M-CSF receptor on the membrane of the osteoclast precursors. This renders cells more susceptible to M-CSF and thereby increases their proliferation. The outcome is an increased pool of osteoclast precursors as seen in *hTNF-α*-tg mice [58].

Ovariectomy is the standard experimental model for postmenopausal osteoporosis. Similarly to *hTNF-α*-tg mice, injection of IL-1 receptor antagonist or TNF-α-binding protein that neutralizes TNF-α was shown to protect ovariectomized rat against bone loss by inhibiting osteoclast formation [59]. In addition, mice lacking TNF-α or TNFR1 but not those lacking TNFR2 are protected against ovariectomy-induced bone loss [60]. In that case, T cells seem to be the major source of TNF-α, as shown by the increased capacity of T cells isolated from ovariectomized mice to secrete TNF-α. In agreement, nude mice are protected from ovariectomy-induced bone loss [61].

Conclusion

TNF-α is a nonessential regulator of bone remodeling. It is a central player in pathological bone loss as it decreases osteoblast activity and survival and increases osteoclast differentiation, recruitment and activity. All functions of TNF-α in bone appear to be dependent on TNFR1.

References

1 Karsenty G: Transcriptional control of skeletogenesis. Annu Rev Genomics Hum Genet 2008;9:183–196.

2 Karsenty G, Wagner EF: Reaching a genetic and molecular understanding of skeletal development. Dev Cell 2002;2:389–406.

3 Karin M, Gallagher E: TNFR signaling: ubiquitin-conjugated TRAFfic signals control stop-and-go for MAPK signaling complexes. Immunol Rev 2009; 228:225–240.

4 Kruglov AA, Kuchmiy A, Grivennikov SI, Tumanov AV, Kuprash DV, Nedospasov SA: Physiological functions of tumor necrosis factor and the consequences of its pathologic overexpression or blockade: mouse models. Cytokine Growth Factor Rev 2008;19:231–244.

5 Azuma Y, Kaji K, Katogi R, Takeshita S, Kudo A: Tumor necrosis factor-alpha induces differentiation of and bone resorption by osteoclasts. J Biol Chem 2000;275:4858–4864.

6 Bu R, Borysenko CW, Li Y, Cao L, Sabokbar A, Blair HC: Expression and function of TNF-family proteins and receptors in human osteoblasts. Bone 2003;33:760–770.

7 Nanes MS: Tumor necrosis factor-alpha: molecular and cellular mechanisms in skeletal pathology. Gene 2003;321:1–15.

8 Teitelbaum SL, Ross FP: Genetic regulation of osteoclast development and function. Nat Rev Genet 2003;4:638–649.

9 Miyamoto T, Ohneda O, Arai F, Iwamoto K, Okada S, Takagi K, Anderson DM, Suda T: Bifurcation of osteoclasts and dendritic cells from common progenitors. Blood 2001;98:2544–2554.

10 Kim N, Odgren PR, Kim DK, Marks SC Jr, Choi Y: Diverse roles of the tumor necrosis factor family member TRANCE in skeletal physiology revealed by TRANCE deficiency and partial rescue by a lymphocyte-expressed TRANCE transgene. Proc Natl Acad Sci USA 2000;97:10905–10910.

11 Kong YY, Yoshida H, Sarosi I, Tan HL, Timms E, Capparelli C, Morony S, Oliveira-dos-Santos AJ, Van G, Itie A, Khoo W, Wakeham A, Dunstan CR, Lacey DL, Mak TW, Boyle WJ, Penninger JM: OPGL is a key regulator of osteoclastogenesis, lymphocyte development and lymph-node organogenesis. Nature 1999;397:315–323.

12 Dougall WC, Glaccum M, Charrier K, Rohrbach K, Brasel K, De Smedt T, Daro E, Smith J, Tometsko ME, Maliszewski CR, Armstrong A, Shen V, Bain S, Cosman D, Anderson D, Morrissey PJ, Peschon JJ, Schuh J: RANK is essential for osteoclast and lymph node development. Genes Dev 1999;13:2412–2424.

13 Lomaga MA, Yeh WC, Sarosi I, Duncan GS, Furlonger C, Ho A, Morony S, Capparelli C, Van G, Kaufman S, van der Heiden A, Itie A, Wakeham A, Khoo W, Sasaki T, Cao Z, Penninger JM, Paige CJ, Lacey DL, Dunstan CR, Boyle WJ, Goeddel DV, Mak TW: TRAF6 deficiency results in osteopetrosis and defective interleukin-1, CD40, and LPS signaling. Genes Dev 1999;13:1015–1024.

14 Naito A, Azuma S, Tanaka S, Miyazaki T, Takaki S, Takatsu K, Nakao K, Nakamura K, Katsuki M, Yamamoto T, Inoue J: Severe osteopetrosis, defective interleukin-1 signalling and lymph node organogenesis in TRAF6-deficient mice. Genes Cells 1999;4:353–362.

15 Feng X: Regulatory roles and molecular signaling of TNF family members in osteoclasts. Gene 2005;350:1–13.

16 Abu-Amer Y, Erdmann J, Alexopoulou L, Kollias G, Ross FP, Teitelbaum SL: Tumor necrosis factor receptors types 1 and 2 differentially regulate osteoclastogenesis. J Biol Chem 2000;275:27307–27310.

17 Kobayashi K, Takahashi N, Jimi E, Udagawa N, Takami M, Kotake S, Nakagawa N, Kinosaki M, Yamaguchi K, Shima N, Yasuda H, Morinaga T, Higashio K, Martin TJ, Suda T: Tumor necrosis factor alpha stimulates osteoclast differentiation by a mechanism independent of the ODF/RANKL-RANK interaction. J Exp Med 2000;191:275–286.

18 Lader CS, Flanagan AM: Prostaglandin E2, interleukin 1alpha, and tumor necrosis factor-alpha increase human osteoclast formation and bone resorption in vitro. Endocrinology 1998;139:3157–3164.

19 Lam J, Takeshita S, Barker JE, Kanagawa O, Ross FP, Teitelbaum SL: TNF-alpha induces osteoclastogenesis by direct stimulation of macrophages exposed to permissive levels of RANK ligand. J Clin Invest 2000;106:1481–1488.

20 Zhang YH, Heulsmann A, Tondravi MM, Mukherjee A, Abu-Amer Y: Tumor necrosis factor-alpha (TNF) stimulates RANKL-induced osteoclastogenesis via coupling of TNF type 1 receptor and RANK signaling pathways. J Biol Chem 2001;276:563–568.

21 Nakao A, Fukushima H, Kajiya H, Ozeki S, Okabe K: RANKL-stimulated TNFalpha production in osteoclast precursor cells promotes osteoclastogenesis by modulating RANK signaling pathways. Biochem Biophys Res Commun 2007;357:945–950.

22 Kim N, Kadono Y, Takami M, Lee J, Lee SH, Okada F, Kim JH, Kobayashi T, Odgren PR, Nakano H, Yeh WC, Lee SK, Lorenzo JA, Choi Y: Osteoclast differentiation independent of the TRANCE-RANK-TRAF6 axis. J Exp Med 2005;202:589–595.

23 Li J, Sarosi I, Yan XQ, Morony S, Capparelli C, Tan HL, McCabe S, Elliott R, Scully S, Van G, Kaufman S, Juan SC, Sun Y, Tarpley J, Martin L, Christensen K, McCabe J, Kostenuik P, Hsu H, Fletcher F, Dunstan CR, Lacey DL, Boyle WJ: RANK is the intrinsic hematopoietic cell surface receptor that controls osteoclastogenesis and regulation of bone mass and calcium metabolism. Proc Natl Acad Sci USA 2000;97:1566–1571.

24 Lee SE, Chung WJ, Kwak HB, Chung CH, Kwack KB, Lee ZH, Kim HH: Tumor necrosis factor-alpha supports the survival of osteoclasts through the activation of Akt and ERK. J Biol Chem 2001; 276: 49343–49349.

25 Wong BR, Besser D, Kim N, Arron JR, Vologodskaia M, Hanafusa H, Choi Y: TRANCE, a TNF family member, activates Akt/PKB through a signaling complex involving TRAF6 and c-Src. Mol Cell 1999;4:1041–1049.

26 Nakamura I, Kadono Y, Takayanagi H, Jimi E, Miyazaki T, Oda H, Nakamura K, Tanaka S, Rodan GA, Duong le T: IL-1 regulates cytoskeletal organization in osteoclasts via TNF receptor-associated factor 6/c-Src complex. J Immunol 2002;168:5103–5109.

27 Aggarwal BB: Signalling pathways of the TNF superfamily: a double-edged sword. Nat Rev Immunol 2003;3:745–756.

28 Chung JY, Park YC, Ye H, Wu H: All TRAFs are not created equal: common and distinct molecular mechanisms of TRAF-mediated signal transduction. J Cell Sci 2002;115:679–688.

29 Kaji K, Katogi R, Azuma Y, Naito A, Inoue JI, Kudo A: Tumor necrosis factor alpha-induced osteoclastogenesis requires tumor necrosis factor receptor-associated factor 6. J Bone Miner Res 2001;16: 1593–1599.

30 Kanazawa K, Azuma Y, Nakano H, Kudo A: TRAF5 functions in both RANKL- and TNFalpha-induced osteoclastogenesis. J Bone Miner Res 2003;18:443–450.

31 Kanazawa K, Kudo A: TRAF2 is essential for TNF-alpha-induced osteoclastogenesis. J Bone Miner Res 2005;20:840–847.

32 Kadono Y, Okada F, Perchonock C, Jang HD, Lee SY, Kim N, Choi Y: Strength of TRAF6 signalling determines osteoclastogenesis. EMBO Rep 2005; 6: 171–176.

33 Gilbert LC, Rubin J, Nanes MS: The p55 TNF receptor mediates TNF inhibition of osteoblast differentiation independently of apoptosis. Am J Physiol Endocrinol Metab 2005;288:E1011–E1018.

34 Yousfi M, Lasmoles F, El Ghouzzi V, Marie PJ: Twist haploinsufficiency in Saethre-Chotzen syndrome induces calvarial osteoblast apoptosis due to increased TNFalpha expression and caspase-2 activation. Hum Mol Genet 2002;11:359–369.

35 Gilbert L, He X, Farmer P, Boden S, Kozlowski M, Rubin J, Nanes MS: Inhibition of osteoblast differentiation by tumor necrosis factor-alpha. Endocrinology 2000;141:3956–3964.

36 Lu X, Gilbert L, He X, Rubin J, Nanes MS: Transcriptional regulation of the osterix (Osx, Sp7) promoter by tumor necrosis factor identifies disparate effects of mitogen-activated protein kinase and NF kappa B pathways. J Biol Chem 2006;281:6297–6306.

37 Li Y, Li A, Strait K, Zhang H, Nanes MS, Weitzmann MN: Endogenous TNFalpha lowers maximum peak bone mass and inhibits osteoblastic Smad activation through NF-kappaB. J Bone Miner Res 2007;22:646–655.

38 Gilbert L, He X, Farmer P, Rubin J, Drissi H, van Wijnen AJ, Lian JB, Stein GS, Nanes MS: Expression of the osteoblast differentiation factor RUNX2 (Cbfa1/AML3/Pebp2alpha A) is inhibited by tumor necrosis factor-alpha. J Biol Chem 2002;277:2695–2701.

39 Kaneki H, Guo R, Chen D, Yao Z, Schwarz EM, Zhang YE, Boyce BF, Xing L: Tumor necrosis factor promotes Runx2 degradation through up-regulation of Smurf1 and Smurf2 in osteoblasts. J Biol Chem 2006;281:4326–4333.

40 Chang J, Wang Z, Tang E, Fan Z, McCauley L, Franceschi R, Guan K, Krebsbach PH, Wang CY: Inhibition of osteoblastic bone formation by nuclear factor-kappaB. Nat Med 2009;15:682–689.

41 Eferl R, Hoebertz A, Schilling AF, Rath M, Karreth F, Kenner L, Amling M, Wagner EF: The Fos-related antigen Fra-1 is an activator of bone matrix formation. EMBO J 2004;23:2789–2799.

42 Jochum W, David JP, Elliott C, Wutz A, Plenk H Jr, Matsuo K, Wagner EF: Increased bone formation and osteosclerosis in mice overexpressing the transcription factor Fra-1. Nat Med 2000;6:980–984.

43 Baron R, Rawadi G, Roman-Roman S: Wnt signaling: a key regulator of bone mass. Curr Top Dev Biol 2006;76:103–127.

44 Kawano Y, Kypta R: Secreted antagonists of the Wnt signalling pathway. J Cell Sci 2003;116:2627–2634.

45 Baron R, Rawadi G: Targeting the Wnt/beta-catenin pathway to regulate bone formation in the adult skeleton. Endocrinology 2007;148:2635–2643.

46 Diarra D, Stolina M, Polzer K, Zwerina J, Ominsky MS, Dwyer D, Korb A, Smolen J, Hoffmann M, Scheinecker C, van der Heide D, Landewe R, Lacey D, Richards WG, Schett G: Dickkopf-1 is a master regulator of joint remodeling. Nat Med 2007;13:156–163.

47 Vincent C, Findlay DM, Welldon KJ, Wijenayaka AR, Zheng TS, Haynes DR, Fazzalari NL, Evdokiou A, Atkins GJ: The proinflammatory cytokines TNF-related weak inducer of apoptosis (TWEAK) and TNFalpha induce the mitogen activated protein kinase (MAPK)-dependent expression of sclerostin in human osteoblasts. J Bone Miner Res 2009;24: 1434–1449.

48 Thomson BM, Mundy GR, Chambers TJ: Tumor necrosis factors alpha and beta induce osteoblastic cells to stimulate osteoclastic bone resorption. J Immunol 1987;138:775–779.

49 Kaplan DL, Eielson CM, Horowitz MC, Insogna KL, Weir EC: Tumor necrosis factor-alpha induces transcription of the colony-stimulating factor-1 gene in murine osteoblasts. J Cell Physiol 1996;168:199–208.

50 Hofbauer LC, Lacey DL, Dunstan CR, Spelsberg TC, Riggs BL, Khosla S: Interleukin-1beta and tumor necrosis factor-alpha, but not interleukin-6, stimulate osteoprotegerin ligand gene expression in human osteoblastic cells. Bone 1999;25:255–259.

51 Abu-Amer Y, Ross FP, Edwards J, Teitelbaum SL: Lipopolysaccharide-stimulated osteoclastogenesis is mediated by tumor necrosis factor via its P55 receptor. J Clin Invest 1997;100:1557–1565.

52 Keffer J, Probert L, Cazlaris H, Georgopoulos S, Kaslaris E, Kioussis D, Kollias G: Transgenic mice expressing human tumour necrosis factor: a predictive genetic model of arthritis. EMBO J 1991;10: 4025–4031.

53 Schett G, Redlich K, Hayer S, Zwerina J, Bolon B, Dunstan C, Gortz B, Schulz A, Bergmeister H, Kollias G, Steiner G, Smolen JS: Osteoprotegerin protects against generalized bone loss in tumor necrosis factor-transgenic mice. Arthritis Rheum 2003;48:2042–2051.

54 Redlich K, Hayer S, Maier A, Dunstan CR, Tohidast-Akrad M, Lang S, Turk B, Pietschmann P, Woloszczuk W, Haralambous S, Kollias G, Steiner G, Smolen JS, Schett G: Tumor necrosis factor alpha-mediated joint destruction is inhibited by targeting osteoclasts with osteoprotegerin. Arthritis Rheum 2002;46:785–792.

55 Redlich K, Hayer S, Ricci R, David JP, Tohidast-Akrad M, Kollias G, Steiner G, Smolen JS, Wagner EF, Schett G: Osteoclasts are essential for TNF-alpha-mediated joint destruction. J Clin Invest 2002;110:1419–1427.

56 Polzer K, Joosten L, Gasser J, Distler JH, Ruiz G, Baum W, Redlich K, Bobacz K, Smolen JS, van den Berg W, Schett G, Zwerina J: IL-1 is essential for systemic inflammatory bone loss. Ann Rheum Dis 2009, Epub ahead of print.

57 Zwerina J, Hayer S, Tohidast-Akrad M, Bergmeister H, Redlich K, Feige U, Dunstan C, Kollias G, Steiner G, Smolen J, Schett G: Single and combined inhibition of tumor necrosis factor, interleukin-1, and RANKL pathways in tumor necrosis factor-induced arthritis: effects on synovial inflammation, bone erosion, and cartilage destruction. Arthritis Rheum 2004;50:277–290.

58 Yao Z, Li P, Zhang Q, Schwarz EM, Keng P, Arbini A, Boyce BF, Xing L: Tumor necrosis factor-alpha increases circulating osteoclast precursor numbers by promoting their proliferation and differentiation in the bone marrow through up-regulation of c-Fms expression. J Biol Chem 2006;281:11846–11855.

59 Kimble RB, Matayoshi AB, Vannice JL, Kung VT, Williams C, Pacifici R: Simultaneous block of interleukin-1 and tumor necrosis factor is required to completely prevent bone loss in the early postovariectomy period. Endocrinology 1995;136:3054–3061.

60 Roggia C, Gao Y, Cenci S, Weitzmann MN, Toraldo G, Isaia G, Pacifici R: Up-regulation of TNF-producing T cells in the bone marrow: a key mechanism by which estrogen deficiency induces bone loss in vivo. Proc Natl Acad Sci USA 2001;98:13960–13965.

61 Cenci S, Weitzmann MN, Roggia C, Namba N, Novack D, Woodring J, Pacifici R: Estrogen deficiency induces bone loss by enhancing T-cell production of TNF-alpha. J Clin Invest 2000;106: 1229–1237.

Dr. Jean-Pierre David
Rheumatology and Immunology, Department of Internal Medicine 3
University of Erlangen-Nuremberg, Krankenhausstrasse 12
DE–91054 Erlangen (Germany)
Tel. +49 9131 853 6445, Fax +49 9131 853 4770, E-Mail jean-pierre.david@uk-erlangen.de

Kollias G, Sfikakis PP (eds): TNF Pathophysiology. Molecular and Cellular Mechanisms.
Curr Dir Autoimmun. Basel, Karger, 2010, vol 11, pp 145–156

TNF-α and Obesity

T. Tzanavari · P. Giannogonas · Katia P. Karalis

Division of Developmental Biology, Biomedical Research Foundation, Academy of Athens, Athens, Greece

Abstract

Obesity, an epidemic of our times with rates rising to alarming levels, is associated with comorbidities including cardiovascular diseases, arthritis, certain cancers, and degenerative diseases of the brain and other organs. Importantly, obesity is a leading cause of insulin resistance and type 2 diabetes. As emerging evidence has shown over the last decade, inflammation is one of the critical processes associated with the development of insulin resistance, diabetes and related diseases, and obesity is now considered as a state of chronic low-grade inflammation. Adipose tissue, apart from its classical role as an energy storage depot, is also a major endocrine organ secreting many factors, whose local and circulating levels are affected by the degree of adiposity. Obesity leads to infiltration of the expanded adipose tissue by macrophages and increased levels in proinflammatory cytokines. The first indication for increased cytokine release in obesity was provided by the identification of increased expression of TNF-α, a proinflammatory cytokine, in the adipose tissue of obese mice in the early 1990s. TNF-α is expressed in and secreted by adipose tissue, its levels correlating with the degree of adiposity and the associated insulin resistance. Targeting TNF-α and/or its receptors has been suggested as a promising treatment for insulin resistance and type 2 diabetes. This review will summarize the available knowledge on the role of TNF-α in obesity and related processes and the potential implications of the above in the development of new therapeutic approaches for obesity and insulin resistance. Recent data from clinical studies will also be described together with late findings on the pathogenesis of obesity and insulin resistance.

TNF-α, a peptide isolated by its cytotoxic effects in tumor cells [1–6], has been shown to be identical to cachectin [7], a molecule that causes body weight loss in vivo and suppressed lipogenesis in vitro [8], mainly through suppression of lipoprotein lipase and induction of proteolysis through the ubiquitin-proteasome proteolytic pathway [9]. The subsequently identified association of this same molecule with obesity [4], an anabolic state, was a major surprise that was confirmed and provided the basis for a nicely evolving theory connecting obesity with inflammation and insulin resistance [4]. Common features between extreme metabolic conditions, such as obesity and cachexia or lipodystrophy, include hyperlipidemia and insulin resistance. As elegantly shown, the connection of TNF-α with cachexia and obesity is primarily related to its effects on lipid metabolism [4].

TNF-α and the Adipose Tissue

TNF-α Expression in Adipose Tissue

TNF-α was first identified in adipose tissue of rodents in 1993, where it was shown to be markedly induced (mRNA and protein levels) in obese animal models [4]. The biological significance of this finding was highlighted by the profound effect of TNF-α neutralization in insulin-induced glucose uptake, suggesting the association of TNF-α with insulin resistance in obese states. In white adipose tissue, TNF-α is expressed both in mature adipocytes and the stromal-vascular cells. In humans, the secretion of TNF-α is mainly by cells of the stromal-vascular and matrix fractions, including the macrophages, its levels being higher in subcutaneous than visceral adipose tissue [10, 11]. Similarly, TNF-α is expressed in preadipocytes; however, its levels increase with the degree of differentiation [12]. TNF-α is synthesized as a 26-kDa transmembrane protein, which following enzymatic processing gives rise to a biologically active 17-kDa soluble form that exerts its effects via type I (p55) and II (p75) TNF-α receptors [11]. Both types of TNF-α receptor are also expressed within adipocytes, with TNF-α expression being correlated with the expression of type II, while type I expression is higher in omental than subcutaneous adipocytes. The expression levels of both receptors are altered in fat and muscle of diabetic obese rats, while insulin-sensitizing agents and food restriction normalize their expression [13]. The above findings provided the initial evidence that TNF-α production by the adipose tissue is associated with the level of obesity, hyperinsulinemia and the corresponding insulin resistance. TNF-α can also induce several other inflammatory cytokines [14].

In obesity, the levels of transmembrane TNF-α in adipocytes are increased without apparent changes in the TNF-α-converting enzyme (TACE), which is responsible for the processing of this molecule. A putative explanation for the above could be the significantly increased expression of the transmembrane form in mature adipocytes compared to its levels in preadipocytes [15]. Furthermore, lipolysis of mature adipocytes by TNF-α resulted in similar reversion of the adipocyte phenotype [12]. TNF-α also acts as a potent inhibitor of adipocyte differentiation, as shown by the suppression of specific markers of fat cell differentiation, such as peroxisome proliferator-activated receptor-γ2, lipoprotein lipase, glycerol-3-phosphate dehydrogenase and glucose transporter type 4 [16]. In adipose tissue, TNF-α suppresses genes involved in the uptake and storage of nonesterified fatty acids and glucose, as well as transcription factors involved in adipogenesis and lipogenesis. Furthermore, it affects the expression of several adipocyte-secreted factors, including interleukin (IL)-6, monocyte chemoattractant protein (MCP)-1 and nerve growth factor, whose levels are substantially increased with the administration of TNF-α in vitro [14, 17]. Finally, in the obese state, TNF-α also contributes to the elevated plasminogen activator inhibitor-1 levels associated with obesity and acute inflammatory conditions [18].

Although, as discussed above, several studies have shown elevated levels of TNF-α in adipose tissue in obesity, the mechanisms leading to this increase remain elusive. It was recently demonstrated that TNF-α can autoregulate positively its own biosynthesis in the adipose tissue, thus contributing to the maintenance of elevated TNF-α levels in obesity. When applied to 3T3-L1 adipocytes, TNF-α autoamplification occurred via the protein kinase C signaling pathway and the transcription nuclear factor-κB (NF-κB). When injected with TNF-α, C57BL/6J mice upregulated TNF-α mRNA levels in adipocytes and in adipose tissue depots. Additionally, hyperinsulinemia potentiated the TNF-α-mediated autoamplification in adipose tissues in a synergistic and dose-dependent manner [19].

TNF-α Function – Genetic Models

In order to unravel the functional role of TNF-α, several studies using TNF-α transgenic mice have been conducted [20–22]. Genetic deficiency in TNF-α signaling reduces adipose TNF-α mRNA levels in *ob/ob* male mice [19]. Mice with partial or complete TNF-α deficiency have been compared with wild-type mice, under normal or calorie-rich diet, for adiposity, insulin sensitivity, and circulating levels of glucose and triglycerides. Transgenic mice have reduced plasma fasting glucose levels, insulin levels, plasma triglycerides and improved glucose tolerance, when fed gold thioglucose that induces massive obesity. The responses of transgenic mice fed a high-fat diet are of similar nature, although of a much lower magnitude [20]. Recent studies have shown that mice deficient in TACE and, thus with significantly lower ability to process TNF-α, are lean and protected from dietary obesity and the corresponding insulin resistance [23]. The effect of TACE deficiency is dose dependent, since heterozygous mice are also protected, although to a lesser extent [24]. In both cases, the phenotype is associated with increased metabolic activity, as shown by increased expression of uncoupling protein-1 and glucose transporter type 4 compared to the wild-type mice. Furthermore, transgenic mice overexpressing transmembrane TNF-α have increased adiposity as indicated by adipocyte area and blood vessel density [25]. The lipolytic action of TNF-α could account for the reduced adiposity detected in mice expressing exclusively transmembrane TNF-α. Lack of protection of the above mice from diet-induced insulin resistance is in further support of the importance of this factor in the obesity-associated insulin resistance and the development of type 2 diabetes [26].

As previously mentioned, TNF-α elicits cellular responses via its two receptors. Complete lack of TNF-α function through targeted mutations in both receptors results in significant improvement of insulin sensitivity in dietary, chemical, or genetic models of rodent obesity [21]. The effects of TNF-α released in obese states, as described above, are receptor-specific. This has been demonstrated by studies in p55 and p75 knockout mice, with most studies supporting the critical role of p55 in these processes [20, 27]. In support of the above studies, in $p55^{-/-}$ mice, recombinant

TNF receptor-Fc fusion protein (TNFR1BP-Fc) prevented the onset of dietary obesity and enhanced peripheral insulin sensitivity [28]. Absence of both TNF receptors had similar effects to deficiency in p55 alone in reducing brown adipocyte apoptosis [29]. Studies injecting TNF-α into lean mice lacking individual TNF-α receptors indicated that TNF-α auto-amplification in adipose tissues is mediated primarily via the p55 receptor, whereas the p75 receptor appeared to augment this response [19]. Other studies in *ob/ob* mice lacking p55 or p75 showed significant improvement in insulin sensitivity, in the former and not in the latter, although deficiency in both receptors had a stronger effect [27], as also suggested by the antidiabetic effects of blockade of the function of both TNF receptors [20]. Finally, a provocative hypothesis was suggested by a very recent study, showing similar insulin resistance and increased adiposity in mice with genetic deficiency in both TNF-α receptors [30].

Mechanisms of TNF-α Action

The underlying reason for the TNF-α-mediated insulin resistance was revealed by demonstration of the effect of TNF-α neutralization on the compromised tyrosine kinase activity of insulin receptor (IR) in the muscle and fat of obese rats. Increased IR tyrosine kinase activity following treatment with a TNF soluble receptor-IgG fusion protein was associated with normalization of circulating glucose, free fatty acid and insulin levels [31]. The molecular pathway mediating the decreased tyrosine kinase activity of IR was elucidated by serine phosphorylation of IRS-1 following TNF-α treatment of murine adipocytes [32]. Serine phosphorylated IRS-1 can function as an inhibitor of the tyrosine kinase activity of the IR. Thus, TNF-α is a mediator of insulin resistance in obesity by modulating IR signaling in target tissues via different pathways [33]. SOCS (suppressor of cytokine signaling) proteins are inhibitors of cytokine signaling involved in negative feedback loops. SOCS-3 expression is increased in the adipose tissue of obese mice while more profoundly increased in transgenic *ob/ob* mice lacking both TNF-α receptors, an effect achieved also in fat by TNF-α administration. SOCS-3 antagonizes insulin-induced IRS-1 tyrosine phosphorylation while it is activated by TNF-α itself, providing a likely factor critically involved in the development of obesity-associated insulin resistance [34].

Obesity is considered an inflammatory state as, in addition to TNF-α, it is characterized by increased production of other proinflammatory cytokines, such as MCP-1, IL6, and chemokine receptors, all expressed in amounts proportional to the increase in adipose mass [35]. Obesity is also characterized by infiltration of the adipose tissue by macrophages, shown to be the predominant source of TNF-α [10]. The cross-talk between adipocytes and macrophages has been demonstrated using relevant in vitro systems that confirmed the importance of macrophages in inducing inflammatory markers in adipocytes [36, 37]. The understanding of the importance of the recruitment and polarization of macrophages in the adipose tissue in obesity has been greatly facilitated by the use of transgenic mice. In terms of macrophage infiltration, it has been shown that mice with genetically engineered deficiency in MCP1 and thus minimal macrophage infiltration of the adipose tissue are protected from diet-induced obesity and development of insulin resistance [38]. Similar results were obtained by genetic deletion of the chemokine receptor CXCR2 that binds MCP1 [39]. However, it was suggested in a recent study that MCP-1 may actually restrain macrophage accumulation in adipose tissue, while being one of the pathways involved in the recruitment of monocytes in the tissue during diet-induced obesity [40].

In addition to the infiltrating macrophages, adipose tissue contains the so-called resident macrophages shown to belong to the M2, or alternatively activated type of macrophages [41]. It seems that in obesity there is a transition of the M2 to the M1 type, or otherwise the classically activated macrophages [42]. The two types differ in their cytokine profile, with M2 producing primarily the anti-inflammatory cytokine IL-10, and M1 proinflammatory cytokines such as IFN-γ, TNF-α, IL-6 and IL-1 [43]. A recent study proposing a protective role of TNF receptors in the development of obesity showed reduced numbers of activated macrophages in the adipose tissue of transgenic mice lacking both TNF receptors [30]. This study suggests that the exact role of the TNF system in the transition of macrophages during obesity and the associated insulin resistance deserves further investigation. Proinflammatory cytokines secreted by the adipose tissue macrophages may act in paracrine, and possibly endocrine, mode to modulate insulin signaling and thus, contribute to systemic insulin resistance. Activation of the proinflammatory cytokines in the adipose tissue follows the induction of IκB kinase (IKK) and NF-κB shown to be associated with the development of insulin resistance [35]. Activation of this pathway has provided the scientific rationale for the use of salicylates in the treatment of insulin resistance and type 2 diabetes [35, 44]. Transgenic models targeting the IKK and thus NF-κB have provided critical insights into the role of this specific pathway in the above states, as well as on the tissue specificity of these responses [45]. Another kinase with critical role in obesity-induced insulin resistance is c-Jun N-terminal kinase (JNK) and more specifically JNK1 [46]. It seems that transgenic mice deficient in JNK1 are protected from development of obesity and the associated increase in TNF-α and development of insulin resistance [46]. Recent studies demonstrated that obesity results in endoplasmic reticulum (ER) stress [31], and suggested this as a main mechanism underlying the obesity-induced JNK activation, inflammatory responses, and peripheral insulin resistance [47]. As shown, intracerebroventricular administration of agents blocking induction of ER stress in mice was able to reduce obesity-induced central leptin resistance [48].

Recently, identification of a pathway leading to NF-κB activation in obesity has been made through the use of Toll-like receptor *(Tlr)* $4^{-/-}$ mice [49, 50]. It seems that TLR4 binds and is activated by fatty acids, resulting in activation of NF-κB and proinflammatory cytokines. Several studies have shown that the above mice are protected from development of dietary obesity, adipose tissue inflammation and insulin resistance, while they have increased levels of adiponectin. The levels of this hormone are reversely associated with obesity [51], insulin resistance [52] and TNF-α levels in mice and humans, while inhibition of its expression by TNF-α has been described [53]. These findings suggest that inflammatory changes in adipose tissue through NF-κB activation are promoted by saturated fatty acids released during lipolysis from hypertrophic adipocytes in obesity, that serve as naturally occurring ligands for TLR4 [54, 55]. These findings highlighted the significance of innate immune activation in the development of obesity and insulin resistance. However, serum TNF-α levels measured in obese *Tlr4*$^{-/-}$ or mutant mice have not provided conclusive results [56] suggesting the multifactorial regulation of the circulating levels of this cytokine in obesity.

TNF-α – A Possible Target for Insulin Resistance Treatment

Human Clinical Studies

As discussed above, studies in animal models identified TNF-α as a key component in obesity-linked insulin resistance through inhibition of IR signaling and glucose transport in insulin-sensitive tissues. TNF-α and its two receptors are expressed in human adipose tissue. In vivo studies were thus designed to further explore its significance in human obesity. However, these have given conflicting results regarding the relationship between circulating TNF-α levels and insulin sensitivity.

Initial studies indicated that in humans, plasma concentrations of both soluble TNF receptors correlate with the degree of adiposity and insulin resistance. However, the reported adipose tissue overexpression of TNF-α does not seem to be reflected in elevated plasma concentrations, suggesting a primarily local role of the cytokine [57, 58]. An indication that in humans TNF-α could also be involved in the development of obesity-associated insulin resistance came from a study in women with severe obesity and high levels of TNF-α. In addition, levels of mRNA of both TNF receptors were increased, with expression of type II being higher in the omental compared to the subcutaneous abdominal depot [59]. Finally, dysregulated proteolysis of TNF-α and IL-6R by TACE/TIMP3 is an important factor for the development of skeletal muscle insulin resistance in obese type 2 diabetes patients through autocrine and/or paracrine mechanisms [60]. TNF-α is a crucial contributor to adipokine dysregulation [61]. Accordingly, adipocytes of obese subjects overproduce adipokines in response to direct exposure to TNF-α. This hyperresponsiveness is mediated by TNFR1 and hyperactivation of the NF-κB pathway. Correspondingly, NF-κB activity is increased in adipocytes of obese subjects and correlates with adipocyte size, adipokine expression, and in vivo insulin resistance. Eventually, adipokine overexpression in adipocytes of obese subjects is prevented by NF-κB inhibitors [62].

A strong inverse correlation between adipose TNF-α secretion and maximum insulin-stimulated glucose transport in adipocytes independent of fat cell volume, age, and BMI was also found in humans [63]. TNF-α induced insulin resistance on glucose uptake in human visceral but not in subcutaneous adipocytes, suggesting depot-specific effects of TNF-α on glucose uptake. Furthermore, activation of JNK1/2 and insulin resistance in glucose uptake could be reversed by LXR agonists [64]. Consistent with findings in rodents [65], an induction of leptin levels by TNF-α was found in humans [66], providing a putative explanation for the role of these factors in the pathophysiology of obesity and cachexia. Levels of MCP-1 [67] and IL-8, a potent proinflammatory cytokine in humans, are increased in obese subjects, and are related to the fat mass and the circulating levels of TNF-α. Increase in circulating IL-8 may be one of the factors linking obesity with greater cardiovascular risk [68].

The reported decrease in serum TNF-α levels and increase in both TNF soluble receptors after treatment of obese subjects with weight-reducing regimens may be a counter-regulatory mechanism to prevent excessive weight loss [69]. Plasma soluble TNF-α receptor (sTNFR) 2 levels are more closely related to abdominal adipose tissue accumulation than to total adiposity [70]. When obese insulin-resistant patients were given a single i.v. injection of a recombinant TNF receptor-Fc fusion protein, no improvement in insulin sensitivity was observed [71]. The circulating concentration of DS-TNFR2 was shown to be inversely linked to metabolic disorders, hinting at a possible anti-inflammatory role [72]. In a more recent study trying to further understand the link between obesity, insulin resistance and type 2 diabetes and increased TNF-α action, subjects were given a biologically active form of sTNFR2 produced by differential splicing (DS-TNFR2) which antagonized TNF-α biological activity.

Therapeutic Potential

The recognition of the role of low-grade chronic inflammation in the pathophysiology of obesity and the metabolic syndrome has made the inflammatory related cytokines, including TNF-α, IL-6 and the soluble TNF-α receptors, potential targets for drug therapy [73]. The primary focus in these states is the improvement of insulin sensitivity. A reasonable approach is to reduce TNF-α levels by using anti-TNF-α antibodies, such as infliximab [74] and adalimumab [75], and targeting the soluble receptor complex. Alternatively, an indirect method, where TNF-α would be targeted through its regulatory factors could prove very successful. As already discussed above, TNF-α activates, but is also regulated by NF-κB. TNF-α also exerts its effects through mitogen-activated protein kinase and Janus-activated kinase/signal transducer and activator of transcription pathways [76]. Thus, additional attempts to target these complex pathways could potentially be of therapeutic value. However, current evidence suggests that neutralizing TNF-α in type 2 diabetic subjects is not sufficient to cause metabolic improvement; and there was no improvement observed in insulin sensitivity in obese insulin-resistant patients administered a single i.v. injection of a recombinant TNF receptor-Fc fusion protein [71]. Despite the initial disappointment regarding the efficacy of inhibitors of the TNF-α system in obese, insulin-resistant patients, more studies are needed to conclude on the biological significance of TNF-α as a contributing factor in common metabolic disturbances in obese patients [77]. Adiponectin antagonizes many effects of TNF-α [53] that, in turn, suppresses adiponectin production. Furthermore, adiponectin secretion from adipocytes is enhanced by thiazolidinediones (TZDs), which also act to antagonize TNF-α effects [16, 78]. Thus, adiponectin could be the common mechanism by which TNF-α promotes, and the TZDs suppress, insulin resistance and inflammation [79], although the insulin-sensitizing action of TZDs is not solely by TNF-α-dependent mechanisms [80]. Unfortunately however, TZDs, due to their adipogenic properties, increase adipose tissue mass, so most likely they should be given in combination with other factors.

Furthermore, although human studies with TNF-α blockers did not prove very successful in the beginning, recent evidence shows that inflammation will provide viable therapeutic targets, most possibly through the use of salicylates, which are typical members of the nonsteroidal anti-inflammatory class of drugs, and IL-1 blockade. Anakinra, the recombinant IL-1 receptor antagonist has been shown to improve glycemia in individuals with type 2 diabetes [81], while salsalate, a salicylate prodrug, has been shown to have glucose-lowering efficacy [82].

Conclusion

The metabolic and immune systems are fundamental for survival while they operate through highly integrated and interdependent pathways. In the past decade, it became apparent that inflammation is critically involved in the pathogenesis of obesity and insulin resistance, as well as the associated morbidities including cardiovascular disease and type 2 diabetes. TNF-α was suggested to mediate obesity-induced inflammation and insulin resistance based on studies in mice and supporting evidence from human studies. Intracellular pathways including JNK, IKKb/NF-κB and factors mediating ER stress were identified as critically involved in obesity-induced insulin resistance. Altered dynamics in immune cells and accumulation in obese adipose tissue, with primary focus on macrophages, have been demonstrated. Finally, recent reports described a critical role for specific T-lymphocyte populations and mast cells, all sites of production and action of TNF-α, in the development of inflammation and insulin resistance in experimental obesity. Initial human studies have not supported a significant therapeutic effect on obesity and insulin resistance of targeting the TNF-α system. Based on the above data and the recent advances in the field, we suggest that it is too early to conclude on the potential of therapeutic interventions targeting TNF-α in humans. More studies are needed that will take advantage of new investigational approaches that factor in multiple pathways and molecules involved in complex conditions, in order to identify specific drug targets for obesity and insulin resistance [83].

References

1 Das UN: Is obesity an inflammatory condition? Nutrition 2001;17:953–966.

2 Festa A, D'Agostino R Jr, Williams K, Karter AJ, Mayer-Davis EJ, Tracy RP, Haffner SM: The relation of body fat mass and distribution to markers of chronic inflammation. Int J Obes Relat Metab Disord 2001;25:1407–1415.

3 Trayhurn P, Wood IS: Adipokines: inflammation and the pleiotropic role of white adipose tissue. Br J Nutr 2004;92:347–355.

4 Hotamisligil GS, Shargill NS, Spiegelman BM: Adipose expression of tumor necrosis factor-alpha: direct role in obesity-linked insulin resistance. Science 1993;259:87–91.

5 Carswell EA, Old LJ, Kassel RL, Green S, Fiore N, Williamson B: An endotoxin-induced serum factor that causes necrosis of tumors. Proc Natl Acad Sci USA 1975;72:3666–3670.

6 Pennica D, Nedwin GE, Hayflick JS, Seeburg PH, Derynck R, Palladino MA, Kohr WJ, Aggarwal BB, Goeddel DV: Human tumour necrosis factor: precursor structure, expression and homology to lymphotoxin. Nature 1984;312:724–729.

7 Cerami A, Beutler B: Cachectin: the dark side of tumor necrosis factor. Cold Spring Harb Symp Quant Biol 1986;51:625–629.

8 Cerami A, Ikeda Y, Le Trang N, Hotez PJ, Beutler B: Weight loss associated with an endotoxin-induced mediator from peritoneal macrophages: the role of cachectin (tumor necrosis factor). Immunol Lett 1985;11:173–177.

9 Tracey KJ, Morgello S, Koplin B, Fahey TJ 3rd, Fox J, Aledo A, Manogue KR, Cerami A: Metabolic effects of cachectin/tumor necrosis factor are modified by site of production. Cachectin/tumor necrosis factor-secreting tumor in skeletal muscle induces chronic cachexia, while implantation in brain induces predominantly acute anorexia. J Clin Invest 1990;86: 2014–2024.

10 Weisberg SP, McCann D, Desai M, Rosenbaum M, Leibel RL, Ferrante AW Jr: Obesity is associated with macrophage accumulation in adipose tissue. J Clin Invest 2003;112:1796–1808.

11 Fain JN, Bahouth SW, Madan AK: TNFalpha release by the nonfat cells of human adipose tissue. Int J Obes Relat Metab Disord 2004;28:616–622.

12 Hube F, Hauner H: The role of TNF-alpha in human adipose tissue: prevention of weight gain at the expense of insulin resistance? Horm Metab Res 1999;31:626–631.

13 Hofmann C, Lorenz K, Braithwaite SS, Colca JR, Palazuk BJ, Hotamisligil GS, Spiegelman BM: Altered gene expression for tumor necrosis factor-alpha and its receptors during drug and dietary modulation of insulin resistance. Endocrinology 1994;134:264–270.

14 Wang B, Trayhurn P: Acute and prolonged effects of TNF-alpha on the expression and secretion of inflammation-related adipokines by human adipocytes differentiated in culture. Pflugers Arch 2006; 452:418–427.

15 Xu H, Uysal KT, Becherer JD, Arner P, Hotamisligil GS: Altered tumor necrosis factor-alpha (TNF-alpha) processing in adipocytes and increased expression of transmembrane TNF-alpha in obesity. Diabetes 2002;51:1876–1883.

16 Peraldi P, Xu M, Spiegelman BM: Thiazolidinediones block tumor necrosis factor-alpha-induced inhibition of insulin signaling. J Clin Invest 1997;100:1863–1869.

17 Ruan H, Hacohen N, Golub TR, Van Parijs L, Lodish HF: Tumor necrosis factor-alpha suppresses adipocyte-specific genes and activates expression of preadipocyte genes in 3T3-L1 adipocytes: nuclear factor-kappaB activation by TNF-alpha is obligatory. Diabetes 2002;51:1319–1336.

18 Pandey M, Tuncman G, Hotamisligil GS, Samad F: Divergent roles for p55 and p75 TNF-alpha receptors in the induction of plasminogen activator inhibitor-1. Am J Pathol 2003;162:933–941.

19 Neels JG, Pandey M, Hotamisligil GS, Samad F: Autoamplification of tumor necrosis factor-alpha: a potential mechanism for the maintenance of elevated tumor necrosis factor-alpha in male but not female obese mice. Am J Pathol 2006;168:435–444.

20 Schreyer SA, Chua SC Jr, LeBoeuf RC: Obesity and diabetes in TNF-alpha receptor- deficient mice. J Clin Invest 1998;102:402–411.

21 Uysal KT, Wiesbrock SM, Marino MW, Hotamisligil GS: Protection from obesity-induced insulin resistance in mice lacking TNF-alpha function. Nature 1997;389:610–614.

22 Ventre J, Doebber T, Wu M, MacNaul K, Stevens K, Pasparakis M, Kollias G, Moller DE: Targeted disruption of the tumor necrosis factor-alpha gene: metabolic consequences in obese and nonobese mice. Diabetes 1997;46:1526–1531.

23 Gelling RW, Yan W, Al-Noori S, Pardini A, Morton GJ, Ogimoto K, Schwartz MW, Dempsey PJ: Deficiency of TNFalpha converting enzyme (TACE/ADAM17) causes a lean, hypermetabolic phenotype in mice. Endocrinology 2008;149:6053–6064.

24 Serino M, Menghini R, Fiorentino L, Amoruso R, Mauriello A, Lauro D, Sbraccia P, Hribal ML, Lauro R, Federici M: Mice heterozygous for tumor necrosis factor-alpha converting enzyme are protected from obesity-induced insulin resistance and diabetes. Diabetes 2007;56:2541–2546.

25 Voros G, Maquoi E, Collen D, Lijnen HR: Influence of membrane-bound tumor necrosis factor (TNF)-alpha on obesity and glucose metabolism. J Thromb Haemost 2004;2:507–513.

26 Xu H, Hirosumi J, Uysal KT, Guler AD, Hotamisligil GS: Exclusive action of transmembrane TNF alpha in adipose tissue leads to reduced adipose mass and local but not systemic insulin resistance. Endocrinology 2002;143:1502–1511.

27 Uysal KT, Wiesbrock SM, Hotamisligil GS: Functional analysis of tumor necrosis factor (TNF) receptors in TNF-alpha-mediated insulin resistance in genetic obesity. Endocrinology 1998;139:4832–4838.

28 Liang H, Yin B, Zhang H, Zhang S, Zeng Q, Wang J, Jiang X, Yuan L, Wang CY, Li Z: Blockade of tumor necrosis factor (TNF) receptor type 1-mediated TNF-alpha signaling protected Wistar rats from diet-induced obesity and insulin resistance. Endocrinology 2008;149:2943–2951.

29 Nisoli E, Briscini L, Giordano A, Tonello C, Wiesbrock SM, Uysal KT, Cinti S, Carruba MO, Hotamisligil GS: Tumor necrosis factor alpha mediates apoptosis of brown adipocytes and defective brown adipocyte function in obesity. Proc Natl Acad Sci USA 2000;97:8033–8038.

30 Pamir N, McMillen TS, Kaiyala KJ, Schwartz MW, LeBoeuf RC: Receptors for tumor necrosis factor-alpha play a protective role against obesity and alter adipose tissue macrophage status. Endocrinology 2009;150:4124–4134.

31 Hotamisligil GS, Arner P, Caro JF, Atkinson RL, Spiegelman BM: Increased adipose tissue expression of tumor necrosis factor-alpha in human obesity and insulin resistance. J Clin Invest 1995; 95: 2409–2415.

32 Hotamisligil GS, Peraldi P, Budavari A, Ellis R, White MF, Spiegelman BM: IRS-1-mediated inhibition of insulin receptor tyrosine kinase activity in TNF-alpha- and obesity-induced insulin resistance. Science 1996;271:665–668.

33 Cheung AT, Ree D, Kolls JK, Fuselier J, Coy DH, Bryer-Ash M: An in vivo model for elucidation of the mechanism of tumor necrosis factor-alpha (TNF-alpha)-induced insulin resistance: evidence for differential regulation of insulin signaling by TNF-alpha. Endocrinology 1998;139:4928–4935.

34 Emanuelli B, Peraldi P, Filloux C, Chavey C, Freidinger K, Hilton DJ, Hotamisligil GS, Van Obberghen E: SOCS-3 inhibits insulin signaling and is up-regulated in response to tumor necrosis factor-alpha in the adipose tissue of obese mice. J Biol Chem 2001;276:47944–47949.

35 Shoelson SE, Lee J, Yuan M: Inflammation and the IKK beta/I kappa B/NF-kappa B axis in obesity- and diet-induced insulin resistance. Int J Obes Relat Metab Disord 2003;27(Suppl 3):S49–S52.

36 Bouloumie A, Curat CA, Sengenes C, Lolmede K, Miranville A, Busse R: Role of macrophage tissue infiltration in metabolic diseases. Curr Opin Clin Nutr Metab Care 2005;8:347–354.

37 Xu H, Barnes GT, Yang Q, Tan G, Yang D, Chou CJ, Sole J, Nichols A, Ross JS, Tartaglia LA, et al: Chronic inflammation in fat plays a crucial role in the development of obesity-related insulin resistance. J Clin Invest 2003;112:1821–1830.

38 Kanda H, Tateya S, Tamori Y, Kotani K, Hiasa K, Kitazawa R, Kitazawa S, Miyachi H, Maeda S, Egashira K, et al: MCP-1 contributes to macrophage infiltration into adipose tissue, insulin resistance, and hepatic steatosis in obesity. J Clin Invest 2006; 116:1494–1505.

39 Weisberg SP, Hunter D, Huber R, Lemieux J, Slaymaker S, Vaddi K, Charo I, Leibel RL, Ferrante AW Jr: CCR2 modulates inflammatory and metabolic effects of high-fat feeding. J Clin Invest 2006; 116:115–124.

40 Kirk EA, Sagawa ZK, McDonald TO, O'Brien KD, Heinecke JW: Monocyte chemoattractant protein deficiency fails to restrain macrophage infiltration into adipose tissue. Diabetes 2008;57:1254–1261.

41 Zeyda M, Stulnig TM: Adipose tissue macrophages. Immunol Lett 2007;112:61–67.

42 Lumeng CN, Bodzin JL, Saltiel AR: Obesity induces a phenotypic switch in adipose tissue macrophage polarization. J Clin Invest 2007;117:175–184.

43 Gordon S: Alternative activation of macrophages. Nat Rev Immunol 2003;3:23–35.

44 Kim JK, Kim YJ, Fillmore JJ, Chen Y, Moore I, Lee J, Yuan M, Li ZW, Karin M, Perret P, et al: Prevention of fat-induced insulin resistance by salicylate. J Clin Invest 2001;108:437–446.

45 Arkan MC, Hevener AL, Greten FR, Maeda S, Li ZW, Long JM, Wynshaw-Boris A, Poli G, Olefsky J, Karin M: IKK-beta links inflammation to obesity-induced insulin resistance. Nat Med 2005;11:191–198.

46 Hirosumi J, Tuncman G, Chang L, Gorgun CZ, Uysal KT, Maeda K, Karin M, Hotamisligil GS: A central role for JNK in obesity and insulin resistance. Nature 2002;420:333–336.

47 Ozcan U, Cao Q, Yilmaz E, Lee AH, Iwakoshi NN, Ozdelen E, Tuncman G, Gorgun C, Glimcher LH, Hotamisligil GS: Endoplasmic reticulum stress links obesity, insulin action, and type 2 diabetes. Science 2004;306:457–461.
48 Ozcan L, Ergin AS, Lu A, Chung J, Sarkar S, Nie D, Myers MG Jr, Ozcan U: Endoplasmic reticulum stress plays a central role in development of leptin resistance. Cell Metab 2009;9:35–51.
49 Shi H, Kokoeva MV, Inouye K, Tzameli I, Yin H, Flier JS: TLR4 links innate immunity and fatty acid-induced insulin resistance. J Clin Invest 2006;116: 3015–3025.
50 Kim JK: Fat uses a TOLL-road to connect inflammation and diabetes. Cell Metab 2006;4:417–419.
51 Arita Y, Kihara S, Ouchi N, Takahashi M, Maeda K, Miyagawa J, Hotta K, Shimomura I, Nakamura T, Miyaoka K, et al: Paradoxical decrease of an adipose-specific protein, adiponectin, in obesity. Biochem Biophys Res Commun 1999;257:79–83.
52 Hotta K, Funahashi T, Arita Y, Takahashi M, Matsuda M, Okamoto Y, Iwahashi H, Kuriyama H, Ouchi N, Maeda K, et al: Plasma concentrations of a novel, adipose-specific protein, adiponectin, in type 2 diabetic patients. Arterioscler Thromb Vasc Biol 2000;20:1595–1599.
53 Masaki T, Chiba S, Tatsukawa H, Yasuda T, Noguchi H, Seike M, Yoshimatsu H: Adiponectin protects LPS-induced liver injury through modulation of TNF-alpha in KK-Ay obese mice. Hepatology 2004; 40:177–184.
54 Song MJ, Kim KH, Yoon JM, Kim JB: Activation of Toll-like receptor 4 is associated with insulin resistance in adipocytes. Biochem Biophys Res Commun 2006;346:739–745.
55 Suganami T, Tanimoto-Koyama K, Nishida J, Itoh M, Yuan X, Mizuarai S, Kotani H, Yamaoka S, Miyake K, Aoe S, et al: Role of the Toll-like receptor 4/NF-kappaB pathway in saturated fatty acid-induced inflammatory changes in the interaction between adipocytes and macrophages. Arterioscler Thromb Vasc Biol 2007;27:84–91.
56 Tsukumo DM, Carvalho-Filho MA, Carvalheira JB, Prada PO, Hirabara SM, Schenka AA, Araujo EP, Vassallo J, Curi R, Velloso LA, et al: Loss-of-function mutation in Toll-like receptor 4 prevents diet-induced obesity and insulin resistance. Diabetes 2007;56:1986–1998.
57 Hauner H, Bender M, Haastert B, Hube F: Plasma concentrations of soluble TNF-alpha receptors in obese subjects. Int J Obes Relat Metab Disord 1998; 22:1239–1243.
58 Mohamed-Ali V, Goodrick S, Bulmer K, Holly JM, Yudkin JS, Coppack SW: Production of soluble tumor necrosis factor receptors by human subcutaneous adipose tissue in vivo. Am J Physiol 1999; 277:E971–E975.
59 Hube F, Birgel M, Lee YM, Hauner H: Expression pattern of tumour necrosis factor receptors in subcutaneous and omental human adipose tissue: role of obesity and non-insulin-dependent diabetes mellitus. Eur J Clin Invest 1999;29:672–678.
60 Monroy A, Kamath S, Chavez AO, Centonze VE, Veerasamy M, Barrentine A, Wewer JJ, Coletta DK, Jenkinson C, Jhingan RM, et al: Impaired regulation of the TNF-alpha converting enzyme/tissue inhibitor of metalloproteinase 3 proteolytic system in skeletal muscle of obese type 2 diabetic patients: a new mechanism of insulin resistance in humans. Diabetologia 2009;52:2169–2181.
61 Lo J, Bernstein LE, Canavan B, Torriani M, Jackson MB, Ahima RS, Grinspoon SK: Effects of TNF-alpha neutralization on adipocytokines and skeletal muscle adiposity in the metabolic syndrome. Am J Physiol Endocrinol Metab 2007;293:E102–E109.
62 Maury E, Noel L, Detry R, Brichard SM: In vitro hyperresponsiveness to tumor necrosis factor-alpha contributes to adipokine dysregulation in omental adipocytes of obese subjects. J Clin Endocrinol Metab 2009;94:1393–1400.
63 Lofgren P, van Harmelen V, Reynisdottir S, Naslund E, Ryden M, Rossner S, Arner P: Secretion of tumor necrosis factor-alpha shows a strong relationship to insulin-stimulated glucose transport in human adipose tissue. Diabetes 2000;49:688–692.
64 Fernandez-Veledo S, Vila-Bedmar R, Nieto-Vazquez I, Lorenzo M: c-Jun N-terminal kinase 1/2 activation by tumor necrosis factor-alpha induces insulin resistance in human visceral but not subcutaneous adipocytes: reversal by liver X receptor agonists. J Clin Endocrinol Metab 2009;94:3583–3593.
65 Sarraf P, Frederich RC, Turner EM, Ma G, Jaskowiak NT, Rivet DJ 3rd, Flier JS, Lowell BB, Fraker DL, Alexander HR: Multiple cytokines and acute inflammation raise mouse leptin levels: potential role in inflammatory anorexia. J Exp Med 1997;185:171–175.
66 Mantzoros CS, Moschos S, Avramopoulos I, Kaklamani V, Liolios A, Doulgerakis DE, Griveas I, Katsilambros N, Flier JS: Leptin concentrations in relation to body mass index and the tumor necrosis factor-alpha system in humans. J Clin Endocrinol Metab 1997;82:3408–3413.
67 Chacon MR, Fernandez-Real JM, Richart C, Megia A, Gomez JM, Miranda M, Caubet E, Pastor R, Masdevall C, Vilarrasa N, et al: Monocyte chemoattractant protein-1 in obesity and type 2 diabetes. Insulin sensitivity study. Obesity (Silver Spring) 2007;15:664–672.

68 Straczkowski M, Dzienis-Straczkowska S, Stepien A, Kowalska I, Szelachowska M, Kinalska I: Plasma interleukin-8 concentrations are increased in obese subjects and related to fat mass and tumor necrosis factor-alpha system. J Clin Endocrinol Metab 2002; 87:4602–4606.
69 Zahorska-Markiewicz B, Janowska J, Olszanecka-Glinianowicz M, Zurakowski A: Serum concentrations of TNF-alpha and soluble TNF-alpha receptors in obesity. Int J Obes Relat Metab Disord 2000;24: 1392–1395.
70 Cartier A, Cote M, Bergeron J, Almeras N, Tremblay A, Lemieux I, Despres JP: Plasma soluble tumour necrosis factor-alpha receptor 2 is elevated in obesity: specific contribution of visceral adiposity. Clin Endocrinol (Oxf) 2009, Epub ahead of print.
71 Paquot N, Castillo MJ, Lefebvre PJ, Scheen AJ: No increased insulin sensitivity after a single intravenous administration of a recombinant human tumor necrosis factor receptor: Fc fusion protein in obese insulin-resistant patients. J Clin Endocrinol Metab 2000;85:1316–1319.
72 Fernandez-Real JM, Botas-Cervero P, Lainez B, Ricart W, Delgado E: An alternatively spliced soluble TNF-alpha receptor is associated with metabolic disorders: a replication study. Clin Immunol 2006; 121:236–241.
73 Mavridis G, Souliou E, Diza E, Symeonidis G, Pastore F, Vassiliou AM, Karamitsos D: Inflammatory cytokines in insulin-treated patients with type 2 diabetes. Nutr Metab Cardiovasc Dis 2008; 18:471–476.
74 Yazdani-Biuki B, Stelzl H, Brezinschek HP, Hermann J, Mueller T, Krippl P, Graninger W, Wascher TC: Improvement of insulin sensitivity in insulin resistant subjects during prolonged treatment with the anti-TNF-alpha antibody infliximab. Eur J Clin Invest 2004;34:641–642.
75 Sarzi-Puttini P, Atzeni F, Scholmerich J, Cutolo M, Straub RH: Anti-TNF antibody treatment improves glucocorticoid induced insulin-like growth factor 1 (IGF1) resistance without influencing myoglobin and IGF1 binding proteins 1 and 3. Ann Rheum Dis 2006;65:301–305.
76 Chen F: Is NF-kappaB a culprit in type 2 diabetes? Biochem Biophys Res Commun 2005;332:1–3.
77 Moller DE: Potential role of TNF-alpha in the pathogenesis of insulin resistance and type 2 diabetes. Trends Endocrinol Metab 2000;11:212–217.
78 Miles PD, Romeo OM, Higo K, Cohen A, Rafaat K, Olefsky JM: TNF-alpha-induced insulin resistance in vivo and its prevention by troglitazone. Diabetes 1997;46:1678–1683.
79 Whitehead JP, Richards AA, Hickman IJ, Macdonald GA, Prins JB: Adiponectin – a key adipokine in the metabolic syndrome. Diabetes Obes Metab 2006;8: 264–280.
80 Wellen KE, Uysal KT, Wiesbrock S, Yang Q, Chen H, Hotamisligil GS: Interaction of tumor necrosis factor-alpha- and thiazolidinedione-regulated pathways in obesity. Endocrinology 2004;145:2214–2220.
81 Larsen CM, Faulenbach M, Vaag A, Volund A, Ehses JA, Seifert B, Mandrup-Poulsen T, Donath MY: Interleukin-1-receptor antagonist in type 2 diabetes mellitus. N Engl J Med 2007;356:1517–1526.
82 Goldfine AB, Silver R, Aldhahi W, Cai D, Tatro E, Lee J, Shoelson SE: Use of salsalate to target inflammation in the treatment of insulin resistance and type 2 diabetes. Clin Transl Sci 2008;1:36–43.
83 Shoelson SE, Goldfine AB: Getting away from glucose: fanning the flames of obesity-induced inflammation. Nat Med 2009;15:373–374.

Dr. Katia P. Karalis
Division of Developmental Biology, Biomedical Research Foundation, Academy of Athens
4 Soranou Ephessiou
GR–11527 Athens (Greece)
Tel. +30 210 6597465, Fax +30 210 6597545, E-Mail kkarali@bioacademy.gr

Kollias G, Sfikakis PP (eds): TNF Pathophysiology. Molecular and Cellular Mechanisms.
Curr Dir Autoimmun. Basel, Karger, 2010, vol 11, pp 157–179

TNF in Host Resistance to Tuberculosis Infection

Valerie F.J. Quesniaux[a] · Muazzam Jacobs[b,c] · Nasiema Allie[b] · Sergei Grivennikov[a,d] · Sergei A. Nedospasov[d,e] · Irene Garcia[f] · Maria L. Olleros[f] · Yuriy Shebzukhov[e] · Dmitry Kuprash[d] · Virginie Vasseur[a] · Stephanie Rose[a] · Nathalie Court[a] · Rachel Vacher[a] · Bernhard Ryffel[a]

[a]Molecular Immunology and Embryology UMR6218, Orleans University and CNRS, Orleans, France; [b]Institute of Infectious Disease and Molecular Medicine, University of Cape Town, [c]National Health Laboratory Service, Cape Town, South Africa; [d]Laboratory of Molecular Immunology, Engelhardt Institute of Molecular Biology, Russian Academy of Sciences, Moscow, Russia; [e]German Rheumatism Research Centre (DRFZ), Berlin, Germany; [f]Department of Pathology and Immunology, CMU, University of Geneva, Geneva, Switzerland

Abstract

TNF is essential to control *Mycobacterium tuberculosis* infection and cannot be replaced by other proinflammatory cytokines. Overproduction of TNF may cause immunopathology, while defective TNF production results in uncontrolled infection. The critical role of TNF in the control of tuberculosis has been illustrated recently by primary and reactivation of latent infection in some patients under pharmacological anti-TNF therapy for rheumatoid arthritis or Crohn's disease. In this review, we discuss results of recent studies aimed at better understanding of molecular, cellular and kinetic aspects of TNF-mediated regulation of host-mycobacteria interactions. In particular, recent data using either mutant mice expressing solely membrane TNF or specific inhibitor sparing membrane TNF demonstrated that membrane TNF is sufficient to control acute *M. tuberculosis* infection. This is opening the way to selective TNF neutralization that might retain the desired anti-inflammatory effect but reduce the infectious risk.

Tuberculosis (TB) infection is a major public health problem caused by *Mycobacterium tuberculosis* (M.tb). The present estimate is that one third of the world population harbors M.tb in a latent form (http://www.who.int), which may be reactivated when the host immune response is suppressed such as in HIV infection [1]. Only 10% of the population which has been in contact with the pathogen develop overt clinical symptoms, while roughly 90% of the infected persons contain the infection. A recent quantification of bacterial growth and death rates showed that M.tb replicates throughout the course of chronic TB infection in mice and is restrained by the host immune system [2]. Unraveling the host immune response during primary and chronic/latent infection is therefore a major challenge. Prominent mechanisms of the host leading to protective immunity controlling TB and reactivation of infection are associated with T cells, macrophages, interferon-γ (IFN-γ), TNF, interleukin-12 (IL-12), nitric oxide (NO), reactive oxygen and reactive nitrogen intermediates (RNIs), as reviewed in references [3–6]. While IL-23 and IL-17 contribute to host resistance [7], they do not seem essential to control acute TB infection [8].

Upon phagocytosis by macrophages, M.tb activates various pattern recognition receptors and stimulates the production TNF, IL-12, RNI as well as the expression of costimulatory molecules. This normally leads to activation of T and NK cells, and IFN-γ production augmenting the microbiocidal activity of the phagocytes [3, 4]. An essential role for IL-1 pathway in the control of acute M.tb infection has also been documented [9, 10], indicating that several proinflammatory cytokines produced during TB are nonredundant.

A simplified view of how M.tb activates antigen-presenting cells and induces T cell activation is depicted in figure 1. A concerted action of chemokines and cytokines leads to a focal accumulation of macrophages containing a few intracellular bacilli, which escaped the initial killing, surrounded by activated T cells forming the typical granulomas of M.tb infection [11]. T cell depletion and inhibition or neutralization of several mediators at different stages of TB infection leads to rapid disease progression, which may be accompanied by granuloma disruption, bacterial growth and dissemination, leading to death. Due to its multiple in vivo activities, excessive TNF may also cause a distinctive set of pathological effects in TB infection, including hyperinflammation, caseous necrosis and cachexia, all of which are correlated with elevated TNF levels [12, 13]. The occurrence of TB reactivation under TNF neutralizing therapy has shed new light on the role of TNF in the control of latent TB infection. This review focuses on the protective role of TNF in the immune response to M.tb infection.

TNF Family

TNF is the founder member of cytokine TNF-like superfamily [for review see 14–16]. TNF is expressed by many different cell types including macrophages, dendritic cells, CD4+ and CD8+ T cells, B cells, but also by other cells such as adipocytes, keratinocytes, mammary and colon epithelium, osteoblasts, or mast cells. TNF is first synthesized as a homotrimeric 26-kDa membrane-bound protein (tmTNF). After proteolytic cleavage by TNF-α-converting enzyme, 17-kDa soluble TNF is released. Levels of circulating TNF in healthy individuals are nearly undetectable; however, they increase substantially in pathological situations [17, 18]. Lymphotoxin-α is a member of TNF superfamily and structurally the closest TNF relative. It exists as a soluble homotrimer (LTα3) or forms a membrane-bound heterotrimeric complex with the anchor LTβ [for review see 19, 20].

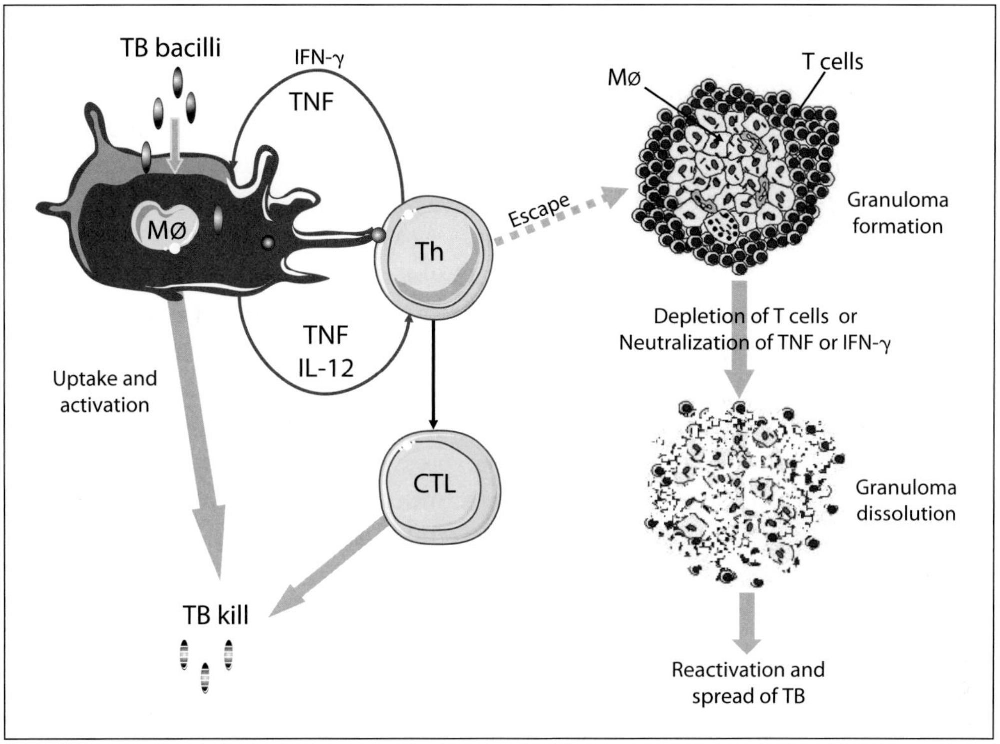

Fig. 1. Macrophage and T cell activation, killing of TB bacilli and granuloma formation. Macrophages are activated by TB bacilli and produce cytokines and T cell activation. Activated macrophages are mycobactericidal, but a few bacilli escape. The cell activation induces lymphocyte recruitment orchestrated by chemokines leading to the formation of granulomas which contain the bacilli. Antibody neutralization of TNF or IFN-γ or T cell depletion results in dissolution of the granuloma structure, rescue of surviving bacilli with dissemination of infection. Mø = Macrophage; Th = T helper; CTL = cytotoxic T lymphocyte.

TNF, LTα and LTβ genes are tightly clustered within 12 kb inside the major histocompatibility complex locus on murine chromosome 17 and human chromosome 6 [21, 22]. Membrane-bound as well as soluble TNF interact with two receptors, TNFR1 (p55 in mouse, p60 in humans, CD120a) and TNFR2 (p75/p80, CD120b). TNFR1, the high-affinity receptor for soluble TNF, is constitutively expressed in nearly all tissues and cell types. TNFR1 contains a protein module called 'death domain' which is essential for induction of apoptosis, as well as for other nonapoptotic functions [14]. The expression of TNFR2 is more restricted to lymphoid tissues [23]. Soluble LTα3 also binds and activates both TNFR1 and TNFR2, whereas membrane-bound LTαβ exerts its unique functions through the engagement of LTβR [for review see 15].

Receptor ligation initiates signals through a complex cascade to activate the nuclear factor NF-κB, JNK-AP1 and p38 signaling axis resulting in activation of TNF-dependent program of gene expression [for review see 24]. Both TNFR1 and

TNFR2 are constitutively shed in substantial amounts in vivo, and soluble TNF receptor shedding is likely to play an important role in regulating TNF activity under physiologic conditions [25]. Macrophage infection by M.tb was shown to induce release of soluble TNFR2 that formed inactive TNF-TNFR2 complexes and reduced TNF bioactivity [26]. In vivo infection with *Mycobacterium bovis* BCG was shown to upregulate soluble TNFR1 and TNFR2 release in the circulation following the release of TNF [27].

Another interesting mode of action that may account for some of TNF functions is the induction of reverse signaling through tmTNF [28]. Bidirectional or reverse signaling has been suggested for a number of transmembrane members of the TNF superfamily including CD40L [29], CD95L (FasL; [30]) and mTNF itself. Reverse signaling through transmembrane TNF has been shown to involve protein kinase C pathway and to induce E-selectin (CD62E) expression on activated human CD4+ T cells [31, 32], as well as for TNF release by monocytes [33]. The molecular mechanisms involved and intracellular pathways activated by reverse signaling remain largely uncharacterized.

Thus, tmTNF, soluble TNF and soluble LTα3 appear to mediate both overlapping and distinct physiological responses in vivo. Their relative roles in inflammatory models and in host defense have not been fully unraveled, in large part due to the limitations in physiologically relevant in vivo models. Membrane-bound TNF mediates cellular responses such as apoptosis [34], proliferation, B cell activation, and some inflammatory responses. To date, the main evidence for an in vivo role for tmTNF has come from genetically modified mice expressing uncleavable membrane-bound TNF [35–38]. While the role of TNF in controlling TB has been extensively studied using a panel of available mouse models [39–42], the role of LTα3 had to be implicated indirectly from the comparative phenotypes of mice deficient in LTα vs. LTβ or TNFR1/TNFR2 vs. TNF and therefore remained much less defined.

TNF Genetic Mouse Models

A detailed understanding of the relations between different members of the TNF family is essential to appreciate the power and the limitations of the available genetic mouse models. Various models to study the role of TNF in vivo have been developed, and the most powerful tools are transgenic and gene-deficient mice. Several transgenic and gene knock-in (KI) mice expressing either human or mouse TNF systemically or in a tissue-specific manner are available, and they are relevant to the conditions of systemic TNF overproduction or local inflammation such as arthritis, colitis or chronic CNS inflammation [43]. In TB research so far the following transgenic mice and gene knockout (KO) mice have been characterized: soluble TNFR1 transgenic, soluble TNFR2 transgenic and mice deficient for TNF/LTα, TNF, LTα, LTβ, TNF/LTβ, TNF/LTα/LTβ, LIGHT, TNF-R1 and R2 and LTβR mice [44–50]. To analyze the specific role of membrane TNF, two recent models which express a functional, normally regulated but uncleavable membrane-bound TNF were reported [37, 38], and a transgenic mouse model expressing membrane TNF in TNF/LTα KO mice [36]. Conversely, the role of the soluble TNFR1 in controlling M.tb infection can now be envisaged using mice expressing a nonsheddable p55 TNFR1 ($TNFR1^{\Delta NS}$) KI [51].

The generation of a novel panel of mice with cell-specific TNF deficiencies [52] and LTβ KO mice [53] allowed the investigation of in vivo functions of TNF or surface LTL produced by distinct cell types of the immune system such as macrophages/neutrophils or lymphocytes, adding yet another powerful tool to dissect TNF cytokine family functions in a more specific way.

Nonredundant Role of TNF to Control Mycobacterial Infection

Macrophages, DC and epithelial cells are among the first cells encountering M.tb bacilli in the airway. Phagocytosis induces the transcriptional machinery resulting in the secretion of several proinflammatory cytokines, chemokines, expression of costimulatory molecules and effector molecules including NO, which has mycobactericidal activity (fig. 1). Mycobacterial proteins are degraded and presented by class II proteins to the T cell receptor inducing clonal activation of CD4 T cells. IFN-γ derived from T cells and NK or NKT cells is a potent activator of APCs, enhancing the killing of M.tb and presentation of mycobacterial peptide to T cells. The concerted action of cytokines and chemokines leads to accumulation of activated macrophages containing a few surviving bacilli surrounded by activated T cells, which constitutes the typical mycobacterial granuloma (fig. 1). Other cell types may participate in this process and include neutrophils, eosinophils, NK, NKT and mast cells and possibly γδ-T cells [3, 4, 6, 7].

Infection with the vaccine strain *M. bovis* BCG is well controlled in normal C57Bl/6 mice. However, the control of *M. bovis* BCG infection is TNF dependent as mice treated with anti-TNF antibodies showed impaired granuloma formation and increased bacillus content [54]. Transgenic mice expressing soluble TNFR1-Fc fusion protein neutralizing TNF and LTα succumbed to BCG infection [44, 55]. Using the first available TNF-LTα double-deficient mice [56], we showed that TNF and/or LTα signaling is required to activate cell of the immune system [50]. TNF-LTα double-deficient mice display high susceptibility and succumb to BCG infection between 8 and 10 weeks. The granuloma response was severely impaired with reduced T cell recruitment and macrophages expressed reduced inducible NO synthase (NOS2), a key mediator of antibacterial defense [50]. We and others further compared the susceptibility of single TNF- and LTα-deficient mice, and showed that both single gene-deficient mice succumbed to BCG infection, suggesting that both TNF and LTα are necessary and non-redundant to control BCG infection [57]. Reintroduction of LTα as a transgene into TNF-LTα double-deficient mice prolonged survival but failed to restore resistance to BCG [57].

Although *M. bovis* BCG is an attenuated strain, the absence of TNF or TNF signaling induced a phenotype essentially similar to an infection with virulent M.tb. Indeed, mice deficient in TNF [39–42], or TNF-R1 [47], or mice treated with soluble TNFR1 or TNFR2 to neutralize TNF [44, 45, 58] have poorly formed granulomas with extensive regions of necrosis and neutrophilic infiltration of the alveoli, and an inability to control mycobacterial replication upon infection with virulent M.tb strains. Bean et al. [39] found comparable MHC class II and inducible NOS expression, serum nitrite levels, and normal activation of T cells and macrophages, while the organization of granulomas was clearly defective and not compensated by LTα. TNF was not required for granuloma formation, but rather for maintaining granuloma integrity indirectly by restricting mycobacterial growth within macrophages and preventing their necrosis in *Mycobacteruium marinum*-infected zebrafish [59]. Similarly, in a murine model of *M. bovis* BCG infection, established hepatic granuloma showed a profound decrease in size and in their population of noninfected macrophages within 2–4 days of anti-TNF treatment [60].

As observed in BCG infection studies, both TNF and LTα seemed necessary to control infection with virulent H37Rv strain of M.tb [41, 49]. However, the very close mutual proximity of genes coding for TNF, LTα and LTβ on mouse chromosome 17 raises the issue of collateral gene damage in mouse models employing targeted modifications of TNF/LT genomic locus. For example, independently generated mouse strains with TNF deficiency behave identically in a number of infection and stress models but demonstrate discrepant phenotypes with regard to the development of Peyer's patches, apparently due to differences in the configuration of the targeted locus [61]. Based on published reports, both removal of a regulatory element controlling transcription of the LT genes and their compensatory upregulation by the actively transcribed neo-resistance cassette can be envisioned. Since LT expression essential for the development of Peyer's patches has to be cell type specific and may be subject to autoregulatory feedback loops, resolution of these discrepancies proved to be a technically challenging task.

Another example of collateral gene damage, probably more relevant to TB research, is dysregulation of TNF expression in the 'conventional' LTα KO mice. Recently generated 'neo-cassette-free' $LT\alpha^{\Delta/\Delta}$ mice were fully capable of producing TNF at normal levels, whereas 'conventional' LTα KO animals displayed significant decrease in TNF synthesis in several critical types of leukocytes both in vitro and in vivo [62]. In the 'conventional' LTα KO mice, TNF deficiency could be corrected by transgenic TNF expression [63]. In agreement with the results of TNF promoter studies, the deficiency appears to be restricted to macrophages and neutrophils [62]. Defective TNF production has been noted, to various extent, by several published reports utilizing 'conventional' LTα KO mice [57, 64]. Once again, cell type-specific collateral damage to transcriptional initiation may be difficult to unambiguously discriminate from physiological mutual regulation of two closely related cytokines sharing some of their receptors. Nevertheless, any conclusions indicating an independent protective role of soluble LTα in intracellular infections based on experiments with conventional LTα KO mice should be taken with certain caution. Our unpublished data indicate that LTα might have a less essential role than anticipated for the control of acute M.tb infection, and the phenotype observed in 'conventional' LTα KO might indeed result at least in part from additional defects such as reduced TNF expression.

TNF Controls Hyper-Inflammatory Response

Although TNF has been mainly considered as a major proinflammatory cytokine, from accumulating studies it appears that TNF mediates both pro- and anti-inflammatory activities which are necessary first for a rapid recruitment of cells to infected sites and then to attenuate this process in order to limit lesions and tissue injury. Evidence was obtained with TNF deficient mice for anti-inflammatory activities of TNF in autoimmune-mediated demyelination [65]. Upon injection with heat-killed *Corynebacterium parvum*, $TNF^{-/-}$ mice showed very little response in the early phase, but later, they developed a strong lethal inflammatory response, whereas wild-type mice exhibited a prompt inflammatory response that resolved [66]. Similarly, infections with M.tb and *M. bovis* BCG in mice unable to use TNF resulted in a delay or very little response during early infection, whereas at late infection, exacerbated inflammatory reaction and disorganized granulomas were observed; [67–69] Mohan, 2001 No. 115; Guler, 2005 No. 1709; Florido, 2007 No. 1748; Zganiacz, 2004 No. 887; Flynn, 1995 No. 119; Jacobs, 2000 No. 17. This exaggerated inflammatory response following TNF deprivation was also observed when the bacterial load in infected organs was low [45]. The pathologies and tissue destruction observed in infected mice were attributed to an excess of IL-12 and IFN-γ production, suggesting that TNF acts as a negative regulator of Th1 immune responses [42]. Infection of $TNFR1^{-/-}$ mice with *Mycobacterium avium* also resulted in granuloma disintegration that was lethal and dependent on IL-12 expression in association with an excess of T cells. In this report, treatment with anti-IL-12 antibodies led to resolution of the exacerbated response in $TNFR1^{-/-}$ mice similar to that observed in wild-type mice [46]. A recent study has shown that granuloma disintegration observed in *M. avium*-infected $TNF^{-/-}$ mice was associated with upregulation of TRAIL, another member of the large TNF family of ligands [70]. These data support the anti-inflammatory in vivo role of TNF in mycobacterial infections which seems to be predominant in granuloma resolution and is associated with a TNF-regulated control of IL-12 and IFN-γ expression. Alternatively, in vitro macrophage infection with M.tb induced TNF but the TNF bioactivity was reduced due to the release of soluble TNFR2 and the formation of inactive TNF-TNFR2 complexes in an IL-10-dependent way [26]. These studies and our unpublished results suggest that TNF has a regulatory role in Th1 cytokine expression preventing a detrimental type 1 immune response. Therefore, complete absence of TNF results in an uncontrolled Th1 cytokine response.

Correcting Experimental TNF Deficiency

Multiple injection of soluble recombinant TNF systemically in vivo did not result in any improvements in sick or infected TNF KO animals or anti-TNF-treated animals [54], indicating that TNF should be present locally. We thus reconstituted TNF deficiency by infecting the TNF-deficient host with recombinant BCG expressing TNF [12]. Indeed, in TNF-deficient mice infected with low doses of BCG expressing TNF, bacillary growth was controlled, granulomas were small and well differentiated, and the mice survived, unlike TNF-deficient mice infected with the wild-type BCG [12]. Therefore, local and not systemic production of TNF at the site of infection enabled a normal response controlling infection. However, infection with high inocula of BCG-expressing TNF induced severe inflammation in the lungs and spleen and earlier death despite a more rapid bacterial clearance. The relative amount of TNF at the site of infection seems to determine whether the cytokine is protective or destructive [12]. It has since been shown that reconstitution of TNF in the host by adenoviral gene transfer improved survival of TNF-deficient mice [42].

Cell-Specific Response and TNF Control of Tuberculosis

The T cell response is critical to control mycobacterial infection. Indeed, antibody-mediated depletion of CD4 T cells in immunocompetent B6 mice leads to uncontrolled infection similar to what is observed in T cell-deficient mice [71]. Depletion of CD4 cells may also lead to reactivation of silent, chronic TB infection, despite almost normal levels of IFN-γ [72]. One of the explanations would be that the TNF produced by CD4 cells is critical for host resistance. Antigen-specific CD8 T cell responses to culture filtrate protein-10 (CFP10) were documented both in human volunteers and in M.tb-infected mice, where CFP10-specific T cells were detected as early as week 3 after infection and reached 30% of CD8 T cells in the lung with long persistence [73]. T cell subsets induced by M.tb infection include Th1 and Th17 cells, but the role of Th17 is still unclear as only the absence of Th1 cells but not of Th17 alters the protective response [74]. In vaccinated animals, however, absence of memory Th17 cells results in loss of accelerated memory Th1 response and protection [8]. Thus, Th1 and Th17 responses seem to cross-regulate each other during mycobacterial infection [8, 74].

Other cells involved in the control of infection, include γδ-T cells and NK cells. γδ-T cells, are recruited into the lung [75] and produce large amounts of IL-17 and may contribute to the host protection [7, 76]. NK cells are associated with early resistance against intracellular pathogens and potent producers of IFN-γ. Aerosol M.tb infection increased NK cell recruitment and activation, and IFN-γ secretion. However, in vivo depletion of NK cells using a lytic antibody had no influence on M.tb clearance. Therefore, NK cells appear to have a minimal role in the host resistance to M.tb [77].

By contrast, NKT cells may play a role in M.tb infection control. Activation of NKT cells by α-galactosylceramide in vivo augmented host resistance to M.tb in mice, which may also be mediated in part by the production of IFN-γ [78]. Moreover, activation of CD1 restricted human T cells increased killing, probably via granulolysin [79]. Indeed, some mycobacterial antigens can be presented to NKT cells in a context of CD1 nonclassical MHC, including mycobacterial PIM [80, 81], thereby mediating interaction of NKT cell with infected cells.

Mast cells are abundant in the lung and interact directly with a wide variety of infectious agents, including M.tb, triggering the release of histamine and β-hexosaminidase, TNF and IL-6, the latter being critically involved in antimycobacterial resistance. M.tb appears to interact with CD48 on mast cells inducing histamine release, which is inhibited by anti-CD48 antibodies. Therefore, M.tb and its antigens recognize and activate mast cells [82]. Recent studies using mast cell degranulation revealed reduced M.tb induced inflammation and reduce host resistance [83]. Further investigations in mast cell-deficient mice are necessary to define the role of mast cells in host response to M.tb infection.

Using mixed radiation bone marrow chimera, we demonstrated that TNF derived from hematopoietic cells rather than stromal cells of mesenchymal origin are essential for a normal host response to BCG infection [84].

Further, using T cell vs. macrophage/neutrophil-specific TNF-deficient mice we are currently analyzing the relative contribution of TNF originating from the different cell types in the control of M.tb infection.

Molecular Mechanisms of Mycobacterial Killing/Resistance

Activation of macrophages and dendritic cells by M.tb induces several proinflammatory cytokines including TNF, LTα and IL-12, and expression of costimulatory molecules that enhance antigen presentation and activation of T cells. Activated T cells produce TNF, IFN-γ and LTα inducing further activation of macrophage and likely other cells including stromal cells. Activated macrophages express NOS2, producing NO and RNIs, which are critical for killing and inhibiting growth of virulent M.tb and BCG [27, 85, 86].

Mycobacteria may inhibit phagosome maturation and fusion with lysosomes, thereby escaping killing [87–90]. Activated macrophages recruit T cells to form granulomas, which contain bacterial growth. The granuloma is a dynamic structure, which requires a permanent signal from activated T cells and macrophages [91]. Any perturbation of this signaling such as neutralization of TNF causes dissolution of granulomas [54] and allows reactivation and spread of infection (fig. 1). Activated T cells not only provide help, but acquire cytotoxic functions, which eradicate bacilli, although the relative contribution of CD4 versus CD8 cells to control TB infection is not fully established.

In order to better understand the effect of TNF on intracellular replication of mycobacteria, we investigated the growth of the vaccine strain BCG in TNF-deficient macrophages. BCG infection resulted in logarithmic growth of the intracellular bacilli, while recombinant BCG-expressing TNF led to bacillary killing associated with production of NO. Therefore, TNF contributes to the expression of NOS2 and to bacterial growth inhibition indirectly [92].

IFN-γ has been shown to be an essential component of immunity to TB. It activates infected host macrophages to directly inhibit the replication of M.tb [3]. Although IFN-γ-inducible NOS2 is considered the principal effector mechanism, other pathways exist. M.tb has developed several mechanisms to escape eradication including inhibition of phagosome maturation [93]. Mycobacteria blocking Ca^{2+} signaling and phagosome maturation in human macrophages or inhibiting sphingosine kinase may allow the escape from eradication in the phagocyte [94–96]. Role of autophagy and ensuing inhibition of phagolysosome formation [97] may be considered, as well as Coronin-1 inhibition as an alternative pathway to prevent phagosome maturation [98]. Autophagy can be induced by IFN-γ and by the immunity-related guanosine triphosphatases (GTPases) or LRG-47 (Irgm-1) which is a member of the IFN-γ-induced 47-kDa GTPase family [99]. LRG-47/Irgm-1 has been linked to autophagosome and autolysosome formation and killing of mycobacteria [100, 101].

Defensins such as cathelicidin (LL37) have an important anti-mycobacterium activity in human macrophages. Liu et al. [102] have reported that activation of Toll-like receptors (TLRs) upregulates the expression of the vitamin D receptor and the vitamin D-1-hydrolase generating $1{,}25(OD)_2D3$, the active form of vitamin D, and leading to induction of the microbicidal peptide cathelicidin and killing of intracellular M.tb in human macrophages. Granulysin contained in CD8 T and NK T cell granules as well as the perforin/granzyme system contribute to elimination of infected macrophages and mycobacteria [79, 103]. Exogenous ATP induces the killing of intracellular M.tb and *M. bovis* BCG in macrophages and involves the purinergic P2X7 receptor which is regulated by IFN-γ [104–107].

Mycobacteria induce apoptosis of macrophages and cause the release of apoptotic vesicles that carry mycobacterial antigens to uninfected antigen-presenting cells, including dendritic cells which are indispensable for subsequent antigen cross-presentation through MHC-I and CD1b. This new pathway for presentation of antigens from a phagosome-contained pathogen illustrated the functional significance of infection-induced apoptosis in the activation of CD8 T cells specific for both protein and glycolipid antigens in TB [108].

Induction of TNF and other proinflammatory cytokines is mediated through several mycobacterial motives triggering different pattern recognition receptors, including TLR2, TLR4 or TLR9. However, while the control of acute TB infection was severely compromised in the absence of MyD88 [109, 110], TLR2, TLR4 and/or TLR9 do not seem essential for the control of acute TB infection but may interfere in the control of chronic infection [111–113]. MyD88 pathway may thus contribute rather through IL-1R signaling to control acute TB [9]. TLR/MyD88-dependent signaling is also required for phagosome maturation [114].

In summary, TNF participates in resistance to mycobacteria in the following ways: (1) activation of macrophages, (2) induction of chemokines and cell recruitment, (3) activation of T cells, (4) killing by macrophages, T and other cells and (5) regulation of apoptosis and signals from TLR/MyD88/IL-1R pathway that contribute to the host response. Since separating the effects of these different TNF functions in vivo is presently difficult or impossible, a computational model was applied to understand specific roles of TNF in control of TB in a single granuloma. The model predicted that macrophage activation is a key effector mechanism for controlling bacterial growth within the granuloma, TNF and bacterial numbers represent strong contributing factors to granuloma structure, and TNF-dependent apoptosis may reduce inflammation at the cost of impaired mycobacterial clearance [115].

Role of Other Members of the TNF Family

In order to dissect the respective roles of soluble LTα3 and membrane-bound LTα-LTβ in the host response to aerosol M.tb infection, and to avoid the complication presented by the absence of secondary lymphoid organs in LT-deficient mice, Roach et al. [49] prepared bone marrow chimeric mice. LTα-deficient chimeras, which lack both secreted LTα3 and membrane-bound LTαLTβ, were highly susceptible and succumbed 5 weeks after infection, while LTβ-deficient chimeras, which lacked only the membrane-bound LTβ, controlled the infection similar to wild-type chimeric mice. T cell responses to mycobacterial antigens and macrophage responses in LTα-deficient chimeras were equivalent to those of wild-type chimeras, but granuloma formation was abnormal with perivascular and peribronchial location of T cells. These studies thus suggested that secreted LTα3 is essential for the organization of functional granulomas and control of pulmonary TB [49]. These data should probably be reevaluated now, in light of the potential defective TNF expression in these 'conventional' KO, as opposed to the more recent 'neo-cassette-free' LTα KO (see above). The role of the LTαβ-LTβR pathway in the control of mycobacterial infection was further studied. Treatment of BCG-infected mice with LTβR-Ig resulted in reduction of iNOS activity and increased bacterial growth [116]. Mice deficient in either LTα or LTβ, which form the LTαβ heterotrimeric ligand of LTβR, showed reduced resistance to M.tb infection, and LTβR KO had increased bacterial load with widespread pulmonary necrosis 35 days after infection, although they expressed normal levels of TNF and IFN-γ and recruited similar numbers of T cells in the pulmonary granulomatous lesions as compared to wild-type mice [117]. Furthermore, inhibition of the LTαβ pathway with soluble LTβR-Fc fusion proteins also compromised immunity against mycobacterial infections [116]. By contrast, LIGHT-deficient mice proved to be resistant to M.tb infection [117]. In conclusion, several members of the TNF family are critically involved in the host response to M.tb infection.

Membrane TNF Biological Activity Controls Acute *Mycobacterium tuberculosis* Infection

Although a key role of TNF in controlling intracellular bacterial infections is uncontested, it is only recently that the specific function of membrane TNF has been appreciated. Membrane TNF is cleaved by the metalloproteinase-disintegrin TNF-α-converting enzyme [118] into secreted, soluble trimeric TNF. Several functions of membrane TNF have been described, such as cytotoxicity, polyclonal activation of B cells, induction of IL-10 by monocytes, ICAM-1 expression on endothelial cells and liver toxicity [34, 37, 119, 120]. The transgenic expression of membrane TNF suggested an in vivo role of membrane TNF [35]. Olleros et al. [67, 69] investigated the resistance to mycobacterial infection in transgenic mice expressing a membrane TNF (Δ12-10; Δ-2-+1; one substitution +11) under the control of proximal TNF promoter and on a TNF-LTα-deficient background. In this model, membrane, TNF had a fully protective effect against *M. bovis* BCG but only partial protective effect against M.tb infection. The recent generation of mice with functional, normally regulated and expressed membrane-bound TNF represents a major advance and allowed interesting insights into the role of membrane TNF in lymphoid structure development and inflammation. KI mice expressing the uncleavable KΔ1-9, K11E TNF [37] and TNF-deficient mice [66] were compared in their resistance to mycobacterial infection. As previously reported for membrane TNF transgenic mice, we and others demonstrated that membrane TNF has important biological functions and substitutes soluble TNF to a large extent [121–124]. Membrane TNF KI mice survived a M.tb aerosol infection for 3 months, were able to recruit and activate macrophages and T cells, generate granuloma and partially control mycobacterial infection in the early stage, unlike complete TNF-deficient mice [122, 123]. However, during the chronic phase of infection, membrane TNF KI mice demonstrated reduced bacterial clearance and succumbed to infection [122]. In another model of targeted mutagenesis in mice, the shedding of membrane TNF was prevented by deleting its cleavage site [38]. Mice expressing noncleavable and regulated Δ1-12 TNF allele partially controlled *M. bovis* BCG infection, with recruitment of activated T cells and macrophages and granuloma formation, while mice with complete TNF deficiency succumbed [125]. It was confirmed that membrane TNF conferred partial protection against virulent *M.tb* infection, and intercrossing these mice with TNF-R1 or TNF-R2 KO mice showed that tmTNFxTNFR2 KO mice were very sensitive, essentially as much as TNF KO mice, while tmTNFxTNFR1 KO mice behaved more like tmTNF mice, suggesting that the protective effect of membrane TNF against acute *M.tb* infection is mediated through TNF-R2 signaling [125].

Therefore, data from the genetic mouse models suggest that membrane-expressed TNF is sufficient and soluble TNF may be dispensable to control the first phase of acute TB infection. However, during the chronic phase membrane TNF alone is not sufficient and soluble TNF seems to be required to control chronic TB infection. The reason for the progressive loss of infectious control is unclear. As previously discussed, soluble TNF may be required to negatively control the Th1 type cytokines. This TNF function may become important during the chronic phase of infection by regulating excess production of IL-12 and IFN-γ by DC and T cells.

TNF in Reactivation of Tuberculosis Infection

Clinical TB in humans may be due to a primary infection or reactivation of latent controlled infection. Secondary immunosuppression due to HIV/AIDS is the most common cause of M.tb reactivation. In the recent years, over a million patients received TNF-neutralizing therapy for the treatment of severe rheumatoid arthritis, Crohn's disease or severe psoriasis. The most common complication of TNF blockade has been the emergence of opportunistic infection and TB. Both reactivation of latent TB and increased susceptibility to new TB in patients without a clinical history of active TB infection was observed [126]. In some patients, TNF-neutralizing antibody, infliximab, or soluble TNFR2-IgG1 Fc fusion protein, etanercept, yielded reactivation of latent TB within 12 weeks and overt clinical disease [127–129], often with extrapulmonary disease manifestations (disseminated infection in lymph node, peritoneum and pleura). The frequency of TB in association with infliximab therapy was higher than the reported frequency of other opportunistic infections associated with this drug [129]. Reactivation of latent TB and primary infection in patients treated with TNF inhibitors are still difficult to clearly define in many cases. Anti-TNF antibody may be more associated with latent TB reactivation than etanercept. The majority of etanercept-associated cases of TB appeared late (90% after 90 days of treatment), suggesting that these cases may have occurred as a result of the inability to control new M.tb infection while 43% of infliximab associated cases of TB occurred during the first 90 days of treatment, indicating that they likely represent reactivation of latent infection [126, 130]. The reactivation of latent TB under TNF-blocking therapy indicates that the normal immune system is able to control, but not able to eradicate, a primary infection, and that TNF plays a role in the long-term containment of residual M.tb in tissues.

In order to study the factors leading to reactivation of chronic or chemotherapy-controlled latent infection, several experimental models have been developed [131]. In the Cornell model, after an intravenous administration of M.tb H37Rv and treatment with pyrazinamide and isoniazide for 12 weeks, mice appear to have cleared the bacilli from organs, but a substantial proportion of animals spontaneously reactivate with acute disease upon cessation of chemotherapy. Since the original publication of the Cornell model, a few variations have been reported [132, 133]. In the low-dose model, infection is exclusively controlled by the host in the absence of chemotherapy [133]. Although considered to better reflect the human host response, bacterial numbers in the organs of these mice remain high during the chronic persistent

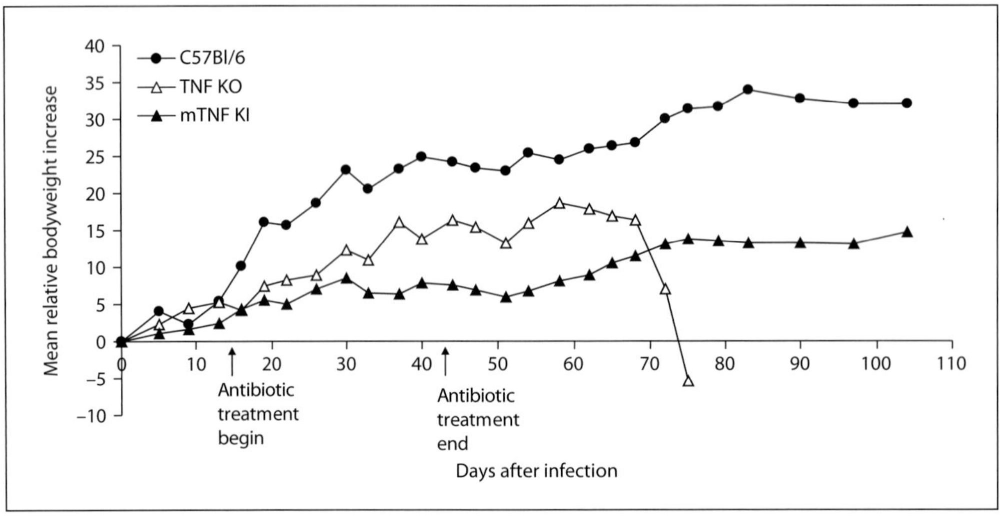

Fig. 2. Comparison of susceptibility of membrane TNF KI and conventional TNF KO reactivating chronic M.tb infection. Wild-type, TNF KO and mTNF KI mice were infected with M.tb (ca 100 CFU i.n.) and treated for 4 weeks with rifampicin and isoniazide (days 14–42) to control the infection. TNF KO mice started to die 6 weeks after the end of the antibiotic treatment, while all mTNF KI mice survived as wild-type mice. In parallel groups infected with M.tb but not treated with antibiotics, TNF KO had to be euthanized at 3 weeks, while 6 out of 8 mTNF KI survived with no marked bodyweight loss.

phase of infection. To date, these models have yielded significant information on the immune effector mechanisms participating in latent or chronic persistent and reactivated TB.

We established the first aerosol infection model of drug-induced latent and reactivated murine TB using rifampicin and isoniazide [132, 134]. In this model, latency was defined as almost undetectable levels of bacilli in mouse organs for a prolonged period of time. Reactivation of infection could be achieved by inhibiting NOS activity by aminoguanidine [132]. Using this model, we showed that a 4 weeks' rifampicin and isoniazide administration cleared infection as assessed by viable bacterial accounts in the organs in both wild-type and TNF-deficient mice. Upon cessation of therapy, massive spontaneous reactivation of M.tb infection occurred within several weeks in TNF-deficient mice with necrotic pneumonia and death, while wild-type mice displayed mild subclinical reactivation [134]. This model allows us to study the role of TNF neutralization in a reactivating infection in the presence of an established specific adaptive immune response.

The role of soluble versus membrane TNF was then studied in this model (fig. 2 and unpubl. data). Although TNF KO mice rapidly lost weight and had to be terminated within 6 weeks after the end of the antibiotic treatment with uncontrolled infection, mTNF KI mice survived as wild-type mice. Therefore, membrane TNF suffices to provide some control of the M.tb infection after reduction of the bacterial burden by an antibiotic treatment, while complete absence of TNF results in rapid progression of the infection.

Pharmacological TNF Neutralization and Tuberculosis Control

The experimental models of TB reactivation described above allow us to test the potential risk of diverse TNF-neutralizing therapies to induce reactivation of TB. Administration of neutralizing TNF antibody but not of soluble TNF receptor was able to reactivate experimental latent infection [135]. TNF neutralization resulted in marked disorganization of the tuberculous granuloma and enhanced expression of specific proinflammatory molecules [68]. A computational approach suggested that TNF bioavailability following anti-TNF therapy is the primary factor for causing reactivation of latent infection and that even very low level of soluble TNF is essential for infection control [136].

Novel approaches to experimentally block soluble TNF are being tested in murine models of TB. One approach is to compete for natural TNF by the use of dominant negative mutant TNF (DN-TNF; see fig. 3) reported to block soluble TNF while sparing membrane TNF [137]. In vivo, DN-TNF attenuated arthritis without suppressing innate immunity to *Listeria monocytogenes* [138]. Similarly, DN-TNF protected mice from acute liver inflammation, without compromising host control of *M. bovis* BCG and M.tb infections [139]. This was in contrast to TNFR2-IgG1 etanercept that inhibits murine soluble and membrane TNF as well as LTα, which severely compromised the host response to M.tb infection [139]. Another novel approach is an active immunization selectively targeting soluble TNF. Vaccination with a virus-like particle linked to a TNF N-terminal peptide resulted in high titers of autoantibodies against soluble TNF. It protected mice from arthritis without inducing reactivation of latent TB [140], while immunization against the entire TNF molecule yielded enhanced reactivation of latent TB. This difference was attributed to recognition of only soluble TNF vs. recognition of both transmembrane and soluble TNF by the elicited antibodies. Thus, specifically targeting soluble TNF has the potential to be effective against inflammatory disorders while overcoming the risk of opportunistic infections known to be associated with the currently available TNF antagonists.

Conclusions and Perspectives

TNF is an essential mediator for the integrity of microbiocidal granulomas and the control of M.tb infection. Experimental TB infection of gene-deficient mice has demonstrated the nonredundant contribution of several proinflammatory

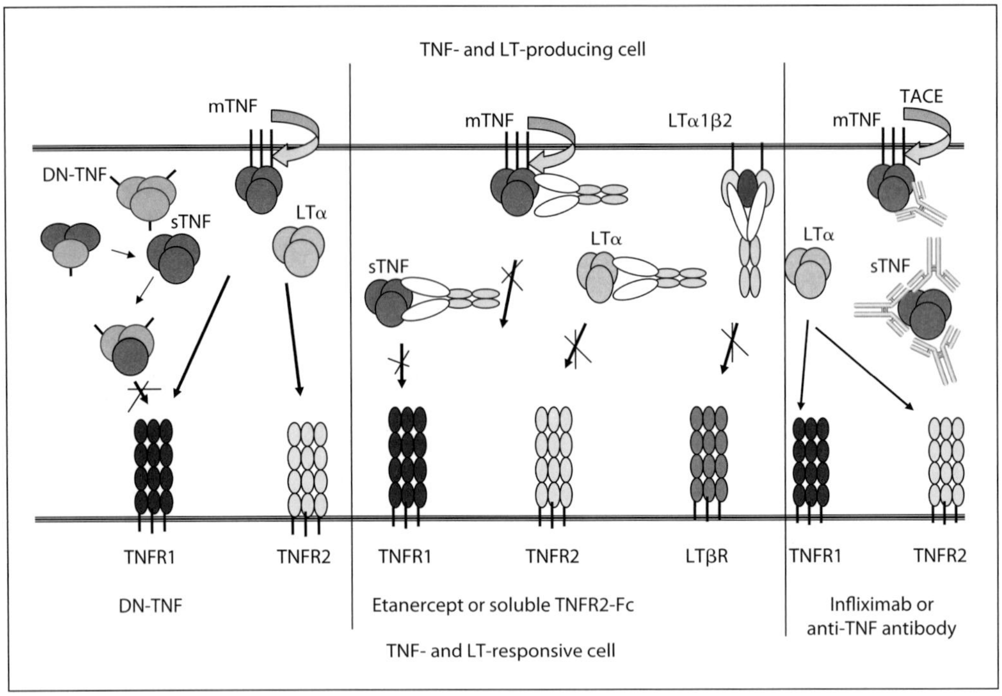

Fig. 3. Mechanisms of action of dominant-negative TNF (DN-TNF) biologics, soluble TNFR2-Fc (etanercept), or anti-TNF antibodies. Left: DN-TNF, a mutated form of human solTNF with disrupted receptor-binding interfaces eliminates solTNF by a subunit exchange mechanism, but is unable to interact with tmTNF and LTα. Center: solTNF, tmTNF, LTα and LTαβ can be neutralized by etanercept, inhibiting interaction with the corresponding receptors. Right: Monoclonal antibodies directed against human TNF neutralize both membrane-bound and soluble TNF, while sparing LTα. Thus, DN-TNF (XENP1595) inhibits solTNF receptor signaling without suppressing tmTNF- or LTα responses to TNFR1 and TNFR2, mediating inflammatory and immune responses.

cytokines such as TNF, IL-12, IFN-γ or IL-1 to the host response to M.tb infection [9, 131]. An important notion is the fact that latent mycobacterial infection can be reactivated by TNF neutralization. The finding that membrane TNF confers partial protection and abrogates the hyperinflammatory syndrome is significant. Sparing membrane TNF in neutralizing TNF therapy used in rheumatic arthritis or Crohn's disease may diminish the infectious complications and reactivation of latent TB infection.

Acknowledgement

The work was supported by the European FP6 grant TB REACT Contract No. 028190.

References

1 Dye C, Watt CJ, Bleed DM, Hosseini SM, Raviglione MC: Evolution of tuberculosis control and prospects for reducing tuberculosis incidence, prevalence, and deaths globally. JAMA 2005;293:2767–2775.

2 Gill WP, Harik NS, Whiddon MR, Liao RP, Mittler JE, Sherman DR: A replication clock for *Mycobacterium tuberculosis*. Nat Med 2009;15:211–214.

3 Flynn JL, Chan J: Immunology of tuberculosis. Annu Rev Immunol 2001;19:93–129.

4 North RJ, Jung YJ: Immunity to tuberculosis. Annu Rev Immunol 2004;22:599–623.

5 Flynn JL: Immunology of tuberculosis and implications in vaccine development. Tuberculosis (Edinb) 2004;84:93–101.

6 Cooper AM: Cell-mediated immune responses in tuberculosis. Annu Rev Immunol 2009;27:393–422.

7 Umemura M, Yahagi A, Hamada S, Begum MD, Watanabe H, Kawakami K, Suda T, Sudo K, Nakae S, Iwakura Y, Matsuzaki G: IL-17-mediated regulation of innate and acquired immune response against pulmonary *Mycobacterium bovis* bacille Calmette-Guerin infection. J Immunol 2007;178:3786–3796.

8 Khader SA, Bell GK, Pearl JE, Fountain JJ, Rangel-Moreno J, Cilley GE, Shen F, Eaton SM, Gaffen SL, Swain SL, Locksley RM, Haynes L, Randall TD, Cooper AM: IL-23 and IL-17 in the establishment of protective pulmonary CD4+ T cell responses after vaccination and during *Mycobacterium tuberculosis* challenge. Nat Immunol 2007;8:369–377.

9 Fremond CM, Togbe D, Doz E, Rose S, Vasseur V, Maillet I, Jacobs M, Ryffel B, Quesniaux VF: IL-1 receptor-mediated signal is an essential component of MyD88-dependent innate response to *Mycobacterium tuberculosis* infection. J Immunol 2007;179:1178–1189.

10 Master SS, Rampini SK, Davis AS, Keller C, Ehlers S, Springer B, Timmins GS, Sander P, Deretic V: *Mycobacterium tuberculosis* prevents inflammasome activation. Cell Host Microbe 2008;3:224–232.

11 Russell DG: Who puts the tubercle in tuberculosis? Nat Rev Microbiol 2007;5:39–47.

12 Bekker LG, Moreira AL, Bergtold A, Freeman S, Ryffel B, Kaplan G: Immunopathologic effects of tumor necrosis factor alpha in murine mycobacterial infection are dose dependent. Infect Immun 2000;68:6954–6961.

13 Ehlers S, Benini J, Held HD, Roeck C, Alber G, Uhlig S: Alphabeta T cell receptor-positive cells and interferon-gamma, but not inducible nitric oxide synthase, are critical for granuloma necrosis in a mouse model of mycobacteria-induced pulmonary immunopathology. J Exp Med 2001;194:1847–1859.

14 Locksley RM, Killeen N, Lenardo MJ: The TNF and TNF receptor superfamilies: integrating mammalian biology. Cell 2001;104:487–501.

15 Ware CF: Network communications: lymphotoxins, LIGHT, and TNF. Annu Rev Immunol 2005;23:787–819.

16 Watts TH: TNF/TNFR family members in costimulation of T cell responses. Annu Rev Immunol 2005;23:23–68.

17 Spahr L, Giostra E, Frossard JL, Bresson-Hadni S, Rubbia-Brandt L, Hadengue A: Soluble TNF-R1, but not tumor necrosis factor alpha, predicts the 3-month mortality in patients with alcoholic hepatitis. J Hepatol 2004;41:229–234.

18 Edrees AF, Misra SN, Abdou NI: Anti-tumor necrosis factor (TNF) therapy in rheumatoid arthritis: correlation of TNF-alpha serum level with clinical response and benefit from changing dose or frequency of infliximab infusions. Clin Exp Rheumatol 2005;23:469–474.

19 Gommerman JL, Browning JL: Lymphotoxin/light, lymphoid microenvironments and autoimmune disease. Nat Rev Immunol 2003;3:642–655.

20 Fu YX, Chaplin DD: Development and maturation of secondary lymphoid tissues. Annu Rev Immunol 1999;17:399–433.

21 Spies T, Morton CC, Nedospasov SA, Fiers W, Pious D, Strominger JL: Genes for the tumor necrosis factors alpha and beta are linked to the human major histocompatibility complex. Proc Natl Acad Sci USA 1986;83:8699–8702.

22 Nedospasov SA, Hirt B, Shakhov AN, Dobrynin VN, Kawashima E, Accolla RS, Jongeneel CV: The genes for tumor necrosis factor (TNF-alpha) and lymphotoxin (TNF-beta) are tandemly arranged on chromosome 17 of the mouse. Nucleic Acids Res 1986;14:7713–7725.

23 Chan KF, Siegel MR, Lenardo JM: Signaling by the TNF receptor superfamily and T cell homeostasis. Immunity 2000;13:419–422.

24 Grivennikov SI, Kuprash DV, Liu ZG, Nedospasov SA: Intracellular signals and events activated by cytokines of the tumor necrosis factor superfamily: from simple paradigms to complex mechanisms. Int Rev Cytol 2006;252:129–161.

25 Pinckard JK, Sheehan KC, Arthur CD, Schreiber RD: Constitutive shedding of both p55 and p75 murine TNF receptors in vivo. J Immunol 1997;158:3869–3873.

26 Balcewicz-Sablinska MK, Keane J, Kornfeld H, Remold HG: Pathogenic *Mycobacterium tuberculosis* evades apoptosis of host macrophages by release of TNF-R2, resulting in inactivation of TNF-alpha. J Immunol 1998;161:2636–2641.

27 Garcia I, Guler R, Vesin D, Olleros ML, Vassalli P, Chvatchko Y, Jacobs M, Ryffel B: Lethal *Mycobacterium bovis* bacillus Calmette Guerin infection in nitric oxide synthase 2-deficient mice: cell-mediated immunity requires nitric oxide synthase 2. Lab Invest 2000;80:1385–1397.
28 Eissner G, Kolch W, Scheurich P: Ligands working as receptors: reverse signaling by members of the TNF superfamily enhance the plasticity of the immune system. Cytokine Growth Factor Rev 2004; 15:353–366.
29 van Essen D, Kikutani H, Gray D: CD40 ligand-transduced co-stimulation of T cells in the development of helper function. Nature 1995;378:620–623.
30 Suzuki I, Fink PJ: Maximal proliferation of cytotoxic T lymphocytes requires reverse signaling through Fas ligand. J Exp Med 1998;187:123–128.
31 Kirchner S, Boldt S, Kolch W, Haffner S, Kazak S, Janosch P, Holler E, Andreesen R, Eissner G: LPS resistance in monocytic cells caused by reverse signaling through transmembrane TNF (mTNF) is mediated by the MAPK/ERK pathway. J Leukoc Biol 2004;75:324–331.
32 Harashima S, Horiuchi T, Hatta N, Morita C, Higuchi M, Sawabe T, Tsukamoto H, Tahira T, Hayashi K, Fujita S, Niho Y: Outside-to-inside signal through the membrane TNF-alpha induces E-selectin (CD62E) expression on activated human CD4+ T cells. J Immunol 2001;166:130–136.
33 Rossol M, Meusch U, Pierer M, Kaltenhauser S, Hantzschel H, Hauschildt S, Wagner U: Interaction between transmembrane TNF and TNFR1/2 mediates the activation of monocytes by contact with T cells. J Immunol 2007;179:4239–4248.
34 Kriegler M, Perez C, DeFay K, Albert I, Lu SD: A novel form of TNF/cachectin is a cell surface cytotoxic transmembrane protein: ramifications for the complex physiology of TNF. Cell 1988;53:45–53.
35 Akassoglou K, Probert L, Kontogeorgos G, Kollias G: Astrocyte-specific but not neuron-specific transmembrane TNF triggers inflammation and degeneration in the central nervous system of transgenic mice. J Immunol 1997;158:438–445.
36 Mueller C, Corazza N, Trachsel-Loseth S, Eugster H-P, Buhler-Jungo M, Brunner T, Imboden MA: Noncleavable transmembrane mouse tumor necrosis factor-alpha (TNFalpha) mediates effects distinct from those of wild-type TNFalpha in vitro and in vivo. J. Biol. Chem. 1999;274:38112–38118.
37 Ruuls SR, Hoek RM, Ngo VN, McNeil T, Lucian LA, Janatpour MJ, Korner H, Scheerens H, Hessel EM, Cyster JG, McEvoy LM, Sedgwick JD: Membrane-bound TNF supports secondary lymphoid organ structure but is subservient to secreted TNF in driving autoimmune inflammation. Immunity 2001;15: 533–543.
38 Alexopoulou L, Kranidioti K, Xanthoulea S, Denis M, Kotanidou A, Douni E, Blackshear PJ, Kontoyiannis DL, Kollias G: Transmembrane TNF protects mutant mice against intracellular bacterial infections, chronic inflammation and autoimmunity. Eur J Immunol 2006;36:2768–2780.
39 Bean AG, Roach DR, Briscoe H, France MP, Korner H, Sedgwick JD, Britton WJ: Structural deficiencies in granuloma formation in TNF gene-targeted mice underlie the heightened susceptibility to aerosol *Mycobacterium tuberculosis* infection, which is not compensated for by lymphotoxin. J Immunol 1999; 162:3504–3511.
40 Kaneko H, Yamada H, Mizuno S, Udagawa T, Kazumi Y, Sekikawa K, Sugawara I: Role of tumor necrosis factor-alpha in Mycobacterium-induced granuloma formation in tumor necrosis factor-alpha-deficient mice. Lab Invest 1999;79:379–386.
41 Roach DR, Bean AG, Demangel C, France MP, Briscoe H, Britton WJ: TNF regulates chemokine induction essential for cell recruitment, granuloma formation, and clearance of mycobacterial infection. J Immunol 2002;168:4620–4627.
42 Zganiacz A, Santosuosso M, Wang J, Yang T, Chen L, Anzulovic M, Alexander S, Gicquel B, Wan Y, Bramson J, Inman M, Xing Z: TNF-alpha is a critical negative regulator of type 1 immune activation during intracellular bacterial infection. J Clin Invest 2004;113:401–413.
43 Douni E, Alexiou M, Kollias G: Genetic engineering in the mouse: tuning TNF/TNFR expression. Methods Mol Med 2004;98:137–170.
44 Garcia I, Miyazaki Y, Marchal G, Lesslauer W, Vassalli P: High sensitivity of transgenic mice expressing soluble TNFR1 fusion protein to mycobacterial infections: synergistic action of TNF and IFN-gamma in the differentiation of protective granulomas. Eur J Immunol 1997;27:3182–3190.
45 Smith S, Liggitt D, Jeromsky E, Tan X, Skerrett SJ, Wilson CB: Local role for tumor necrosis factor alpha in the pulmonary inflammatory response to *Mycobacterium tuberculosis* infection. Infect Immun 2002;70:2082–2089.
46 Ehlers S, Kutsch S, Ehlers EM, Benini J, Pfeffer K: Lethal granuloma disintegration in mycobacteria-infected TNFRp55$^{-/-}$ mice is dependent on T cells and IL-12. J Immunol 2000;165:483–492.
47 Flynn JL, Goldstein MM, Chan J, Triebold KJ, Pfeffer K, Lowenstein CJ, Schreiber R, Mak TW, Bloom BR: Tumor necrosis factor-alpha is required in the protective immune response against *Mycobacterium tuberculosis* in mice. Immunity 1995;2: 561–572.

48 Pfeffer K, Matsuyama T, Kundig TM, Wakeham A, Kishihara K, Shahinian A, Wiegmann K, Ohashi PS, Kronke M, Mak TW: Mice deficient for the 55 kD tumor necrosis factor receptor are resistant to endotoxic shock, yet succumb to *L. monocytogenes* infection. Cell 1993;73:457–467.
49 Roach DR, Briscoe H, Saunders B, France MP, Riminton S, Britton WJ: Secreted lymphotoxin-alpha is essential for the control of an intracellular bacterial infection. J Exp Med 2001;193:239–246.
50 Jacobs M, Brown N, Allie N, Ryffel B: Fatal *Mycobacterium bovis* BCG infection in TNF-LT-alpha-deficient mice. Clin Immunol 2000;94:192–199.
51 Xanthoulea S, Pasparakis M, Kousteni S, Brakebusch C, Wallach D, Bauer J, Lassmann H, Kollias G: Tumor necrosis factor (TNF) receptor shedding controls thresholds of innate immune activation that balance opposing TNF functions in infectious and inflammatory diseases. J Exp Med 2004;200:367–376.
52 Grivennikov SI, Tumanov AV, Liepinsh DJ, Kruglov AA, Marakusha BI, Shakhov AN, Murakami T, Drutskaya LN, Forster I, Clausen BE, Tessarollo L, Ryffel B, Kuprash DV, Nedospasov SA: Distinct and nonredundant in vivo functions of TNF produced by t cells and macrophages/neutrophils: protective and deleterious effects. Immunity 2005;22:93–104.
53 Tumanov A, Kuprash D, Lagarkova M, Grivennikov S, Abe K, Shakhov A, Drutskaya L, Stewart C, Chervonsky A, Nedospasov S: Distinct role of surface lymphotoxin expressed by B cells in the organization of secondary lymphoid tissues. Immunity 2002;17:239–250.
54 Kindler V, Sappino AP, Grau GE, Piguet PF, Vassalli P: The inducing role of tumor necrosis factor in the development of bactericidal granulomas during BCG infection. Cell 1989;56:731–740.
55 Guler R, Olleros ML, Vesin D, Parapanov R, Garcia I: Differential effects of total and partial neutralization of tumor necrosis factor on cell-mediated immunity to *Mycobacterium bovis* BCG infection. Infect Immun 2005;73:3668–3676.
56 Eugster HP, Muller M, Karrer U, Car BD, Schnyder B, Eng VM, Woerly G, Le Hir M, di Padova F, Aguet M, Zinkernagel R, Bluethmann H, Ryffel B: Multiple immune abnormalities in tumor necrosis factor and lymphotoxin-alpha double-deficient mice. Int Immunol 1996;8:23–36.
57 Bopst M, Garcia I, Guler R, Olleros ML, Rulicke T, Muller M, Wyss S, Frei K, Le Hir M, Eugster HP: Differential effects of TNF and LTalpha in the host defense against *M. bovis* BCG. Eur J Immunol 2001;31:1935–1943.
58 Adams LB, Mason CM, Kolls JK, Scollard D, Krahenbuhl JL, Nelson S: Exacerbation of acute and chronic murine tuberculosis by administration of a tumor necrosis factor receptor-expressing adenovirus. J Infect Dis 1995;171:400–405.
59 Clay H, Volkman HE, Ramakrishnan L: Tumor necrosis factor signaling mediates resistance to mycobacteria by inhibiting bacterial growth and macrophage death. Immunity 2008;29:283–294.
60 Egen JG, Rothfuchs AG, Feng CG, Winter N, Sher A, Germain RN: Macrophage and T cell dynamics during the development and disintegration of mycobacterial granulomas. Immunity 2008;28:271–284.
61 Kuprash DV, Tumanov AV, Liepinsh DJ, Koroleva EP, Drutskaya MS, Kruglov AA, Shakhov AN, Southon E, Murphy WJ, Tessarollo L, Grivennikov SI, Nedospasov SA: Novel tumor necrosis factor-knockout mice that lack Peyer's patches. Eur J Immunol 2005;35:1592–1600.
62 Liepinsh DJ, Grivennikov SI, Klarmann KD, Lagarkova MA, Drutskaya MS, Lockett SJ, Tessarollo L, McAuliffe M, Keller JR, Kuprash DV, Nedospasov SA: Novel lymphotoxin alpha (LTalpha) knockout mice with unperturbed tumor necrosis factor expression: reassessing LTalpha biological functions. Mol Cell Biol 2006;26:4214–4225.
63 Alexopoulou L, Pasparakis M, Kollias G: Complementation of lymphotoxin alpha knockout mice with tumor necrosis factor-expressing transgenes rectifies defective splenic structure and function. J Exp Med 1998;188:745–754.
64 Schluter D, Kwok LY, Lutjen S, Soltek S, Hoffmann S, Korner H, Deckert M: Both lymphotoxin-alpha and TNF are crucial for control of *Toxoplasma gondii* in the central nervous system. J Immunol 2003; 170:6172–6182.
65 Liu JQ, Bai XF, Shi FD, Xiao BG, Li HL, Levi M, Mustafa M, Wahren B, Link H: Inhibition of experimental autoimmune encephalomyelitis in Lewis rats by nasal administration of encephalitogenic MBP peptides: synergistic effects of MBP 68–86 and 87–99. Int Immunol 1998;10:1139–1148.
66 Marino MW, Dunn A, Grail D, Inglese M, Noguchi Y, Richards E, Jungbluth A, Wada H, Moore M, Williamson B, Basu S, Old LJ: Characterization of tumor necrosis factor-deficient mice. Proc Natl Acad Sci U S A 1997;94:8093–8098.
67 Olleros ML, Guler R, Corazza N, Vesin D, Eugster HP, Marchal G, Chavarot P, Mueller C, Garcia I: Transmembrane TNF induces an efficient cell-mediated immunity and resistance to *Mycobacterium bovis* bacillus Calmette-Guerin infection in the absence of secreted TNF and lymphotoxin-alpha. J Immunol 2002;168:3394–3401.

68 Chakravarty SD, Zhu G, Tsai MC, Mohan VP, Marino S, Kirschner DE, Huang L, Flynn J, Chan J: Tumor necrosis factor blockade in chronic murine tuberculosis enhances granulomatous inflammation and disorganizes granulomas in the lungs. Infect Immun 2008;76:916–926.
69 Olleros ML, Guler R, Vesin D, Parapanov R, Marchal G, Martinez-Soria E, Corazza N, Pache JC, Mueller C, Garcia I: Contribution of transmembrane tumor necrosis factor to host defense against *Mycobacterium bovis* bacillus Calmette-Guerin and *Mycobacterium tuberculosis* infections. Am J Pathol 2005;166:1109–1120.
70 Florido M, Appelberg R: Characterization of the deregulated immune activation occurring at late stages of mycobacterial infection in TNF-deficient mice. J Immunol 2007;179:7702–7708.
71 Flory CM, Hubbard RD, Collins FM: Effects of in vivo T lymphocyte subset depletion on mycobacterial infections in mice. J Leukoc Biol 1992;51:225–229.
72 Caruso AM, Serbina N, Klein E, Triebold K, Bloom BR, Flynn JL: Mice deficient in CD4 T cells have only transiently diminished levels of IFN-gamma, yet succumb to tuberculosis. J Immunol 1999;162:5407–5416.
73 Kamath AB, Woodworth J, Xiong X, Taylor C, Weng Y, Behar SM: Cytolytic CD8+ T cells recognizing CFP10 are recruited to the lung after *Mycobacterium tuberculosis* infection. J Exp Med 2004;200:1479–1489.
74 Khader SA, Cooper AM: IL-23 and IL-17 in tuberculosis. Cytokine 2008;41:79–83.
75 Lockhart E, Green AM, Flynn JL: IL-17 production is dominated by gammadelta T cells rather than CD4 T cells during *Mycobacterium tuberculosis* infection. J Immunol 2006;177:4662–4669.
76 Hamada S, Umemura M, Shiono T, Tanaka K, Yahagi A, Begum MD, Oshiro K, Okamoto Y, Watanabe H, Kawakami K, Roark C, Born WK, O'Brien R, Ikuta K, Ishikawa H, Nakae S, Iwakura Y, Ohta T, Matsuzaki G: IL-17A produced by gammadelta T cells plays a critical role in innate immunity against *Listeria monocytogenes* infection in the liver. J Immunol 2008;181:3456–3463.
77 Junqueira-Kipnis AP, Kipnis A, Jamieson A, Juarrero MG, Diefenbach A, Raulet DH, Turner J, Orme IM: NK cells respond to pulmonary infection with *Mycobacterium tuberculosis*, but play a minimal role in protection. J Immunol 2003;171:6039–6045.
78 Chackerian A, Alt J, Perera V, Behar SM: Activation of NKT cells protects mice from tuberculosis. Infect Immun 2002;70:6302–6309.
79 Gansert JL, Kiessler V, Engele M, Wittke F, Rollinghoff M, Krensky AM, Porcelli SA, Modlin RL, Stenger S: Human NKT cells express granulysin and exhibit antimycobacterial activity. J Immunol 2003;170:3154–3161.
80 Fischer K, Scotet E, Niemeyer M, Koebernick H, Zerrahn J, Maillet S, Hurwitz R, Kursar M, Bonneville M, Kaufmann SH, Schaible UE: Mycobacterial phosphatidylinositol mannoside is a natural antigen for CD1d-restricted T cells. Proc Natl Acad Sci U S A 2004;101:10685–10690.
81 Zajonc DM, Ainge GD, Painter GF, Severn WB, Wilson IA: Structural characterization of mycobacterial phosphatidylinositol mannoside binding to mouse CD1d. J Immunol 2006;177:4577–4583.
82 Munoz S, Hernandez-Pando R, Abraham SN, Enciso JA: Mast cell activation by *Mycobacterium tuberculosis*: mediator release and role of CD48. J Immunol 2003;170:5590–5596.
83 Carlos D, de Souza Junior DA, de Paula L, Jamur MC, Oliver C, Ramos SG, Silva CL, Faccioli LH: Mast cells modulate pulmonary acute inflammation and host defense in a murine model of tuberculosis. J Infect Dis 2007;196:1361–1368.
84 Jacobs M, Marino MW, Brown N, Abel B, Bekker LG, Quesniaux VJ, Fick L, Ryffel B: Correction of defective host response to *Mycobacterium bovis* BCG infection in TNF-deficient mice by bone marrow transplantation. Lab Invest 2000;80:901–914.
85 MacMicking JD, Nathan C, Hom G, Chartrain N, Fletcher DS, Trumbauer M, Stevens K, Xie QW, Sokol K, Hutchinson N, et al: Altered responses to bacterial infection and endotoxic shock in mice lacking inducible nitric oxide synthase. Cell 1995;81:641–650.
86 Chan J, Xing Y, Magliozzo RS, Bloom BR: Killing of virulent *Mycobacterium tuberculosis* by reactive nitrogen intermediates produced by activated murine macrophages. J Exp Med 1992;175:1111–1122.
87 Sturgill-Koszycki S, Schaible UE, Russell DG: Mycobacterium-containing phagosomes are accessible to early endosomes and reflect a transitional state in normal phagosome biogenesis. EMBO J 1996;15:6960–6968.
88 Russell DG, Dant J, Sturgill-Koszycki S: *Mycobacterium avium*- and *Mycobacterium tuberculosis*-containing vacuoles are dynamic, fusion-competent vesicles that are accessible to glycosphingolipids from the host cell plasmalemma. J Immunol 1996;156:4764–4773.
89 Mwandumba HC, Russell DG, Nyirenda MH, Anderson J, White SA, Molyneux ME, Squire SB: *Mycobacterium tuberculosis* resides in nonacidified vacuoles in endocytically competent alveolar macrophages from patients with tuberculosis and HIV infection. J Immunol 2004;172:4592–4598.
90 Xu S, Cooper A, Sturgill-Koszycki S, van Heyningen T, Chatterjee D, Orme I, Allen P, Russell DG: Intracellular trafficking in *Mycobacterium tuberculosis* and *Mycobacterium avium*-infected macrophages. J Immunol 1994;153:2568–2578.

91 Ehlers S, Kutsch S, Benini J, Cooper A, Hahn C, Gerdes J, Orme I, Martin C, Rietschel ET: NOS2-derived nitric oxide regulates the size, quantity and quality of granuloma formation in *Mycobacterium avium*-infected mice without affecting bacterial loads. Immunology 1999;98:313–323.
92 Bekker LG, Freeman S, Murray PJ, Ryffel B, Kaplan G: TNF-alpha controls intracellular mycobacterial growth by both inducible nitric oxide synthase-dependent and inducible nitric oxide synthase-independent pathways. J Immunol 2001;166: 6728–6734.
93 Flynn JL, Chan J: Immune evasion by *Mycobacterium tuberculosis*: living with the enemy. Curr Opin Immunol 2003;15:450–455.
94 Malik ZA, Thompson CR, Hashimi S, Porter B, Iyer SS, Kusner DJ: Cutting edge: *Mycobacterium tuberculosis* blocks Ca^{2+} signaling and phagosome maturation in human macrophages via specific inhibition of sphingosine kinase. J Immunol 2003;170:2811–2815.
95 Malik ZA, Iyer SS, Kusner DJ: *Mycobacterium tuberculosis* phagosomes exhibit altered calmodulin-dependent signal transduction: contribution to inhibition of phagosome-lysosome fusion and intracellular survival in human macrophages. J Immunol 2001;166:3392–3401.
96 Malik ZA, Denning GM, Kusner DJ: Inhibition of Ca^{2+} signaling by *Mycobacterium tuberculosis* is associated with reduced phagosome-lysosome fusion and increased survival within human macrophages. J Exp Med 2000;191:287–302.
97 Deretic V: Autophagy, an immunologic magic bullet: *Mycobacterium tuberculosis* phagosome maturation block and how to bypass it. Future Microbiol 2008;3:517–524.
98 Jayachandran R, Sundaramurthy V, Combaluzier B, Mueller P, Korf H, Huygen K, Miyazaki T, Albrecht I, Massner J, Pieters J: Survival of mycobacteria in macrophages is mediated by coronin 1-dependent activation of calcineurin. Cell 2007;130:37–50.
99 MacMicking JD, Taylor GA, McKinney JD: Immune control of tuberculosis by IFN-gamma-inducible LRG-47. Science 2003;302:654–659.
100 Singh SB, Davis AS, Taylor GA, Deretic V: Human IRGM induces autophagy to eliminate intracellular mycobacteria. Science 2006;313:1438–1441.
101 MacMicking JD: Immune control of phagosomal bacteria by p47 GTPases. Curr Opin Microbiol 2005;8:74–82.
102 Liu PT, Stenger S, Li H, Wenzel L, Tan BH, Krutzik SR, Ochoa MT, Schauber J, Wu K, Meinken C, Kamen DL, Wagner M, Bals R, Steinmeyer A, Zugel U, Gallo RL, Eisenberg D, Hewison M, Hollis BW, Adams JS, Bloom BR, Modlin RL: Toll-like receptor triggering of a vitamin D-mediated human antimicrobial response. Science 2006;311:1770–1773.
103 Stenger S, Hanson DA, Teitelbaum R, Dewan P, Niazi KR, Froelich CJ, Ganz T, Thoma-Uszynski S, Melian A, Bogdan C, Porcelli SA, Bloom BR, Krensky AM, Modlin RL: An antimicrobial activity of cytolytic T cells mediated by granulysin. Science 1998;282:121–125.
104 Lammas DA, Stober C, Harvey CJ, Kendrick N, Panchalingam S, Kumararatne DS: ATP-induced killing of mycobacteria by human macrophages is mediated by purinergic P2Z(P2X7) receptors. Immunity 1997;7:433–444.
105 Kusner DJ, Barton JA: ATP stimulates human macrophages to kill intracellular virulent *Mycobacterium tuberculosis* via calcium-dependent phagosome-lysosome fusion. J Immunol 2001;167:3308–3315.
106 Fairbairn IP, Stober CB, Kumararatne DS, Lammas DA: ATP-mediated killing of intracellular mycobacteria by macrophages is a P2X(7)-dependent process inducing bacterial death by phagosome-lysosome fusion. J Immunol 2001;167:3300–3307.
107 Humphreys BD, Dubyak GR: Modulation of P2X7 nucleotide receptor expression by pro- and anti-inflammatory stimuli in THP-1 monocytes. J Leukoc Biol 1998;64:265–273.
108 Schaible UE, Winau F, Sieling PA, Fischer K, Collins HL, Hagens K, Modlin RL, Brinkmann V, Kaufmann SH: Apoptosis facilitates antigen presentation to T lymphocytes through MHC-I and CD1 in tuberculosis. Nat Med 2003;9:1039–1046.
109 Feng CG, Scanga CA, Collazo-Custodio CM, Cheever AW, Hieny S, Caspar P, Sher A: Mice lacking myeloid differentiation factor 88 display profound defects in host resistance and immune responses to *Mycobacterium avium* infection not exhibited by Toll-like receptor 2 (TLR2)- and TLR4-deficient animals. J Immunol 2003;171:4758–4764.
110 Fremond CM, Yeremeev V, Nicolle DM, Jacobs M, Quesniaux VF, Ryffel B: Fatal *Mycobacterium tuberculosis* infection despite adaptive immune response in the absence of MyD88. J Clin Invest 2004;114: 1790–1799.
111 Drennan MB, Nicolle D, Quesniaux VJ, Jacobs M, Allie N, Mpagi J, Fremond C, Wagner H, Kirschning C, Ryffel B: Toll-like receptor 2-deficient mice succumb to *Mycobacterium tuberculosis* infection. Am J Pathol 2004;164:49–57.
112 Bafica A, Scanga CA, Feng CG, Leifer C, Cheever A, Sher A: TLR9 regulates Th1 responses and cooperates with TLR2 in mediating optimal resistance to *Mycobacterium tuberculosis*. J Exp Med 2005; 202: 1715–1724.

113 Holscher C, Reiling N, Schaible UE, Holscher A, Bathmann C, Korbel D, Lenz I, Sonntag T, Kroger S, Akira S, Mossmann H, Kirschning CJ, Wagner H, Freudenberg M, Ehlers S: Containment of aerogenic *Mycobacterium tuberculosis* infection in mice does not require MyD88 adaptor function for TLR2, -4 and -9. Eur J Immunol 2008;38:680–694.
114 Blander JM, Medzhitov R: Regulation of phagosome maturation by signals from Toll-like receptors. Science 2004;304:1014–1018.
115 Ray JC, Flynn JL, Kirschner DE: Synergy between individual TNF-dependent functions determines granuloma performance for controlling *Mycobacterium tuberculosis* infection. J Immunol 2009; 182:3706–3717.
116 Lucas R, Tacchini-Cottier F, Guler R, Vesin D, Jemelin S, Olleros ML, Marchal G, Browning JL, Vassalli P, Garcia I: A role for lymphotoxin beta receptor in host defense against *Mycobacterium bovis* BCG infection. Eur J Immunol 1999;29:4002–4010.
117 Ehlers S, Holscher C, Scheu S, Tertilt C, Hehlgans T, Suwinski J, Endres R, Pfeffer K: The lymphotoxin beta receptor is critically involved in controlling infections with the intracellular pathogens *Mycobacterium tuberculosis* and *Listeria monocytogenes*. J Immunol 2003;170:5210–5218.
118 Black RA, Rauch CT, Kozlosky CJ, Peschon JJ, Slack JL, Wolfson MF, Castner BJ, Stocking KL, Reddy P, Srinivasan S, Nelson N, Boiani N, Schooley KA, Gerhart M, Davis R, Fitzner JN, Johnson RS, Paxton RJ, March CJ, Cerretti DP: A metalloproteinase disintegrin that releases tumour-necrosis factor-alpha from cells. Nature 1997;385:729–733.
119 Decker T, Lohmann-Matthes ML, Gifford GE: Cell-associated tumor necrosis factor (TNF) as a killing mechanism of activated cytotoxic macrophages. J Immunol 1987;138:957–962.
120 Grell M, Douni E, Wajant H, Lohden M, Clauss M, Maxeiner B, Georgopoulos S, Lesslauer W, Kollias G, Pfizenmaier K, et al: The transmembrane form of tumor necrosis factor is the prime activating ligand of the 80 kDa tumor necrosis factor receptor. Cell 1995;83:793–802.
121 Torres D, Janot L, Quesniaux VF, Grivennikov SI, Maillet I, Sedgwick JD, Ryffel B, Erard F: Membrane tumor necrosis factor confers partial protection to listeria infection. Am J Pathol 2005;167:1677–1687.
122 Fremond C, Allie N, Dambuza I, Grivennikov SI, Yeremeev V, Quesniaux VF, Jacobs M, Ryffel B: Membrane TNF confers protection to acute mycobacterial infection. Respir Res 2005;6:136.
123 Saunders BM, Tran S, Ruuls S, Sedgwick JD, Briscoe H, Britton WJ: Transmembrane TNF is sufficient to initiate cell migration and granuloma formation and provide acute, but not long-term, control of *Mycobacterium tuberculosis* infection. J Immunol 2005;174:4852–4859.
124 Dambuza I, Allie N, Fick L, Johnston N, Fremond C, Mitchell J, Quesniaux VF, Ryffel B, Jacobs M: Efficacy of membrane TNF mediated host resistance is dependent on mycobacterial virulence. Tuberculosis (Edinb) 2008;88:221–234.
125 Allie N, Alexopoulou L, Quesniaux VJ, Fick L, Kranidioti K, Kollias G, Ryffel B, Jacobs M: Protective role of membrane tumour necrosis factor in the host's resistance to mycobacterial infection. Immunology 2008;125:522–534.
126 Wallis RS: Tumour necrosis factor antagonists: structure, function, and tuberculosis risks. Lancet Infect Dis 2008;8:601–611.
127 Keane J, Gershon S, Wise RP, Mirabile-Levens E, Kasznica J, Schwieterman WD, Siegel JN, Braun MM: Tuberculosis associated with infliximab, a tumor necrosis factor alpha-neutralizing agent. N Engl J Med 2001;345:1098–1104.
128 Mohan AK, Cote TR, Block JA, Manadan AM, Siegel JN, Braun MM: Tuberculosis following the use of etanercept, a tumor necrosis factor inhibitor. Clin Infect Dis 2004;39:295–299.
129 Keane J: TNF-blocking agents and tuberculosis: new drugs illuminate an old topic. Rheumatology (Oxford) 2005;44:714–720.
130 Wallis RS, Broder M, Wong J, Lee A, Hoq L: Reactivation of latent granulomatous infections by infliximab. Clin Infect Dis 2005;(suppl 3):S194–S198.
131 Flynn JL: Lessons from experimental *Mycobacterium tuberculosis* infections. Microbes Infect 2006;8:1179–1188.
132 Botha T, Ryffel B: Reactivation of latent tuberculosis by an inhibitor of inducible nitric oxide synthase in an aerosol murine model. Immunology 2002; 107:350–357.
133 Flynn JL, Chan J: Tuberculosis: latency and reactivation. Infect Immun 2001;69:4195–4201.
134 Botha T, Ryffel B: Reactivation of latent tuberculosis infection in TNF-deficient mice. J Immunol 2003; 171:3110–3118.
135 Plessner HL, Lin PL, Kohno T, Louie JS, Kirschner D, Chan J, Flynn JL: Neutralization of tumor necrosis factor (TNF) by antibody but not TNF receptor fusion molecule exacerbates chronic murine tuberculosis. J Infect Dis 2007;195:1643–1650.

136 Marino S, Sud D, Plessner H, Lin PL, Chan J, Flynn JL, Kirschner DE: Differences in reactivation of tuberculosis induced from anti-TNF treatments are based on bioavailability in granulomatous tissue. PLoS Comput Biol 2007;3:1909–1924.
137 Steed PM, Tansey MG, Zalevsky J, et al: Inactivation of TNF signaling by rationally designed dominant-negative TNF variants. Science 2003;301:1895–1898.
138 Zalevsky J, Secher T, Ezhevsky SA, Janot L, Steed PM, O'Brien C, Eivazi A, Kung J, Nguyen DH, Doberstein SK, Erard F, Ryffel B, Szymkowski DE: Dominant-negative inhibitors of soluble TNF attenuate experimental arthritis without suppressing innate immunity to infection. J Immunol 2007; 179: 1872–1883.
139 Olleros ML, Vesin D, Lambou AF, Janssens JP, Ryffel B, Rose S, Fremond C, Quesniaux VF, Szymkowski DE, Garcia I: Dominant-negative tumor necrosis factor protects from *Mycobacterium bovis* bacillus Calmette-Guerin (BCG) and endotoxin-induced liver injury without compromising host immunity to BCG and *Mycobacterium tuberculosis*. J Infect Dis 2009;199:1053–1063.
140 Spohn G, Guler R, Johansen P, Keller I, Jacobs M, Beck M, Rohner F, Bauer M, Dietmeier K, Kundig TM, Jennings GT, Brombacher F, Bachmann MF: A virus-like particle-based vaccine selectively targeting soluble TNF-alpha protects from arthritis without inducing reactivation of latent tuberculosis. J Immunol 2007;178:7450–7457.

Valerie F.J. Quesniaux, PhD
Molecular Immunology and Embryology UMR6218
Orleans University and CNRS, 3B rue de la Férollerie
FR–45071 Orleans Cedex 2 (France)
Tel. +33 2 38 25 54 38, Fax +33 2 38 25 79 79, E-Mail quesniaux@cnrs-orleans.fr

Kollias G, Sfikakis PP (eds): TNF Pathophysiology. Molecular and Cellular Mechanisms.
Curr Dir Autoimmun. Basel, Karger, 2010, vol 11, pp 180–210

The First Decade of Biologic TNF Antagonists in Clinical Practice: Lessons Learned, Unresolved Issues and Future Directions

Petros P. Sfikakis

First Department of Propedeutic and Internal Medicine, Laikon Hospital, Athens University Medical School, Athens, Greece

Abstract

Results from clinical trials of biologic anti-TNF drugs performed in the late 1990s confirmed the biological relevance of TNF function in the pathogenesis of chronic noninfectious inflammation of joints, skin and gut, which collectively affects 2–3% of the population. Up to April 2009, more than two million patients worldwide have received the first marketed drugs, namely the monoclonal anti-TNF antibodies infliximab and adalimumab and the soluble TNF receptor etanercept. All three are equally effective in rheumatoid arthritis, ankylosing spondylitis, psoriasis and psoriatic arthritis, but, for not clearly defined reasons, only the monoclonal antibodies are effective in inflammatory bowel disease. About 60% of patients who do not benefit from standard nonbiologic treatments for these diseases respond to TNF antagonists. Less than half of responding patients achieve complete remission of disease. Importantly, some of those patients with rheumatoid arthritis in whom long-term anti-TNF therapy induced disease remission remain disease-free after discontinuation of any kind of treatment. There are not yet reliable predictors of which patients will or will not respond on anti-TNF therapy, whereas subsequent loss of an initial clinical response occurs frequently. The spectrum of efficacy anti-TNF therapies widens to include diseases such as systemic vasculitis and sight-threatening uveitis. While paradoxical new adverse effects are recognized, i.e. exacerbation or development of new onset psoriasis, reactivation of latent tuberculosis remains the most important safety issue of anti-TNF therapies. Clinical practice guidelines and consensus statements on the criteria of introduction, duration of treatment and cessation of TNF antagonists, including safety issues, are under constant revision as data from longer periods of patient exposure accumulate. It is hoped that more efficacious drugs that will ideally target the deleterious proinflammatory properties of TNF without compromising its protective role in host defense and (auto)immunity will be available in the near future.

Results from multiple randomized clinical trials of anti-TNF drugs confirmed the biological relevance of TNF function in the pathogenesis of chronic noninfectious inflammation of joints, skin and gut [1]. Since the first license of biologic TNF antagonists for clinical use in 1998, about 2 million patients worldwide have received these drugs for approved indications that include rheumatoid arthritis, inflammatory bowel disease, psoriatic arthritis, juvenile idiopathic arthritis, plaque psoriasis, ankylosing spondylitis and uveoretinitis associated with Adamantiades-Behcet's disease (approved indication in Japan). TNF antagonists have marked a new era in the treatment of these diseases which collectively affect 2–3% of the population. As a class, these drugs have shown remarkable efficacy and acceptable long-term safety profiles [2]. These issues have been the subject of meta-analyses of randomized clinical trials-derived results, as well as of observational studies from national patients' registries reporting postmarketed experience.

Herein, the approved indications and dosages based on trials sponsored by the manufacturers of currently licensed TNF antagonists are presented, whereas mechanisms of action and their differences are briefly discussed. Emphasis is given on efficacy data accumulated over the first decade of their use in clinical practice for the approved indications, as well as for other diseases which are potential targets of TNF blockade. A brief critical review of safety data is also presented. Finally, some examples of how much remains to be learned about the individualized factors that may enhance efficacy and safety of the available TNF antagonists are mentioned.

TNF Antagonists in 2010: Approved Indications and Dosages

According to data on file provided by the pharmaceutical companies marketing the three first licensed TNF antagonists, up to April 2009 about 1,100,000 patients worldwide have been treated with infliximab, 550,000 patients with etanercept and 350,000 patients with adalimumab. Certolizumab pegol and golimumab have only recently been introduced on the market; thus, the available clinical experience is limited. The five currently licensed anti-TNF drugs and their indications are shown in table 1.

Infliximab (Remicade®) is a chimeric IgG1κ monoclonal antibody composed of human constant and murine variable regions. Approved indications and dosages include the following: (a) Rheumatoid arthritis, for reducing signs and symptoms, inhibiting the progression of structural damage, and improving physical function in patients with moderately to severely active disease. Initial dose is 3 mg/kg given as an intravenous infusion followed by additional similar doses at 2 and 6 weeks after the first infusion and every 8 weeks thereafter, in combination with methotrexate. (b) Ankylosing spondylitis, for reducing signs and symptoms in patients with active disease. Initial dose is 5 mg/kg followed by additional similar doses at 2 and 6 weeks after the first infusion and every 6 weeks thereafter. (c) Psoriatic arthritis (dosing similar to ankylosing spondylitis, with or without methotrexate), for reducing signs and symptoms of active arthritis, inhibiting the progression of structural damage, and improving physical function. (d) Plaque psoriasis (dosing similar to ankylosing spondylitis), for the treatment of adult patients with chronic severe (i.e. extensive

Table 1. Licensed TNF antagonists and their approved indications in the USA (FDA), Europe (EMEA) and Japan (July 2009)

Molecule, MW in kDa, type, route of administration	Indications
Infliximab (Remicade®), 149, chimeric humanized mAb, IV	RA, AS, PsA, pPs, CD, UC, ABD uveoretinitis (Japan)
Adalimumab (Humira®), 148, recombinant human mAb, SC	RA, JIA, AS, PsA, pPs, CD
Certolizumab pegol (Cimzia®), 40, recombinant humanized antibody Fab fragment, SC	RA, CD
Golimumab (Simponi®), 147, fully human mAb, SC	RA, AS, PsA
Etanercept (Enbrel®), 150, fusion protein p75 TNF receptor, SC	RA, JIA, AS, PsA, pPs

MW = Molecular weight; mAb = monoclonal antibody; RA = rheumatoid arthritis; AS = ankylosing spondylitis; PsA = psoriatic arthritis; pPs = plaque psoriasis; CD = Crohn's disease; UC = ulcerative colitis; ABD = Adamantiades-Behcet's disease; JIA = juvenile idiopathic arthritis.

and/or disabling) disease who are candidates for systemic therapy and when other systemic therapies are medically less appropriate. (e) Crohn's disease (5 mg/kg at 0, 2 and 6 weeks followed by a maintenance regimen of 5 mg/kg every 8 weeks), for reducing signs and symptoms and inducing and maintaining clinical remission in adult and pediatric patients with moderately to severely active disease who have had an inadequate response to conventional therapy. In addition, infliximab is indicated for reducing the number of draining enterocutaneous and rectovaginal fistulas and maintaining fistula closure in adult patients with fistulizing Crohn's disease, as well as for patients with moderately to severely active ulcerative colitis who have had an inadequate response to conventional therapy. Across all indications, the dose can be adjusted up to 10 mg/kg for patients who have an incomplete response, or treating as often as every 4 weeks.

Adalimumab (Humira®) is a recombinant IgG1 monoclonal antibody specific for human TNF with human-derived heavy and light chain variable regions and human IgG1:K constant regions. Approved indications of adalimumab are the same as of infliximab with the exception of ulcerative colitis. In addition to these indications, adalimumab has been approved for the treatment of patients 4–17 years of age with polyarticular juvenile idiopathic arthritis. The recommended dose of adalimumab is 40 mg given as a subcutaneous injection every other week. Some patients with rheumatoid arthritis not taking concomitant methotrexate may derive additional benefit from increasing the dosing frequency to 40 mg every week. The recommended dose for adult patients with Crohn's disease is 160 mg initially at day 1, followed by 80 mg 2 weeks later, and a maintenance dose of 40 mg every other week, 2 weeks later. The recommended dose for psoriasis is an initial dose of 80 mg, followed by 40 mg given every other week, whereas in juvenile idiopathic arthritis the dose is based on weight (approximately 1 mg/kg every other week).

Certolizumab pegol (Cimzia®) is a recombinant, humanized antibody Fab′ fragment, specific for human TNF, conjugated to polyethylene glycol. Current approved indications of certolizumab include (a) Crohn's disease, for reducing signs and symptoms of and maintaining clinical response in adult patients with moderately to severely active disease who have had an inadequate response to conventional therapy, and (b) rheumatoid arthritis patients with moderately to severely active disease. The recommended dose is 400 mg given as subcutaneous injection initially, and at weeks 2 and 4 followed by 200 mg every other week. In patients who obtain a clinical response, the recommended maintenance regimen is 400 mg every 4 weeks.

Golimumab (Simponi®) is a fully human IgG1k monoclonal antibody specific for human TNF, approved as a 50-mg subcutaneous injection once a month and is indicated for adult patients with (a) rheumatoid arthritis and moderately to severely active disease in combination with methotrexate, (b) psoriatic arthritis alone or in combination with methotrexate, and (c) ankylosing spondylitis.

Etanercept (Enbrel®) is a dimeric fusion protein consisting of the extracellular ligand-binding portion of the human 75-kDa (p75) receptor of TNF, linked to the Fc portion of human IgG1. The Fc component of etanercept contains the C_H2 domain, the C_H3 domain and hinge region, but not the C_H1 domain of IgG1. Approved indications and dosages include the following: (a) Rheumatoid arthritis, for reducing signs and symptoms, inducing major clinical response, inhibiting the progression of structural damage, and improving physical function in patients with moderately to severely active disease. Etanercept is given as a subcutaneous injection of 50 mg per week, with or without methotrexate. (b) Polyarticular juvenile idiopathic arthritis (0.8 mg/kg per week, up to a maximum of 50 mg per week) for reducing signs and symptoms of moderately to severely active in patients aged 2 years and older. (c) Psoriatic arthritis (dosing similar to rheumatoid arthritis), for reducing signs and symptoms, inhibiting the progression of structural damage of active arthritis, and improving physical function, in combination with methotrexate in patients who do not respond adequately to methotrexate alone. (d) Ankylosing spondylitis (dosing similar to rheumatoid arthritis) for reducing signs and symptoms in patients with active disease. (e) Plaque psoriasis, for the treatment of adult patients with chronic moderate to severe disease who are candidates for systemic therapy or phototherapy (50 mg dose given twice weekly for 3 months followed by a reduction to a maintenance dose of 50 mg per week).

Mechanisms of Action and Differences between TNF Antagonists

Multiple TNF-dependent or TNF-mediated biological processes that are involved in the initiation and perpetuation of inflammation are negatively affected in patients treated with these drugs. The main mechanisms of therapeutic action, which are not distinct and may overlap, include the following: reduction of proinflammatory cytokines, chemokines and acute phase proteins, downregulation of adhesion molecule expression, attenuation of vascular permeability and angiogenesis, deactivation of epithelial, endothelial and dendritic cells, myofibroblasts and osteoclasts, increases in circulating regulatory T cells, and diminished recruitment of inflammatory cells from blood to the inflamed tissue [reviewed in detail in references 3, 4]. These mechanisms, and probably those which will be revealed in future studies, operate at different levels in different diseases, as well as in different patients within the same disease spectrum.

TNF antagonists work by directly neutralizing the activity of soluble TNF and preventing its binding to the two TNF receptors. These receptors, namely the p55 or TNFR1 and p75 or TNFR2, are expressed on the membrane of monocytes and T lymphocytes and circulate in the blood in soluble forms. In addition to neutralization of soluble TNF activity, anti-TNF molecules induce complement-dependent cytotoxicity (CDC) and antibody-dependent cytotoxicity (ADCC) by binding to transmembrane TNF expressed in various cell types [5]. Moreover, this binding may affect intracellular signaling, with the end result being either programmed cell death, or suppression of cytokine production, or cell growth arrest [6–9]. In addition to TNF, etanercept neutralizes lymphotoxin, whereas anti-TNF monoclonal antibodies do not; this may impact other in vivo functions [10]. For example, infliximab, adalimumab and certolizumab inhibit the inflammation in Crohn's disease but etanercept does not work in this disease state.

In vitro mechanistic studies of TNF antagonists have yielded variable results due to differences in experimental conditions. Some studies have shown differences among TNF antagonists in their binding to soluble versus membrane TNF, resulting in differential ability to induce cell apoptosis [7, 10–12]. Another study addressing the specificity of anti-TNF binding to the p55 or p75 receptor [13] found that infliximab, adalimumab and certolizumab equally neutralized membrane-bound TNF via the p55 or p75 receptors, but certolizumab was 2-fold more potent in neutralizing soluble TNF signaling via either receptor than infliximab and adalimumab. In a system of NS0 myeloma cells expressing transmembrane TNF, it was shown that etanercept bound fewer TNF molecules than infliximab, adalimumab and certolizumab [5]. However, only etanercept, infliximab and adalimumab were able to affect CDC and ADCC in this system, whereas certolizumab (also approved for Crohn's disease) exhibited neither CDC nor ADCC because it is a Fab′ fragment without an Fc portion. In other experimental systems, infliximab and adalimumab induced CDC more potently than etanercept [12, 14].

On the other hand, as shown in a recent study in a different experimental setting, although the binding properties of adalimumab, infliximab, and etanercept were similar for soluble TNF, and very similar for transmembrane TNF, none of the three agents was able to induce CDC in activated normal human peripheral blood mononuclear cells [15]. Finally, in Jurkat T cells stably expressing an uncleavable form of transmembrane TNF, outside-to-inside (reverse) signal transduction through transmembrane TNF was induced by adalimumab and infliximab, but not by etanercept; this may partly explain their differential efficacy in Crohn's disease [14].

Rheumatoid Arthritis

Rheumatoid arthritis is a chronic, progressive, inflammatory systemic disease affecting about 1% of the population that may reduce life expectancy. It targets primarily the synovial tissues and results in joint destruction leading to substantial disability, functional declines and work disability. Certain disease-modifying antirheumatic drugs (DMARDs) reduce clinical disease activity, improve function and retard disease progression; of them, methotrexate is today considered the mainstay of rheumatoid arthritis treatment. Despite the use of DMARDs as mono- or combination therapy, between 25 and 50% of patients still have clinically active synovitis with progressive structural articular damage.

TNF antagonists given as monotherapy are efficacious to a similar degree as methotrexate in suppressing inflammation and improving clinical outcomes in rheumatoid arthritis (for example, 20, 50, or 70% improvement according to American College of Rheumatology response criteria). However, these drugs are more potent than methotrexate in inhibiting the progression of structural joint damage [16]. Several lines of evidence suggest that the latter effect of TNF antagonists is not necessarily a direct result of the former, since TNF itself is a potent inducer of osteoclastogenesis, and osteoclasts are key mediators of bone loss [1]. Indeed, some patients exhibit radiological improvement on anti-TNF treatments, without achieving clinical response [17].

The combination of TNF antagonists with methotrexate is probably the most powerful therapy we currently have to treat rheumatoid arthritis, although combinations of methotrexate with newer biologic agents can be considered. Multiple randomized trials have convincingly shown that addition of an anti-TNF agent to methotrexate is significantly superior to methotrexate monotherapy in any aspect of the disease. The same happens when an anti-TNF agent is compared with placebo in patients for whom previous DMARD treatment has failed. Disease remission in these trials is achieved by 30–40% of patients, with a placebo response ranging from 6 to 20% [18–32].

Encouraged by the positive results of randomized trials and using an analogy from oncology (two therapeutic phases: remission induction and maintenance), investigators explored the possibility that by treating patients with an anti-TNF agent in early stages (less than 3 years), the disease could be 'eliminated' and long-lasting remission could occur. Along this line, of 120 patients with early rheumatoid arthritis who were initially treated with infliximab in combination with methotrexate (25 mg/week), 56% had persistent low disease activity and discontinued infliximab after a median of 9.9 months, while keeping a median methotrexate dosage of 10 mg/week after 2 years

[33]. At 4 years, 69% of patients treated initially with this combination had not any significant radiographic progression of joint damage. Importantly, 18% of patients were able to discontinue methotrexate and achieve 'drug-free' remission of disease at 4 years [34], emphasizing that 'initial treatment' could mean 'temporary treatment'.

Observational data from routine rheumatology clinical practice settings also highlight the remarkable efficacy of TNF antagonists in rheumatoid arthritis and provide additional information beyond data of randomized trials which reflect a strictly selected proportion of the general patient population. For daily practice patients who are eligible for randomized trials, clinical responses are almost similar to responses reached in these trials. In contrast, for those considered ineligible for the trials (mainly because of lower baseline disease activity, more comorbidities, and lower functional status), the clinical response rates were lower than that in eligible patients, although the majority would still benefit from TNF antagonists [35–37]. On the other hand, data derived from a national registry, which reflect general clinical practice, show a significant retardation of radiographic progression in patients treated with anti-TNF and the magnitude of the improvement seen is similar to results from randomized trials [38]. Moreover, the risk of developing cardiovascular disease, the leading cause of death in rheumatoid arthritis, was lower in Swedish patients treated with TNF antagonists versus those not treated. This is compatible with the hypothesis that chronic inflammation contributes to the development of cardiovascular events in these patients [39]. In addition, a review of the current literature revealed a beneficial effect of anti-TNF on bone density as measured by dual energy radiograph absorptiometry at the lumbar spine and hip, as well as on markers of bone turnover and resorption [40].

Although no head-to-head studies are available, comparison of the results obtained in randomized trials of the five currently licensed TNF antagonists suggests that all are similarly efficacious. Overall, about two thirds of patients respond favorably to TNF antagonists and about one third of them achieve complete clinical remission. However, there are patients who lose efficacy during therapy (secondary failure or acquired therapeutic resistance) [41]. For some of these patients with secondary failure to a given TNF-antagonist, instead of increasing the dose [42], switching to another TNF-antagonist can restore a good clinical response, as suggested from observational data from national registries [43, 44] and from a recently published randomized trial [45].

Juvenile Idiopathic Arthritis

Juvenile idiopathic arthritis affects about 0.1% of children younger than 16 years of age, and is one of the more common chronic diseases in childhood. The term refers to a group of distinct but heterogeneous disorders (seven subtypes) that have chronic inflammatory arthritis in common. Despite long-term treatment, most patients do not achieve remission and more than 30% of them have significant functional limitations in the adult life. Nonsteroidal anti-inflammatory drugs (NSAIDs) are effective in only about 25% of patients, whereas corticosteroids should be avoided due to their deleterious effects on bone and growth. Of the DMARDs used in adults, only methotrexate is licensed for use in children. Methotrexate is the cornerstone of management for most patients, especially for those with the polyarticular form of the disease which resembles adult-onset rheumatoid arthritis.

The use of TNF antagonists has markedly increased our ability to effectively treat these children. In particular, more than 50% of patients with polyarticular juvenile idiopathic arthritis achieve complete disease control, estimated by a 70% improvement according to American College of Rheumatology pediatric response criteria, with TNF antagonists, including patients who failed methotrexate monotherapy. These drugs may also slow radiologic damage and increase bone density [46–49]. Initial trials of infliximab, etanercept and adalimumab have shown almost similar efficacy in polyarticular juvenile idiopathic arthritis. Current clinical experience is larger with etanercept than adalimumab, whereas infliximab is currently not approved by FDA. Higher doses of etanercept seem to be of significant benefit for difficult cases [50], but higher doses of infliximab are associated with increased rates of adverse events [48].

Regarding other subtypes of juvenile idiopathic arthritis, all three drugs appear to be also effective in enthesitis-related arthritis (juvenile spondyloarthropathy), but are significantly less effective in the systemic form of the disease [46, 49, 51, 52]. In the German registry with 322 children treated with etanercept, only 24% of those with the systemic form had a 70% improvement at 12 months, compared to 54% of patients with other subtypes [53].

Ankylosing Spondylitis

Ankylosing spondylitis is a chronic progressive disease characterized by sacroiliitis, synovitis and inflammation at the sites of tendinous insertions into bone (enthesitis), affecting about 0.5% of the population. Chronic spinal inflammation leads to ankylosis, abnormal posture and permanent disability. Extraskeletal manifestations involve mainly the eye (uveitis) and bowel inflammation. Before the introduction of TNF antagonists, empiric pharmacological therapy was dominated by NSAIDs which are effective in relief of axial and peripheral symptoms. However, full-dose indomethacine, the most commonly prescribed NSAID, results in low response rates, i.e. only 6 and 30% of patients achieve clinical improvement by 40 and 20% by standard response criteria, respectively [54]. Corticosteroids and DMARDs have shown little benefit in the treatment of ankylosing spondylitis and, being ineffective for spinal disease in contrast to their proven efficacy in rheumatoid and psoriatic arthritis, are not able to modify disease progression.

Multiple trials of etanercept [55–58], infliximab [59–63], adalimumab [64], and golimumab [65] clearly demonstrated that TNF antagonists significantly improve the signs, symptoms, functional status and quality of life of patients with ankylosing spondylitis. The extent of short-term response appears to be comparable among the four drugs. Clinical improvement is rapid and can be seen as early as 2 weeks after the start of therapy. Overall, improvement by 20% is seen in more than two thirds of patients (with a placebo response in about one fifth), whereas improvement by 40% is seen in more than half of the responders. Beneficial effects of TNF antagonists include large improvement of thoracoabdominal motion during breathing, which is usually severely impaired [66]. Moreover, data from relevant clinical trials and open-label experience suggest that flares of uveitis occurred less frequently under TNF antagonist treatment compared with placebo [67]. Disease flare occurs upon discontinuation of therapy, but good responses are generally restored upon readministration of anti-TNF treatment [68]. There is growing body of evidence that TNF antagonists also attenuate disease progression, as shown by imaging studies.

The beneficial results demonstrated in randomized trials of TNF antagonists are reproduced in the everyday clinical practice. This has led to consensus recommending TNF antagonists in patients with ankylosing spondylitis who fail to respond to NSAID therapy, meaning that virtually all patients should try these drugs at some point. Thus, it is not surprising that TNF antagonists are used in these patients more frequently than in other diseases. For example, in a Norwegian registry and as of 2005, the proportion of patients who have received anti-TNF drugs in ankylosing spondylitis was 54%, versus 23, 22, and 37% in rheumatoid, psoriatic, and juvenile arthritis, respectively [69]. Results from this registry indicate that 1-year retention rate of anti-TNF treatment was 78% versus 65% in rheumatoid arthritis, suggesting that survival of anti-TNF treatment is superior in ankylosing spondylitis. Larger improvements in quality of life in these patients may contribute to the differences in drug survival [70].

Psoriasis and Psoriatic Arthritis

Psoriasis is a common, chronic, inflammatory disease with predominantly skin/nail and joint manifestations affecting approximately 2% of the population. Patients afflicted with moderate to severe psoriasis suffer the aggravation associated with the pain, itchiness, and bleeding of the psoriatic lesions which impacts their daily functions and social well-being. About one fifth of patients with psoriasis develop at some point arthritis involving peripheral joints and axial skeleton. A significant proportion of patients with psoriatic arthritis have bone erosions and peripheral joint destruction. In addition, enthesitis and extensive bone resorption at the distal phalages are common features. Traditional pharmacologic measures in psoriasis and psoriatic arthritis are largely similar to those used in rheumatoid arthritis, including DMARDs (mainly methotrexate and cyclosporine-A), but have produced at best only partial responses.

Multiple randomized trials have shown that TNF antagonists are effective for the treatment of patients with either moderate to severe psoriasis or psoriatic arthritis, producing rapid and significant improvements in joint and skin manifestations and reducing disability. All monoclonal antibodies under trial, namely, infliximab [71–73], adalimumab [74–76] and golimumab [77], as well as the soluble receptor etanercept [78, 79], result in high clinical response rates in the majority of patients, including those who failed methotrexate. Efficacy in peripheral joint disease was roughly similar to the efficacy observed in rheumatoid arthritis. Enthesitis and dactylitis also showed significant improvement and radiographic progression of joint damage slowed. However, no difference between treatment and placebo was found in specific radiologic changes, such as pencil-in-cup change, osteolysis, or periostitis, presumably because these are chronic and fixed types of changes.

It appears that all anti-TNF agents have comparable efficacy, but relevant evidence is more limited than in other diseases [80–81]. Overall percentages of patients who experience 75% improvement in skin involvement from baseline are high. For example, at week 10, 80% of patients with moderate to severe psoriasis treated with infliximab achieved at least a 75% improvement from baseline and 57% achieved at least a 90% improvement, compared with 3 and 1% in the placebo group, respectively; these responses were generally maintained at week 50 [71]. Comparable results, along with significant improvement in health-related quality of life, have been produced in trials assessing adalimumab and etanercept.

The impressive trial results led to extensive use of TNF antagonists in patients with psoriasis refractory to traditional regimens, as well as in patients with active psoriatic arthritis that is not controlled with NSAIDs in the case of axial disease and DMARDs in the case of peripheral arthritis [80–85]. In the Norwegian registry and as of 2005, the proportion of patients with psoriatic arthritis who have received anti-TNF drugs was 22% [69]; 1-year retention rate of anti-TNF treatment was 77%, which is higher than that in rheumatoid arthritis (65%) probably due to greater improvements in quality of life [78]. For these patients, TNF antagonists are currently the best available treatment option.

Inflammatory Bowel Disease: Crohn's Disease and Ulcerative Colitis

Chronic inflammatory bowel disease, a systemic disorder of unknown etiology involving mainly the gastrointestinal tract, is divided in two major groups: Crohn's disease, which primarily affects the small intestine (also known as regional enteritis), and chronic nonspecific ulcerative colitis, an inflammatory reaction primarily involving the colonic mucosa. These disorders are distinguished pathologically in most instances and clinically are characterized by recurrent inflammation of intestinal segments manifested as diarrhea, bloody diarrhea, malabsorption, persistent perianal sepsis and abdominal pain. Extracolonic manifestations include arthritis, ocular inflammation (most often uveitis), liver disease, various skin and mucosal lesions and nutritional and metabolic complications. Inflammatory bowel disease has been traditionally treated with major immunosuppressive regimens, but it often has an unpredictable course and carries significant morbidity and mortality.

The first approved indication of an anti-TNF monoclonal antibody was Crohn's disease. Infliximab in the pivotal randomized trial was given as a single infusion of 5, 10, or 20 mg/kg; patients who did not respond 4 weeks after infusion were given infliximab in an open-label fashion at 10 mg/kg and followed for an additional 12 weeks. Remarkably, 81% of patients given 5 mg/kg had a clinical response at week 4, and 33% of patients treated with infliximab went into remission compared with 4% in the placebo group [86]. Subsequent trials with infliximab [87–89], adalimumab [90–91] and certolizumab [92–94] confirmed the remarkable efficacy of anti-TNF agents in inducing remission in a significant proportion of patients with moderate to severe Crohn's disease. As expected, continuous maintenance therapy results in clinically meaningful improvements in quality of life and maintenance of the quality and quantity of the remission and response. The proportion of patients returning to a normal life was significant greater among those treated with TNF-antagonists compared with those who received placebo.

In a meta-analysis of fourteen randomized trials enrolling 3,995 patients with luminal Crohn's disease, it was found that the mean difference between TNF antagonist and placebo was 11% for induction of remission at week 4. Moreover, the mean difference between TNF antagonist and placebo in maintenance of remission at weeks 20–30 in patients who responded to induction therapy and in patients randomized before induction was 23% [95]. It should be noted that many of the studies on the maintenance therapy of patients with moderate Crohn's disease reveal a placebo rate of about 35%, probably due to concurrent symptoms of irritable bowel syndrome in these patients. In overall analysis, anti-TNF therapy was effective for fistula closure only in maintenance trials after open-label induction. Efficacy of anti-TNF agents other than infliximab in treating fistulizing Crohn's disease requires additional investigations [95].

Up to date, thousands of patients with Crohn's disease, including pediatric patients, have benefited from this treatment. Anti-TNF monoclonal antibodies are the first drugs shown to induce endoscopic and histologic healing in these patients, and this healing has now been established as a new benchmark by the FDA for the development of new pharmaceuticals for Crohn's disease. Extraintestinal manifestations, such as uveitis, episcleritis, and arthritis [96] or pyoderma gangrenosum [97] may also respond to TNF blockade.

In contrast to the proven efficacy of anti-TNF monoclonal antibodies in Crohn's disease, etanercept was not significantly effective compared to placebo at 4 weeks or at 8 weeks of treatment in a randomized trial of 48 patients [98].

Despite the proven efficacy of anti-TNF antibodies in the treatment of Crohn's disease, evidence for efficacy of these drugs in ulcerative colitis is scant. Only infliximab, which is currently the approved TNF antagonist for reducing signs and symptoms, inducing and maintaining clinical remission and mucosal healing, and eliminating corticosteroid use in ulcerative colitis, has been studied. Of the four randomized trials for the treatment of moderate to severe steroid-refractory disease published so far, the first involved 43 patients [99]; infliximab and placebo were not significantly different with respect to clinical and sigmoidoscopic remission or quality of life 2 and 6 weeks after infliximab treatment. In the second study of 45 patients, infliximab was associated with a significantly reduced need for colectomy (29%) compared with placebo (67%) [100]. The two subsequent studies together included 728 patients who were either intolerant of or refractory to immunosuppressant agents or steroids, or steroid dependent [101]. The rate of clinical response at 8 weeks, the primary outcome point, was significantly higher with infliximab (about 60%) compared with placebo (about 30%). Based on these data, infliximab was associated with improved health-related quality of life. It is clear, however, that further trials are needed to assess infliximab's impact on the treatment and progression of ulcerative colitis. On the other hand, real-life clinical practice may have better outcomes than those shown in randomized controlled trials, as suggested by recent observational studies in patients with perhaps milder forms of disease than patients included in the trials [102].

Adamantiades-Behcet's Disease

Adamantiades-Behcet's disease is a chronic, relapsing vasculitic disorder of unknown origin that causes mucocutaneous lesions, ocular inflammation (uveoretinitis), central nervous system manifestations, as well as arthritis and gastrointestinal lesions. This disease is rare in the USA and Northern Europe, but prevalent in the Middle East, Far East and the Mediterranean basin. Uveoretinitis occurs in about 70% of the patients and usually has a chronic relapsing course which, despite major immunosuppressive treatment, can lead to blindness.

In the first study of infliximab in a small group of patients with sight-threatening uveoretinitis, effective suppression of ocular inflammation within the first 24 h following a single infusion of infliximab infusion was noted [103]. Three subsequent independent, open-label, prospective, self-controlled studies on the long-term effects of repetitive infliximab infusions in preventing ocular relapses, maintaining visual acuity, and the ability to taper immunosuppressive therapy in patients who were unresponsive or intolerant to standard immunosuppressive treatment had positive results [104–106]. Given the paucity of effective and fast-acting therapies particularly for patients with eye disease, these results led to infliximab's approval in Japan for the treatment of 'Behcet's disease complicated with refractory uveoretinitis which does not respond to conventional therapies' (Osaka, Japan, January 26, 2007, JCN Newswire). Infliximab is also effective for extraocular manifestations in patients treated for uveoretinitis, as well as in other patients with recalcitrant orogenital ulcers, arthritis, intestinal or central nervous system involvement, and in isolated patients with pulmonary artery aneurysm [107, 108]. Moreover, in a 4-week small but randomized trial in patients with mucocutaneous manifestations, significantly beneficial effects of etanercept versus placebo were reported; no data are at hand from this study on eye involvement [109].

Based on the available evidence, recommendations on the use of TNF-antagonists in Adamantiades-Behcet's disease have been developed from an expert panel [108]. According to these recommendations, TNF antagonists should be used in selected patients with severe disease. Patients with two or more relapses of posterior uveitis per year, low visual acuity due to chronic cystoid macular edema, or active central nervous system disease, and/or selected patients with intestinal inflammation, or arthritic and mucocutaneous manifestations which reduce significantly the quality of life would fit in this category. According to the experience accumulated so far, infliximab seems to be more efficacious than etanercept in disease manifestations other than mucocutaneous or joint involvement, while data on adalimumab are still limited. Moreover, a single infusion of infliximab (5 mg/kg) can be used as a first-line agent for sight-threatening, bilateral posterior eye segment inflammation, when the fast-onset of response is considered to be critical to prevent fixed retinal lesions and thus permanent visual loss. In those patients whose ocular relapses are not controlled by azathioprine and/or cyclosporine, a maintenance therapy with infliximab at the dose of 5 mg/kg every 6–8 weeks could be used for up to 2 years, provided no relapse occurs between intervals.

Other Diseases as Targets of TNF Antagonists: Preliminary Successes and Failures

Animal and human studies, including pilot clinical studies of TNF antagonists with positive results, suggest that TNF plays a major pathogenetic role in a broad spectrum of inflammatory diseases (table 2). For example, a role of TNF in ocular inflammatory conditions has been recently demonstrated in small trials reporting preliminary results on the efficacy of these agents in patients with noninfectious uveitis, regardless of the origin of the disease [reviewed in reference 110]. These reports suggest that TNF antagonists are useful in the treatment of ocular inflammation, not only when it occurs in the course of the approved systemic diseases, but also in idiopathic uveitis or scleritis, birdshot retinochoroiditis, sarcoidosis and Grave's disease ophthalmopathy. Infliximab was also beneficial in small numbers of patients with uveitic [111] and sight-threatening diabetic cystoid macular edema [112], as well as in patients with age-related macular degeneration [113], the most common cause of blindness in patients over the age of 60 years. Most recently, following experiments in rabbits to

Table 2. Off-label use of TNF antagonists in noninfectious inflammatory diseases with variably beneficial results

Systemic diseases
Systemic lupus erythematosus-associated glomerulonephritis
Systemic sclerosis
Dermatomyositis
Sjogren's syndrome
Adult-onset Still's disease
Wegener's granulomatosis
Giant cell arteritis
Sarcoidosis
Organ-specific autoimmune diseases
Diabetes type 1
Autoimmune inner ear disease
Ocular diseases
Idiopathic uveitis or scleritis
Birdshot retinochoroiditis
Grave's disease ophthalmopathy
Idiopathic scleritis
Cystoid macular edema (uveitic and diabetic)
Age-related macular degeneration
Dermatological diseases
Pemphigoid
Hidradenitis suppurativa
Pyoderma gangrenosum
Acne
Aphthous stomatitis
Pityriasis rubra pilaris
Eosinophilic fasciitis
Panniculitis
Other
Chronic obstructive pulmonary disease
Herniated disc
Polymyalgia rheumatica
Idiopathic membranous nephropathy

establish the safety of intraocular delivery of infliximab [114], intravitreal injections of infliximab were beneficial in a small series of patients with age-related macular degeneration refractory to any available treatment option [115].

Several open-label studies reported positive results with either infliximab, adalimumab or etanercept in a variety of dermatologic diseases other than psoriasis, including pemphigoid, hidradenitis suppurativa, pyoderma gangrenosum, acne, aphthous stomatitis, pityriasis rubra pilaris, eosinophilic fasciitis, and panniculitis [reviewed in references 116, 117]. Anti-TNF therapies have also been reported in pilot studies with varied success in patients with proliferative glomerulonephritis associated with systemic lupus erythematosus [118], idiopathic membranous nephropathy [119], children with new onset diabetes type 1 [120], patients with systemic sclerosis [121], dermatomyositis [122–124], adult-onset Still's disease [125, 126], various forms of systemic vasculitis [127, 128], and sarcoidosis [129, 130].

However, several important controlled trials emphasized the crucial role of patient randomization to receive active treatment, or not, in determining whether anti-TNF treatment is effective, even in the face of promising preliminary data. Indeed, some randomized trials performed so far have shown that TNF antagonists are not superior to placebo in Sjogren's syndrome (infliximab) [131], Wegener's granulomatosis (etanercept) [132], autoimmune inner ear disease (etanercept) [133], chronic obstructive pulmonary disease (infliximab) [134], herniated disc (infliximab) [135], while epidural etanercept may hold promise for the treatment of sciatica [136], giant cell arteritis (infliximab) [137], polymyalgia rheumatica (infliximab) [138], and pulmonary sarcoidosis (infliximab) [139]. The results of these trials do not preclude adjunctive anti-TNF treatment in selected patients with these diseases who are refractory to standard therapeutic regimens [132, 137, 140]. On the other hand, despite multiple lines of evidence that TNF is involved in the pathogenesis of diseases such as multiple sclerosis and chronic heart failure, anti-TNF therapy proved deleterious in multiple sclerosis [141, 142] and adversely affected the clinical condition of patients with moderate-to-severe chronic heart failure [143, 144].

Safety Issues

The key safety considerations which emerged during the first years of clinical use of TNF antagonists included infections, autoimmune disease, demyelinating disease, malignancies and congestive heart failure. Overall rates of these conditions in randomized controlled trials were not significantly increased during treatment compared with placebo. However, extension studies, case reports and postmarketing surveillance data highlighted the deleterious effects of blocking the protective role of TNF in host defense in many patients, as well as the clinical importance of attenuating the role of TNF in regulating (auto)antigen-presenting cell function in rare cases (table 3).

In a large, randomized, placebo-controlled trial, the risk of serious infections in patients receiving the approved infliximab dose of 3 mg/kg plus methotrexate for rheumatoid arthritis was similar to that in patients receiving methotrexate alone. Patients receiving an induction regimen of 10 mg/kg infliximab plus methotrexate followed by a 10 mg/kg maintenance regimen had an increased risk of serious infections through week 22 [145], emphasizing the risk which accompanies a more profound inhibition of TNF actions. The German Registry of patients on TNF antagonists reported that after adjusting for differences in the case patient mix, the relative risks

Table 3. Safety issues associated with TNF antagonists

Screening prior to treatment	
	Latent tuberculosis
	Hepatitis B
	Hepatitis C
	Human immunodeficiency virus
Contraindications	
	Advanced congestive heart failure
	Preexisting multiple sclerosis
	Active infection
	Lymphoma
	Solid tumors
	Pregnancy (relative)
Rare adverse reactions	
	Aplastic anemia
	Demyelination
	Psoriasis
	Systemic lupus erythematosus
	Vasculitis

of serious infections were 2.2 for patients receiving etanercept and 2.1 for patients receiving infliximab (pulmonary and skin infections, particularly herpes, were more frequent), compared with controls [146]. In the UK nationwide registry, there were no significant differences in rates of serious infections, defined as those requiring hospitalization and IV antibiotics or resulting in death, between patients treated with TNF antagonists or DMARDs [147].

TNF antagonists might reactivate chronic HBV infection, yet concurrent treatment of infliximab with lamivudine can stabilize HBV disease activity [148]. Caution is needed when considering anti-TNF therapy in patients with chronic HCV infection. It is advised to screen all patients for hepatitis B and C prior to treatment with anti-TNF therapy and in case of a positive result, anti-TNF therapy should be initiated only in combination with hepatitis treatment with antiviral agents and close monitoring of serum aminotransferases and serum viral DNA levels [148]. Limited evidence suggests that treatment with anti-TNF therapy is a viable alternative in HIV patients without advanced disease and associated rheumatic diseases refractory to standard therapy [149], but great caution is needed [150]. Although, fungal infections were not significantly increased in randomized trials, based on current case reports and post-marketing surveillance data, if a patient on anti-TNF therapy develops a fever, fungal infections should be considered.

The important role played by TNF in regulating (auto)antigen-presenting cell function may also be linked with new clinical signs of autoimmune disease arising in some patients treated with TNF antagonists [151]. Development of systemic lupus erythematosus in patients with inflammatory arthritides is rare, whereas withdrawal of anti-TNF therapy usually leads to resolution of symptoms [151, 152]. On the other hand, development of antinuclear antibodies following anti-TNF treatment is a common finding, and even antibodies against double-stranded DNA or cardiolipin often appear [153–155]. Several demyelinating and neurologic events, including exacerbations of preexisting multiple sclerosis, were reported in association with TNF antagonism in patients with inflammatory arthritides; however, the incidence of demyelinating disease does not appear to be increased in patients on anti-TNF therapy when compared to the general population.

There is thus far no firm evidence that these drugs are associated with an increased incidence of solid tumor development, or recurrence rate in those who had solid malignancies previously. Also, although current data suggest a higher rate of lymphomas in patients receiving TNF antagonists relative to the general population, whether TNF antagonists increase the risk of lymphomas in patients with chronic inflammatory conditions receiving conventional immunosuppressive regiments, associated per se with increased lymphoma rates, remains uncertain [156, 157]. A cohort study of 1,152 biologic users and 7,306 methotrexate users reported that a propensity score-adjusted pooled hazard ratio was 1.37 for hematologic malignancies and 0.91 for solid tumors, which does not support the likelihood of increased risk of malignancies by the use of TNF antagonists [158]. The results of randomized controlled trials show that 26% of malignancies occur within 12 weeks from enrollment in patients receiving anti-TNF therapies, suggesting preexistence [159]; thus, current cancer screening procedures in clinical practice may need further study.

Multiple lines of evidence suggest that the first three marketed TNF antagonists have a comparable safety profile, with the exception of tuberculosis reactivation risk which is discussed below. Regarding infliximab, acute allergic reactions were seen in approximately 5% of intravenous infusions, but using appropriate treatment protocols these reactions are effectively treated in nearly all patients. Provided that anti-TNF agents are not used to treat patients with active infection, malignancy, preexisting demyelinating conditions, and heart failure, their safety profile is excellent. For example, in the most recent safety analysis in 19,041 patients exposed to adalimumab in 36 global clinical trials from 1997 to 2007, cumulative rates of serious adverse events remained stable over time. Overall malignancy and mortality rates for adalimumab-treated patients were as expected for the general population [160].

TNF Antagonists and Tuberculosis

Tuberculosis has been the most common serious infection observed with TNF antagonists [157]. Its incidence is influenced by age, concomitant immunosuppressive regimens, low socioeconomic status, and geography [161]. The cumulative incidence of reported cases among patients in the USA receiving these drugs from January 1998 through September 2002 was estimated at 54/100,000 for infliximab and 28/100,000 for etanercept, suggesting that infliximab carries a higher risk than etanercept [161]. Subsequent studies on postmarketing surveillance data have confirmed the increased risk of tuberculosis posed by anti-TNF antibodies compared with soluble TNF receptor, particularly with regard to reactivation of latent infection occurring at greatest frequency within the first 12 weeks of anti-TNF treatment [162, 163]. Various structural and functional differences seem to account for this finding [163–166]. In in vitro studies using therapeutic drug concentrations in whole-blood culture, infliximab and adalimumab reduced the proportion of tuberculosis-responsive CD69+ CD4 cells by 70 and 49%, and suppressed antigen-induced interferon-γ production by 70 and 64%, respectively; in contrast, etanercept produced no significant effect. Adalimumab and etanercept had divergent, concentration-dependent effects on control of intracellular growth of mycobacterium tuberculosis. None of the drugs induced significant levels of apoptosis or necrosis in either monocytes or T cells, suggesting that the tuberculosis risk posed by antibodies may reflect their combined effects on TNF and interferon-γ [165].

Screening of patients who are about to start anti-TNF therapy has dramatically reduced the risk of activating tuberculosis. Screening should include history, physical examination, and purified protein derivative skin tests, whereas it may be appropriate to consider chest films and QuantiFERON-TB Gold test (measures interferon-α production after 16- to 24-hour incubation of whole blood with mycobacterial peptides) in patients when there is a suspicion of a compromised skin test. Several consensus guidelines for screening and treatment of latent tuberculosis in these patients have been proposed by different organizations. Compliance with recommendations to prevent reactivation of latent infection leads is imperative. According to data from the BIOBADASER registry, which evaluated new cases of active tuberculosis in 5,198 patients treated with TNF antagonists after the dissemination of recommendations [167], the probability of developing active tuberculosis was seven times higher when recommendations were not followed. In the case of latent tuberculosis, starting preventive therapy before beginning TNF blocking agents is imperative. TNF antagonists should be immediately discontinued in the setting of active tuberculosis; the time to resume anti-TNF therapy after completion of antituberculosis therapy is still controversial.

Paradoxical Adverse Reactions of TNF Antagonists

The development of psoriasis and psoriasiform lesions in patients treated with either etanercept, infliximab or adalimumab, in particular palmoplantar pustulosis in more than half of all reported cases, was severely questioned when first described [168, 169]. However, psoriasis was subsequently documented in hundreds of such patients

[170, 171]. In a recent prospective study of 9,826 patients, 25 patients with rheumatoid arthritis were described with new onset of psoriasis as a consequence of anti-TNF treatment. Interestingly, in a comparison group of 2,880 patients treated with traditional disease-modifying antirheumatic drugs, there was no evidence of psoriasis [172]. Most of the patients who developed psoriasis were treated with adalinumab, whereas 6 were treated with infliximab and a further 6 with etanercept, suggesting a class effect rather than a drug-specific type of hypersensitivity [172]. This paradoxical phenomenon, when considering the proven effects of anti-TNF agents in psoriasis and psoriatic arthritis, has now been classified as an accepted, even though rare, side effect in patients receiving TNF antagonists.

The mechanisms underlying psoriasis development after anti-TNF treatment are puzzling and may relate to a promotion of activation of autoreactive T cells, leading to tissue damage via autoimmune mechanisms [171]. In the setting of other autoimmune diseases, therapeutic TNF blockade has been shown to have differential effects, improving lupus nephritis in some patients, but inducing typical SLE in others, as discussed above. The development of pustular eruptions may have a different pathogenetic background, compared to the more plaque type psoriasis-like lesions. As proposed recently, palmoplantar pustulosis originates from an inherent dysfunction of the palmoplantar sweat glands and, as shown by expression studies, TNF may possibly play a crucial role in their function [173].

Other paradoxical, but very rare, adverse reactions of anti-TNF therapies include vasculitis, which can be potentially serious [151, 174], and sarcoidosis [175]. Finally, biopsy-proven glomerulonephritis, uveitis and Crohn's disease have been reported in isolated patients treated with TNF antagonists; however, a true association remains uncertain.

Conclusions and Future Directions

The remarkable efficacy of TNF antagonists in various diseases helped us to better understand that common pathways operate in chronic noninfectious inflammatory diseases affecting joints, skin and gut, as well as that these diseases are more closely related than previously recognized. It seems that the anti-TNF monoclonal antibodies infliximab and adalimumab and the soluble TNF receptor etanercept are equally effective treatments for rheumatoid arthritis, juvenile chronic arthritis, ankylosing spondylitis, psoriatic arthritis and psoriasis. In contrast, only the monoclonal antibodies are efficacious in inflammatory bowel disease, and perhaps in uveitis, underscoring the dissimilarities of TNF-mediated pathology between inflammatory disease states. More studies on the different outcomes obtained when particular agents are used, as well as monitoring TNF-mediated biological processes during the course of treatment in different patients will help us better understand the pathophysiology of the diseases currently being treated by these agents and perhaps identify those patients who will benefit the most from TNF inhibition. Such information will offer new insights into the mechanistic interplays among patients and the heterogeneous pathogenic processes operating even within the same disease spectrum.

The introduction of TNF antagonists in clinical practice has transformed the lives of many patients who were refractory to previously available therapies. For example, it is estimated that among the rheumatoid arthritis population in Europe about 30% of patients are currently being treated successfully with TNF antagonists. With increasing clinical experience, these drugs will remain in the coming years the first choice of biologic treatment for patients failing traditional DMARDs, whereas any novel biologic therapy needs to demonstrate superiority to anti-TNF therapy in order to progress up the treatment algorithm. On the other hand, the high cost of TNF antagonists, currently ranging between USD 13,000 and 20,000 per year per patient, limits many patients' access to therapy and fuels the demand for cheaper drugs.

Much remains to be learned about the individualized factors that may enhance the efficacy of TNF antagonists in the approved indications, but also about additional potential indications. It is now clear that about one third of patients suffering from any approved indication do not clinically respond to TNF antagonists, while disease remission is achieved in only a minority of the responders. Whether the combination of anti-TNF with other biologic therapies is of further benefit and safe, remains to be tested. Selection and dosing of TNF antagonist needs to be individualized to achieve a maximal response, and in order to maintain this response the dose may be increased or the frequency between dosing decreased. Much lower than the conventional doses are enough in some patients. In many responders, unless anti-TNF therapy is maintained the manifestations of the disease return. In contrast, some of those patients with early rheumatoid arthritis in whom long-term anti-TNF therapy induced disease remission remain disease-free after discontinuation of any kind of treatment. On the other hand, optimism for targeting TNF in patients with various organ-specific or systemic (auto)immune-mediated diseases provides the foundation and justification for additional randomized controlled trials.

Regarding selection of patients who will likely have a major response, it seems that the main factor influencing therapeutic efficacy is the prior response to DMARD treatment [176]. The problem of treating patients with TNF antagonists very early is to identify the right patient who will really benefit. Since no clinical or laboratory tests are available to predict response to anti-TNF therapies, great need exists for predictive biomarkers. For example, changes in circulating autoantibody titers do not predict treatment outcome [177, 178]. Preliminary results suggest that a combination of multi-step proteomics approach with arthritis antigen arrays, a multiplex cytokine assay, and conventional ELISA methods enables pretreatment classification and prediction of etanercept responders among patients with rheumatoid arthritis [179].

Clinical practice guidelines and consensus statements on the criteria of introduction, duration of treatment and cessation of TNF antagonists, including safety issues, are under constant revision as data from longer periods of patient exposure accumulate. The safety profile is clearly favorable, especially if recommended doses are used. However, more long-term data are needed to develop a consensus on the estimated risk of lymphoma with anti-TNF therapy. Continued surveillance of patients is warranted until the relationship with lymphoma development is fully characterized. Moreover, the balance between effective TNF blockade to control aggressive inflammatory disease states and adequate remaining TNF activity to confer immunoprotection against infections such as tuberculosis is an important issue. More studies on how each of the available TNF antagonists affects the complex interactions of the inflammation cascade and apoptosis, and whether the effects are modulatory or destructive, are needed.

New anti-TNF agents, including monoclonal antibodies, soluble receptors, small molecules disassembling TNF trimers, or inhibiting intracellular signaling resulting in reduced TNF production are currently being developed [180]. Finally, experimental evidence in murine disease models indicates a heterogeneity of TNF receptor usage in autoimmune suppression versus inflammatory tissue damage, and put forward a rationale for a predictably beneficial effect of 'anti-TNF receptor' instead of 'anti-TNF' treatment in human chronic inflammatory and autoimmune conditions [181]. Other studies indicate that transmembrane TNF is capable of exerting antilisterial host defenses while remaining inadequate to mediate arthritogenic functions, and suppresses experimental encephalomyelitis while retaining the autoimmune suppressive properties of wild-type TNF [182]. Along these lines, it is hoped that more efficacious drugs that will ideally target the deleterious proinflammatory biologic effects of TNF without compromising its protective role in host defense and (auto) immunity will be available in the future.

References

1 Sfikakis PP, Kollias G: Tumor necrosis factor biology in experimental and clinical arthritis. Curr Opin Rheumatol 2003;5:380–386.

2 Lin J, Ziring D, Desai S, Kim S, Wong M, Korin Y, Braun J, Reed E, Gjertson D, Singh RR: TNFα blockade in human diseases: an overview of efficacy and safety. Clin Immunol 2008;126:13–30.

3 Tracey D, Klareskog L, Sasso EH, Selfeld JG, Tak PP: Tumor necrosis factor antagonist mechanisms of action. Pharmacol Ther 2008;117:244–279.

4 Wong M, Ziring D, Korin Y, Desai S, Kim S, Lin J, Gjertson D, Braun J, Reed E, Singh RR: TNFa blockade in human diseases: mechanisms and future directions. Clin Immunol 2008;126:121–136.

5 Fossati G, Nesbitt AM: In vitro complement-dependent cytotoxicity and antibody-dependent cellular cytotoxicity by the anti-TNF agents adalimumab, etanercept, infliximab, and certolizumab pegol (CDP870). Am J Gastroenterol 2005;100:S287.

6 Lugering A, Schmidt M, Lugering N, Pauels HG, Domschke W, Kucharzik T: Infliximab induces apoptosis in monocytes from patients with chronic active Crohn's disease by using a caspase-dependent pathway. Gastroenterology 2001;121:1145–1157.

7 Mitoma H, Horiuchi T, Tsukamoto H: Binding activities of infliximab and etanercept to transmembrane tumor necrosis factor-alpha. Gastroenterology 2004;126:934–935; author reply 935–936.

8 Mitoma H, Horiuchi T, Hatta N, Tsukamoto H, Harashima S, Kikuchi Y, Otsuka J, Okamura S, Fujita S, Harada M: Infliximab induces potent anti-inflammatory responses by outside-to-inside signals through transmembrane TNF-alpha. Gastroenterology 2005;128:376–392.

9 Shen C, Van Assche G, Rutgeerts P, Ceuppens JL: Caspase activation and apoptosis induction by adalimumab: demonstration in vitro and in vivo in a chimeric mouse model. Inflamm Bowel Dis 2006; 12:22–28.

10 Scallon B, Cai A, Solowski N, Rosenberg A, Song XY, Shealy D, Wagner C: Binding and functional comparisons of two types of tumor necrosis factor antagonists. J Pharmacol Exp Ther 2002;301:418–426.
11 Van den Brande JM, Braat H, van den Brink GR, Versteeg HH, Bauer CA, Hoedemaeker I, van Montfrans C, Hommes DW, Peppelenbosch MP, van Deventer SJ: Infliximab but not etanercept induces apoptosis in lamina propria T-lymphocytes from patients with Crohn's disease. Gastroenterology 2003;124:1774–1785.
12 Arora T, Padaki R, Liu L, Hamburger AE, Ellison AR, Stevens SR, Louie JS, Kohno T: Differences in binding and effector functions between classes of TNF antagonists. Cytokine 2009;45:124–131.
13 Gramlick A, Fossati G, Nesbitt AM: Neutralization of soluble and membrane tumor necrosis factor (TNF) by infliximab, adalimumab, or certolizumab pegol using p55 or p75 TNF-receptor-specific bioassays. Gastroenterology 2006;130:697.
14 Mitoma H, Horiuchi T, Tsukamoto H, Tamimoto Y, Kimoto Y, Uchino A, To K, Harashima S, Hatta N, Harada M: Mechanisms for cytotoxic effects of anti-tumor necrosis factor agents on transmembrane tumor necrosis factor alpha-expressing cells: comparison among infliximab, etanercept, and adalimumab. Arthritis Rheum 2008;58:1248–1257.
15 Kaymakcalan Z, Sakorafas P, Bose S, Scesney S, Xiong L, Hanzatian DK, Salfeld J, Sasso EH: Comparisons of affinities, avidities, and complement activation of adalimumab, infliximab, and etanercept in binding to soluble and membrane tumor necrosis factor. Clin Immunol 2009;131:308–316.
16 Donahue KE, Gartlehner G, Jonas DE, Lux LJ, Thieda P, Jonas BL, Hansen RA, Morgan LC, Lohr KN: Systematic review: comparative effectiveness and harms of disease-modifying medications for rheumatoid arthritis. Ann Intern Med 2008;148:124–134.
17 Genovese MC, Bathon JM, Martin RW, Fleischmann RM, Tesser JR, Schiff MH, Keystone EC, Wasko MC, Moreland LW, Weaver AL, Markenson J, Cannon GW, Spencer-Green G, Finck BK: Etanercept versus methotrexate in patients with early rheumatoid arthritis: two-year radiographic and clinical outcomes. Arthritis Rheum 2002;46:1443–1450.
18 Weinblatt M, Kremer J, Bankhurst A, Bulpitt K, Fleischmann R, Fox R, Jackson C, Lange M, Burge D: A trial of etanercept, a recombinant tumor necrosis factor receptor: Fc fusion protein, in patients with rheumatoid arthritis receiving methotrexate. N Engl J Med 1999;340:253–259.
19 Moreland L, Schiff M, Baumgartner S, Tindall E, Fleischmann R, Bulpitt K, Weaver A, Keystone E, Furst D, Mease P, Ruderman E, Horwitz D, Arkfeld D, Garrison L, Burge D, Blosch C, Lange M, McDonnell N, Weinblatt M: Etanercept therapy in rheumatoid arthritis. A randomized, controlled trial. Ann Intern Med 1999;130:478–486.
20 Lipsky PE, van der Heijde DM, St Clair EW, Furst DE, Breedveld FC, Kalden JR, Smolen JS, Weisman M, Emery P, Feldmann M, Harriman GR, Maini RN: Infliximab and methotrexate in the treatment of rheumatoid arthritis. Anti-Tumor Necrosis Factor Trial in Rheumatoid Arthritis with Concomitant Therapy Study Group. N Engl J Med 2000;343:1594–1602.
21 Weinblatt ME, Keystone EC, Furst DE, Moreland LW, Weisman MH, Birbara CA, Teoh LA, Fischkoff SA, Chartash EK: Adalimumab, a fully human anti-tumor necrosis factor alpha monoclonal antibody, for the treatment of rheumatoid arthritis in patients taking concomitant methotrexate: the ARMADA trial. Arthritis Rheum 2003;48:35–45.
22 Keystone EC, Kavanaugh AF, Sharp JT, Tannenbaum H, Hua Y, Teoh LS, Fischkoff SA, Chartash EK: Radiographic, clinical, and functional outcomes of treatment with adalimumab (a human anti-tumor necrosis factor monoclonal antibody) in patients with active rheumatoid arthritis receiving concomitant methotrexate therapy: a randomized, placebo-controlled, 52-week trial. Arthritis Rheum 2004;50:1400–1411.
23 Klareskog L, van der Heijde D, de Jager JP, Gough A, Kalden J, Malaise M, Martin Mola E, Pavelka K, Sany J, Settas L, Wajdula J, Pedersen R, Fatenejad S, Sanda M: Therapeutic effect of the combination of etanercept and methotrexate compared with each treatment alone in patients with rheumatoid arthritis: double-blind randomised controlled trial. Lancet 2004;363:675–681.
24 van de Putte LB, Atkins C, Malaise M, Sany J, Russell AS, van Riel PL, Settas L, Bijlsma JW, Todesco S, Dougados M, Nash P, Emery P, Walter N, Kaul M, Fischkoff S, Kupper H: Efficacy and safety of adalimumab as monotherapy in patients with rheumatoid arthritis for whom previous disease modifying antirheumatic drug treatment has failed. Ann Rheum Dis 2004;63:508–516.
25 St Clair EW, van der Heijde DM, Smolen JS, Maini RN, Bathon JM, Emery P, Keystone E, Schiff M, Kalden JR, Wang B, Dewoody K, Weiss R, Baker D: Combination of infliximab and methotrexate therapy for early rheumatoid arthritis: a randomized, controlled trial. Arthritis Rheum 2004;50:3432–3443.

26 Smolen JS, Van Der Heijde DM, St Clair EW, Emery P, Bathon JM, Keystone E, Maini RN, Kalden JR, Schiff M, Baker D, Han C, Han J, Bala M: Predictors of joint damage in patients with early rheumatoid arthritis treated with high-dose methotrexate with or without concomitant infliximab: results from the ASPIRE trial. Arthritis Rheum 2006;54:702–710.

27 Breedveld FC, Weisman MH, Kavanaugh AF, Cohen SB, Pavelka K, van Vollenhoven R, Sharp J, Perez JL, Spencer-Green GT: The PREMIER study: a multicenter, randomized, double-blind clinical trial of combination therapy with adalimumab plus methotrexate versus methotrexate alone or adalimumab alone in patients with early, aggressive rheumatoid arthritis who had not had previous methotrexate treatment. Arthritis Rheum. 2006;54:26–37.

28 Kay J, Matteson EL, Dasgupta B, Nash P, Durez P, Hall S, Hsia EC, Han J, Wagner C, Xu Z, Visvanathan S, Rahman MU: Golimumab in patients with active rheumatoid arthritis despite treatment with methotrexate: a randomized, double-blind, placebo-controlled, dose-ranging study. Arthritis Rheum 2008; 58:964–975.

29 Keystone E, Heijde D, Mason D Jr, Landewé R, Vollenhoven RV, Combe B, Emery P, Strand V, Mease P, Desai C, Pavelka K: Certolizumab pegol plus methotrexate is significantly more effective than placebo plus methotrexate in active rheumatoid arthritis: findings of a fifty-two-week, phase III, multicenter, randomized, double-blind, placebo-controlled, parallel-group study. Arthritis Rheum 2008;58:3319–3329.

30 Keystone EC, Genovese MC, Klareskog L, Hsia EC, Hall ST, Miranda PC, Pazdur J, Bae SC, Palmer W, Zrubek J, Wiekowski M, Visvanathan S, Wu Z, Rahman MU: Golimumab, a human antibody to tumour necrosis factor alpha given by monthly subcutaneous injections, in active rheumatoid arthritis despite methotrexate therapy: the GO-FORWARD Study. Ann Rheum Dis 2009;68:789–796.

31 Smolen J, Landewé RB, Mease P, Brzezicki J, Mason D, Luijtens K, van Vollenhoven RF, Kavanaugh A, Schiff M, Burmester GR, Strand V, Vencovsky J, van der Heijde D: Efficacy and safety of certolizumab pegol plus methotrexate in active rheumatoid arthritis: the RAPID 2 study. A randomised controlled trial. Ann Rheum Dis 2009;68:797–804.

32 Fleischmann R, Vencovsky J, van Vollenhoven RF, Borenstein D, Box J, Coteur G, Goel N, Brezinschek HP, Innes A, Strand V: Efficacy and safety of certolizumab pegol monotherapy every 4 weeks in patients with rheumatoid arthritis failing previous disease-modifying antirheumatic therapy: the FAST4WARD study. Ann Rheum Dis 2009;68:805–811.

33 van der Bijl AE, Goekoop-Ruiterman YP, de Vries-Bouwstra JK, Ten Wolde S, Han KH, van Krugten MV, Allaart CF, Breedveld FC, Dijkmans BA: Infliximab and methotrexate as induction therapy in patients with early rheumatoid arthritis. Arthritis Rheum 2007;56:2629–2634.

34 van der Kooij SM, Goekoop-Ruiterman YP, de Vries-Bouwstra JK, Güler-Yüksel M, Zwinderman AH, Kerstens PJ, van der Lubbe PA, de Beus WM, Grillet BA, Ronday HK, Huizinga TW, Breedveld FC, Dijkmans BA, Allaart CF: Drug-free remission, functioning and radiographic damage after 4 years of response-driven treatment in patients with recent-onset rheumatoid arthritis. Ann Rheum Dis 2009;68:914–921.

35 Zink A, Strangfeld A, Schneider M, Herzer P, Hierse F, Stoyanova-Scholz M, Wassenberg S, Kapelle A, Listing J: Effectiveness of tumor necrosis factor inhibitors in rheumatoid arthritis in an observational cohort study: comparison of patients according to their eligibility for major randomized clinical trials. Arthritis Rheum 2006;54:3399–3407.

36 Kievit W, Fransen J, Oerlemans AJ, Kuper HH, van der Laar MA, de Rooij DJ, De Gendt CM, Ronday KH, Jansen TL, van Oijen PC, Brus HL, Adang EM, van Riel PL: The efficacy of anti-TNF in rheumatoid arthritis, a comparison between randomised controlled trials and clinical practice. Ann Rheum Dis 2007;66:1473–1478.

37 Greenberg JD, Kishimoto M, Strand V, Cohen SB, Olenginski TP, Harrington T, Kafka SP, Reed G, Kremer JM: Consortium of Rheumatology Researchers of North America Investigators. Tumor necrosis factor antagonist responsiveness in a United States rheumatoid arthritis cohort. Am J Med 2008;121:532–538.

38 Pavelka K, Gatterová J, Tegzová D, Jarosová K, Tomasová Studýnková J, Svobodník A, Svihálek J, Dusek L, Vencovský J: Radiographic progression of rheumatoid arthritis in patients from the Czech National Registry receiving infliximab treatment. Clin Exp Rheumatol 2007;25:540–545.

39 Jacobsson LT, Turesson C, Gülfe A, Kapetanovic MC, Petersson IF, Saxne T, Geborek P: Treatment with tumor necrosis factor blockers is associated with a lower incidence of first cardiovascular events in patients with rheumatoid arthritis. J Rheumatol 2005;32:1213–1218.

40 Barnabe C, Hanley DA: Effect of tumor necrosis factor alpha inhibition on bone density and turnover markers in patients with rheumatoid arthritis and spondyloarthropathy. Semin Arhritis Rheum 2008, Epub ahead of print.

41 Finckh A, Simard JF, Gabay C, Guerne PA: SCQM physicians: evidence for differential acquired drug resistance to anti-tumour necrosis factor agents in rheumatoid arthritis. Ann Rheum Dis 2006;65:746–752.

42 Pavelka K, Jarosova K, Suchy D, Senolt L, Chroust K, Dusek L, Vencovsky J: Increasing the infliximab dose in rheumatoid arthritis patients: a randomized, double blind study failed to confirm its efficacy.Ann Rheum Dis. 2009;68:1285–1289.

43 van Vollenhoven R, Harju A, Brannemark S, Klareskog L: Treatment with infliximab (Remicade) when etanercept (Enbrel) has failed or vice versa: data from the STURE registry showing that switching tumour necrosis factor alpha blockers can make sense. Ann Rheum Dis 2003;62:1195–1198.

44 Wick MC, Ernestam S, Lindblad S, Bratt J, Klareskog L, van Vollenhoven RF: Adalimumab (Humira) restores clinical response in patients with secondary loss of efficacy from infliximab (Remicade) or etanercept (Enbrel): results from the STURE registry at Karolinska University Hospital. Scand J Rheumatol 2005;34:353–358.

45 Smolen JS, Kay J, Doyle MK, Landewé R, Matteson EL, Wollenhaupt J, Gaylis N, Murphy FT, Neal JS, Zhou Y, Visvanathan S, Hsia EC, Rahman MU, for the GO-AFTER study investigators: Golimumab in patients with active rheumatoid arthritis after treatment with tumour necrosis factor alpha inhibitors (GO-AFTER study): a multicentre, randomised, double-blind, placebo-controlled, phase III trial. Lancet 2009;374:210–221.

46 Lovell DJ, Giannini EH, Reiff A, Cawkwell GD, Silverman ED, Nocton JJ, Stein LD, Gedalia A, Ilowite NT, Wallace CA, Whitmore J, Finck BK: Etanercept in children with polyarticular juvenile rheumatoid arthritis. Pediatric Rheumatology Collaborative Study Group. N Engl J Med 2000; 342: 763–769.

47 Lovell DJ, Giannini EH, Reiff A, Jones OY, Schneider R, Olson JC, Stein LD, Gedalia A, Ilowite NT, Wallace CA, Lange M, Finck BK, Burge DJ: Long-term efficacy and safety of etanercept in children with polyarticular-course juvenile rheumatoid arthritis: interim results from an ongoing multicenter, open-label, extended-treatment trial. Arthritis Rheum 2003;48:218–226.

48 Ruperto N, Lovell DJ, Cuttica R, et al, Paediatric Rheumatology International Trials Organisation, Pediatric Rheumatology Collaborative Study Group: A randomized, placebo-controlled trial of infliximab plus methotrexate for the treatment of polyarticular-course juvenile rheumatoid arthritis. Arthritis Rheum 2007;56:3096–3106.

49 Lovell DJ, Ruperto N, Goodman S, Reiff A, Jung L, Jarosova K, Nemcova D, Mouy R, Sandborg C, Bohnsack J, Elewaut D, Foeldvari I, Gerloni V, Rovensky J, Minden K, Vehe RK, Weiner LW, Horneff G, Huppertz HI, Olson NY, Medich JR, Carcereri-De-Prati R, McIlraith MJ, Giannini EH, Martini A, Pediatric Rheumatology Collaborative Study Group, Pediatric Rheumatology International Trials Organisation: Adalimumab with or without methotrexate in juvenile rheumatoid arthritis. N Engl J Med 2008;359:810–820.

50 Horneff G, Ebert A, Fitter S, Minden K, Foeldvari I, Kümmerle-Deschner J, Thon A, Girschick HJ, Weller F, Huppertz HI: Safety and efficacy of once weekly etanercept 0.8 mg/kg in a multicentre 12 week trial in active polyarticular course juvenile idiopathic arthritis. Rheumatology (Oxford) 2009; 48:916–919.

51 Quartier P, Taupin P, Bourdeaut F, Lemelle I, Pillet P, Bost M, Sibilia J, Kone-Paut I, Gandon-Laloum S, LeBideau M, Bader-Meunier B, Mouy R, Debre M, Landais P, Prieur AM: Efficacy of etanercept for the treatment of juvenile idiopathic arthritis according to the onset type. Arthritis Rheum 2003;48:1093–1101.

52 Kimura Y, Pinho P, Walco G, Higgins G, Hummell D, Szer I, Henrickson M, Watcher S, Reiff A: Etanercept treatment in patients with refractory systemic onset juvenile rheumatoid arthritis. J Rheumatol 2005;32:935–942.

53 Horneff G, Schmeling H, Biedermann T, Foeldvari I, Ganser G, Girschick HJ, Hospach T, Huppertz HI, Keitzer R, Kuster RM, Michels H, Moebius D, Rogalski B, Thon A: The German etanercept registry for treatment of juvenile idiopathic arthritis. Ann Rheum Dis 2004;63:1638–1644.

54 Cinar M, Dinc A, Simsek I, Erdem H, Koc B, Pay S, Kilic S: Evaluation of the short-term efficacy of NSAIDs on patients with active ankylosing spondylitis in daily practice: a 3-month, longitudinal, observational study. Rheumatol Int 2009, Epub ahead of print.

55 Gorman JD, Sack KE, Davis JC Jr: Treatment of ankylosing spondylitis by inhibition of tumor necrosis factor alpha. N Engl J Med. 2002;346:1349–1356.

56 Brandt J, Khariouzov A, Listing J, Haibel H, Sorensen H, Grassnickel L, Rudwaleit M, Sieper J, Braun J: Six-month results of a double-blind, placebo-controlled trial of etanercept treatment in patients with active ankylosing spondylitis. Arthritis Rheum 2003;48:1667–1675.

57 Davis JC Jr, van der HD, Braun J, Dougados M, Cush J, Clegg DO, Kivitz A, Fleischmann R, Inman R, Tsuji W: Recombinant human tumor necrosis factor receptor (etanercept) for treating ankylosing spondylitis: a randomized, controlled trial. Arthritis Rheum 2003;48:3230–3236.

58 Calin A, Dijkmans BA, Emery P, Hakala M, Kalden J, Leirisalo-Repo M, Mola EM, Salvarani C, Sanmarti R, Sany J, Sibilia J, Sieper J, van der LS, Veys E, Appel AM, Fatenejad S: Outcomes of a multicentre randomised clinical trial of etanercept to treat ankylosing spondylitis. Ann Rheum Dis 2004;63: 1594–1600.

59 Braun J, Brandt J, Listing J, Zink A, Alten R, Golder W, Gromnica-Ihle E, Kellner H, Krause A, Schneider M, Sorensen H, Zeidler H, Thriene W, Sieper J: Treatment of active ankylosing spondylitis with infliximab: a randomised controlled multicentre trial. Lancet 2002;359:1187–1193.

60 Braun J, Brandt J, Listing J, Zink A, Alten R, Burmester G, Golder W, Gromnica-Ihle E, Kellner H, Schneider M, Sorensen H, Zeidler H, Reddig J, Sieper J: Long-term efficacy and safety of infliximab in the treatment of ankylosing spondylitis: an open, observational, extension study of a three-month, randomized, placebo-controlled trial. Arthritis Rheum 2003;48:2224–2233.

61 Braun J, Brandt J, Listing J, Zink A, Alten R, Burmester G, Gromnica-Ihle E, Kellner H, Schneider M, Sorensen H, Zeidler H, Sieper J: Two year maintenance of efficacy and safety of infliximab in the treatment of ankylosing spondylitis. Ann Rheum Dis 2005;64:229–234.

62 Braun J, Baraliakos X, Brandt J, Listing J, Zink A, Alten R, Burmester G, Gromnica-Ihle E, Kellner H, Schneider M, Sorensen H, Zeidler H, Sieper J: Persistent clinical response to the anti-TNF-alpha antibody infliximab in patients with ankylosing spondylitis over 3 years. Rheumatology (Oxford) 2005;44:670–676.

63 van der Heijde D, Dijkmans B, Geusens P, Sieper J, DeWoody K, Williamson P, Braun J: Efficacy and safety of infliximab in patients with ankylosing spondylitis: results of a randomized, placebo-controlled trial (ASSERT). Arthritis Rheum 2005;52: 582–591.

64 van der Heijde D, Kivitz A, Schiff MH, Sieper J, Dijkmans BA, Braun J, Dougados M, Reveille JD, Wong RL, Kupper H, Davis JC Jr: Efficacy and safety of adalimumab in patients with ankylosing spondylitis: results of a multicenter, randomized, double-blind, placebo-controlled trial. Arthritis Rheum 2006;54:2136–2146.

65 Inman RD, Davis JC Jr, Heijde D, Diekman L, Sieper J, Kim SI, Mack M, Han J, Visvanathan S, Xu Z, Hsu B, Beutler A, Braun J: Efficacy and safety of golimumab in patients with ankylosing spondylitis: results of a randomized, double-blind, placebo-controlled, phase III trial. Arthritis Rheum 2008;58: 3402–3412.

66 Tzelepis GE, Kalliakosta G, Tzioufas AG, Sfikakis PP, Mandros C, Boki KA, Roussos C, Moutsopoulos HM: Thoracoabdominal motion in ankylosing spondylitis: association with standardised clinical measures and response to therapy. Ann Rheum Dis 2009;68:966–971.

67 Braun J, Baraliakos X, Listing J, Sieper J: Decreased incidence of anterior uveitis in patients with ankylosing spondylitis treated with the anti-tumor necrosis factor agents infliximab and etanercept. Arthritis Rheum 2005;52:2447–2451.

68 Baraliakos X, Listing J, Rudwaleit M, Brandt J, Alten R, Burmester G, Gromnica-Ihle E, Haibel H, Schewe S, Schneider M, Sorensen H, Zeidler H, Visvanathan S, Sieper J, Braun J: Safety and efficacy of readministration of infliximab after long-term continuous therapy and withdrawal in patients with ankylosing spondylitis. J Rheumatol 2007;34:510–515.

69 Kvien TK, Heiberg, Lie E, Kaufmann C, Mikkelsen K, Nordvåg BY, Rødevand E: A Norwegian DMARD register: prescriptions of DMARDs and biological agents to patients with inflammatory rheumatic diseases. Clin Exp Rheumatol 2005;23(suppl 39): S188–S194.

70 Heiberg MS, Koldingsnes W, Mikkelsen K, Rødevand E, Kaufmann C, Mowinckel P, Kvien TK: The comparative one-year performance of anti-tumor necrosis factor alpha drugs in patients with rheumatoid arthritis, psoriatic arthritis, and ankylosing spondylitis: results from a longitudinal, observational, multicenter study. Arthritis Rheum 2008; 59:234–240.

71 Reich K, Nestle FO, Papp K, Ortonne JP, Evans R, Guzzo C, Li S, Dooley LT, Griffiths CE: Infliximab induction and maintenance therapy for moderate-to-severe psoriasis: a phase III, multicentre, double-blind trial. Lancet 2005;366:1367–1374.

72 Feldman SR, Gordon KB, Bala M, Evans R, Li S, Dooley LT, Guzzo C, Patel K, Menter A, Gottlieb AB: Infliximab treatment results in significant improvement in the quality of life of patients with severe psoriasis: a double-blind placebo-controlled trial. Br J Dermatol 2005;152:954–960.

73 Antoni C, Krueger GG, de VK, Birbara C, Beutler A, Guzzo C, Zhou B, Dooley LT, Kavanaugh A: Infliximab improves signs and symptoms of psoriatic arthritis: results of the IMPACT 2 trial. Ann Rheum Dis 2005;64:1150–1157.

74 Mease PJ, Gladman DD, Ritchlin CT, Ruderman EM, Steinfeld SD, Choy EH, Sharp JT, Ory PA, Perdok RJ, Weinberg MA: Adalimumab for the treatment of patients with moderately to severely active psoriatic arthritis: results of a double-blind, randomized, placebo-controlled trial. Arthritis Rheum 2005;52:3279–3289.

75 Gordon KB, Langley RG, Leonardi C, Toth D, Menter MA, Kang S, Heffernan M, Miller B, Hamlin R, Lim L, Zhong J, Hoffman R, Okun MM: Clinical response to adalimumab treatment in patients with moderate to severe psoriasis: double-blind, randomized controlled trial and open-label extension study. J Am Acad Dermatol 2006;55:598–606.

76 Gladman DD, Mease PJ, Ritchlin CT, Choy EH, Sharp JT, Ory PA, Perdok RJ, Sasso EH: Adalimumab for long-term treatment of psoriatic arthritis: forty-eight week data from the adalimumab effectiveness in psoriatic arthritis trial. Arthritis Rheum 2007;56:476–488.

77 Kavanaugh A, McInnes I, Mease P, Krueger GG, Gladman D, Gomez-Reino J, Papp K, Zrubek J, Mudivarthy S, Mack M, Visvanathan S, Beutler A: Golimumab, a new human tumor necrosis factor alpha antibody, administered every four weeks as a subcutaneous injection in psoriatic arthritis: twenty-four-week efficacy and safety results of a randomized, placebo-controlled study. Arthritis Rheum 2009;60:976–986.

78 Mease PJ, Goffe BS, Metz J, VanderStoep A, Finck B, Burge DJ: Etanercept in the treatment of psoriatic arthritis and psoriasis: a randomised trial. Lancet 2000;356:385–390.

79 Mease PJ, Kivitz AJ, Burch FX, Siegel EL, Cohen SB, Ory P, Salonen D, Rubenstein J, Sharp JT, Tsuji W: Etanercept treatment of psoriatic arthritis: safety, efficacy, and effect on disease progression. Arthritis Rheum 2004;50:2264–2272.

80 Gladman DD: Adalimumab, etanercept and infliximab are equally effective treatments for patients with psoriatic arthritis. Nat Clin Pract Rheumatol 2008;4:510–511.

81 Ritchlin CT, Kavanaugh A, Gladman DD, Mease PJ, Helliwell P, Boehncke WH, de Vlam K, Fiorentino D, Fitzgerald O, Gottlieb AB, McHugh N, Nash PT, Qureshi A, Soriano ER, Taylor WJ: Treatment recommendations for psoriatic arthritis. Ann Rheum Dis 2008, Epub ahead of print.

82 Kyle S, Chandler D, Griffiths CE, Helliwell P, Lewis J, McInnes I, Oliver S, Symmons D, McHugh N: Guideline for anti-TNF-alpha therapy in psoriatic arthritis. Rheumatology (Oxford) 2005;44:390–397.

83 Boehncke WH, Brasie RA, Barker J, Chimenti S, Dauden E, de Rie M, Dubertret L, Giannetti A, Katsambas A, Kragballe K, Naeyaert JM, Ortonne JP, Peyri J, Prinz JC, Saurat JH, Strohal R, van de Kerkhof P, Sterry W: Recommendations for the use of etanercept in psoriasis: a European dermatology expert group consensus. J Eur Acad Dermatol Venereol 2006;20:988–998.

84 Menter A, Gottlieb A, Feldman SR, Van Voorhees AS, Leonardi CL, Gordon KB, Lebwohl M, Koo JY, Elmets CA, Korman NJ, Beutner KR, Bhushan R: Guidelines of care for the management of psoriasis and psoriatic arthritis: Section 1. Overview of psoriasis and guidelines of care for the treatment of psoriasis with biologics. J Am Acad Dermatol 2008;58:826–850.

85 Paller AS, Siegfried EC, Langley RG, Gottlieb AB, Pariser D, Landells I, Hebert AA, Eichenfield LF, Patel V, Creamer K, Jahreis A: Etanercept treatment for children and adolescents with plaque psoriasis. N Engl J Med 2008;358:241–251.

86 Targan S, Hanauer S, van Deventer S, Mayer L, Present D, Braakman T, DeWoody K, Schaible T, Rutgeerts P: A short-term study of chimeric monoclonal antibody cA2 to tumor necrosis factor alpha for Crohn's disease. Crohn's Disease cA2 Study Group. N Engl J Med 1997;337:1029–1035.

87 Lichtenstein GR, Yan S, Bala M, Blank M, Sands B: Infliximab maintenance treatment reduces hospitalizations, surgeries, and procedures in fistulizing Crohn's disease. Gastroenterology 2005;128:862–869.

88 Present DH, Rutgeerts P, Targan S, Hanauer SB, Mayer L, van Hogez RA, Podolsky DK, Sands BE, Braakman T, DeWoody KL, Schaible TF, van Deventer SJ: Infliximab for the treatment of fistulas in patients with Crohn's disease. N Engl J Med 1999;340:1398–1405.

89 Hyams JS, Crandall W, Kugathasan S, Heuschkel R, Griffiths A, Cohen S, Markowitz J, Olsen A, Johanns J, Baldassano R: A randomized, multicenter, open-label study to evaluate the safety and efficacy of infliximab in pediatric patients with moderate-to-severe Crohn's disease (abstract). J Pediatr Gastroenterol Nutr 2005;41:539.

90 Papadakis KA, Shaye OA, Vasiliauskas EA, Ippoliti A, Dubinsky MC, Birt J, Paavola J, Lee SK, Price J, Targan SR, Abreu MT: Safety and efficacy of adalimumab (D2E7) in Crohn's disease patients with an attenuated response to infliximab. Am J Gastroenterol 2005;100:75–79.

91 Sandborn WJ, Hanauer SB, Rutgeerts P, Fedorak RN, Lukas M, MacIntosh DG, Panaccione R, Wolf D, Kent JD, Bittle B, Li L, Pollack PF: Adalimumab for maintenance treatment of Crohn's disease: results of the CLASSIC II trial. Gut 2007;56:1232–1239.

92 Sandborn WJ, Feagan BG, Stoinov S, Honiball PJ, Rutgeerts P, Mason D, Bloomfield R, Schreiber S, PRECISE 1 Study Investigators: Certolizumab pegol for the treatment of Crohn's disease. N Engl J Med 2007;357:228–238.

93 Schreiber S, Khaliq-Kareemi M, Lawrance IC, Thomsen OØ, Hanauer SB, McColm J, Bloomfield R, Sandborn WJ, PRECISE 2 Study Investigators: Maintenance therapy with certolizumab pegol for Crohn's disease. N Engl J Med 2007;357:239–250.

94 Feagan BG, Coteur G, Tan S, Keininger DL, Schreiber S: Clinically meaningful improvement in health-related quality of life in a randomized controlled trial of certolizumab pegol maintenance therapy for Crohn's disease. Am J Gastroenterol 2009, Epub ahead of print.

95 Peyrin-Biroulet L, Deltenre P, de Suray N, Branche J, Sandborn WJ, Colombel JF: Efficacy and safety of tumor necrosis factor antagonists in Crohn's disease: meta-analysis of placebo-controlled trials. Clin Gastroenterol Hepatol 2008;6:644–653.

96 Rispo A, Scarpa R, Di Girolamo E, Cozzolino A, Lembo G, Atteno M, De Falco T, Lo Presti M, Castiglione F: Infliximab in the treatment of extra-intestinal manifestations of Crohn's disease. Scand J Rheumatol 2005;34:387–391.

97 Brooklyn TN, Dunnill MG, Shetty A, Bowden JJ, Williams JD, Griffiths CE, Forbes A, Greenwood R, Probert CS: Infliximab for the treatment of pyoderma gangrenosum: a randomised, double blind, placebo controlled trial. Gut 2006;55:505–509.

98 Sandborn WJ, Hanauer SB, Katz S, Safdi M, Wolf DG, Baerg RD, Tremaine WJ, Johnson T, Diehl NN, Zinsmeister AR: Etanercept for active Crohn's disease: a randomized, double-blind, placebo-controlled trial. Gastroenterology 2001;121:1088–1094.

99 Probert CS, Hearing SD, Schreiber S, Kühbacher T, Ghosh S, Arnott ID, Forbes A: Infliximab in moderately severe glucocorticoid resistant ulcerative colitis: a randomised controlled trial. Gut 2003;52: 998–1002.

100 Järnerot G, Hertervig E, Friis-Liby I, Blomquist L, Karlén P, Grännö C, Vilien M, Ström M, Danielsson A, Verbaan H, Hellström PM, Magnuson A, Curman B: Infliximab as rescue therapy in severe to moderately severe ulcerative colitis: a randomized, placebo-controlled study. Gastroenterology 2005;128: 1805–1811.

101 Rutgeerts P, Sandborn WJ, Feagan BG, Reinisch W, Olson A, Johanns J, Travers S, Rachmilewitz D, Hanauer SB, Lichtenstein GR, de Villiers WJ, Present D, Sands BE, Colombel JF: Infliximab for induction and maintenance therapy for ulcerative colitis. N Engl J Med 2005;353:2462–2476; erratum in N Engl J Med 2006;354:2200.

102 Gonzalez-Lama Y, Fernandez-Blanco I, Lopez-SanRoman A, Taxonera C, Casis B, Tabernero S, Bermejo F, Martinez-Silva F, Mendoza JL, Martinez-Montiel P, Carneros JA, Sanchez F, Maté J, Gisbert JP: Group for the Study of Inflammatory Bowel Diseases from Madrid. Open-label infliximab therapy in ulcerative colitis: a multicenter survey of results and predictors of response. Hepatogastroenterology 2008;55:1609–1614.

103 Sfikakis PP, Theodossiadis PG, Katsiari CG, Kaklamanis P, Markomichelakis NN: Effect of Infliximab on sight-threatening panuveitis in Behcet's disease. Lancet 2001;358:295–296.

104 Sfikakis PP, Kaklamanis P, Elezoglou A, Katsilambros N, Theodossiadis PG, Papaefthimiou S, Markomichelakis NN: Infliximab for recurrent sight-threatening ocular inflammation in Adamantiades-Behcet's disease. Ann Intern Med 2004; 140:404–406.

105 Ohno S, Nakamura S, Hori S, Shimakawa M, Kawashima H, Mochizuki M, Sugita S, Ueno S, Yoshizaki K, Inaba G: Efficacy, safety, and pharmacokinetics of multiple administration of infliximab in Behcet's disease with refractory uveoretinitis. J Rheumatol 2004;31:1362–1368.

106 Tugal-Tutkun I, Mudun A, Urgancioglu M, Kamali S, Kasapoglu E, Inanc M, Gül A: Efficacy of Infliximab in the treatment of uveitis that is refractory to treatment with the combination of azathioprine, cyclosporine, and corticosteroids in Behcet's disease: an open-label trial. Arthritis Rheum 2005; 52:2478–2484.

107 Sfikakis PP: Behcet's disease: a new target for anti-tumor necrosis factor treatment. Ann Rheum Dis 2002;61(suppl 2):ii51–ii53.

108 Sfikakis PP, Markomichelakis N, Alpsoy E, Assaad-Khalil S, Bodaghi B, Gul A, Ohno S, Pipitone N, Schirmer M, Stanford M, Wechsler B, Zouboulis C, Kaklamanis P, Yazici H: Anti-TNF therapy in the management of Behcet's disease – review and basis for recommendations. Rheumatology (Oxford). 2007;46:736–741.

109 Melikoglu M, Fresko I, Mat C, Ozyazgan Y, Gogus F, Yurdakul S, Hamuryudan V, Yazici H: Short-term trial of etanercept in Behcet's disease: a double blind, placebo controlled study. J Rheumatol 2005;32:98–105.

110 Theodossiadis PG, Markomichelakis NN, Sfikakis PP: Tumor necrosis factor antagonists: preliminary evidence for an emerging approach in the treatment of ocular inflammation. Retina J Ret Vit Dis 2007;27:399–413.

111 Markomichelakis N, Theodossiadis PG, Pantelia E, Papaefthimiou S, Theodossiadis GP, Sfiakakis PP: Infliximab for chronic cystoid macular edema associated with uveitis. Am J Ophthalmol 2004;138:648–650.

112 Sfikakis PP, Markomichelakis N, Theodossiadis GP, Grigoropoulos V, Katsilambros N, Theodossiadis PG: Regression of sight-threatening macular edema in type 2 diabetes following treatment with the anti-tumor necrosis factor monoclonal antibody Infliximab. Diabetes Care 2005;28:445–447.

113 Markomichelakis N, Theodossiadis PG, Sfikakis PP: Regression of neovascular age-related macular degeneration following Infliximab therapy. Am J Ophthalmol 2005;139:537–540.

114 Theodossiadis PG, Liarakos VS, Sfikakis PP, Charonis A, Agrogiannis G, Kavantzas N, Vergados IA: Intravitreal administration of the anti-TNF monoclonal antibody Infliximab in the rabbit. Graefes Arch Clin Exp Ophtalmol 2009;247:273–281.

115 Theodossiadis PG, Liarakos VS, Sfikakis PP, Vergados IA, Theodossiadis GP: Intravitreal administration of the anti-TNF agent Infliximab for neovascular age-related macular degeneration. Am J Ophthalmol 2009;147:825–830.

116 Alexis AF, Strober BE: Off-label dermatologic uses of anti-TNF-a therapies. J Cutan Med Surg 2005; 9:296–302.

117 Rigopoulos D, Korfitis C, Gregoriou S, Katsambas AD: Infliximab in dermatological treatment: beyond psoriasis. Expert Opin Biol Ther 2008;8:123–133.

118 Aringer M, Graninger WB, Steiner G, Smolen JS: Safety and efficacy of tumor necrosis factor alpha blockade in systemic lupus erythematosus: an open-label study. Arthritis Rheum 2004;50:3161–3169.

119 Lionaki S, Siamopoulos K, Theodorou I, Papadimitraki E, Bertsias G, Boumpas D, Boletis J: Inhibition of tumour necrosis factor alpha in idiopathic membranous nephropathy: a pilot study. Nephrol Dial Transplant 2009;24:2144–2150.

120 Mastrandrea L, Yu J, Behrens T, Buchlis J, Albini C, Fourtner S, Quattrin T: Etanercept treatment in children with new-onset type 1 diabetes: pilot randomized, placebo-controlled, double-blind study. Diabetes Care 2009;32:1244–1249.

121 Denton CP, Engelhart M, Tvede N, Wilson H, Khan K, Shiwen X, Carreira PE, Diaz Gonzalez F, Black CM, van den Hoogen FH: An open-label pilot study of infliximab therapy in diffuse cutaneous systemic sclerosis. Ann Rheum Dis 2008, Epub ahead of print.

122 Efthimiou P, Schwartzman S, Kagen LJ: Possible role for tumour necrosis factor inhibitors in the treatment of resistant dermatomyositis and polymyositis: a retrospective study of eight patients. Ann Rheum Dis 2006;65:1233–1236.

123 Dastmalchi M, Grundtman C, Alexanderson H, Mavragani CP, Einarsdottir H, Helmers SB, Elvin K, Crow MK, Nennesmo I, Lundberg IE: A high incidence of disease flares in an open pilot study of infliximab in patients with refractory inflammatory myopathies. Ann Rheum Dis. 2008;67:1670–1677.

124 Riley P, McCann LJ, Maillard SM, Woo P, Murray KJ, Pilkington CA: Effectiveness of infliximab in the treatment of refractory juvenile dermatomyositis with calcinosis. Rheumatology (Oxford) 2008;47: 877–880.

125 Kokkinos A, Iliopoulos A, Greka P, Efthymiou A, Katsilambros N, Sfikakis PP: Successful treatment of refractory adult-onset Still's disease with infliximab. A prospective, non-comparative series of four patients. Clin Rheumatol 2004;23:45–49.

126 Fautrel B, Sibilia J, Mariette X, Combe B; Club Rhumatismes et Inflammation: Tumour necrosis factor alpha blocking agents in refractory adult Still's disease: an observational study of 20 cases. Ann Rheum Dis 2005;64:262–266.

127 Saint Marcoux B, De Bandt M, CRI (Club Rhumatismes et Inflammation): Vasculitides induced by TNFalpha antagonists: a study in 39 patients in France. Joint Bone Spine 2006;73:710–713.

128 Lamprecht P, Till A, Steinmann J, Aries PM, Gross WL: Current state of biologicals in the management of systemic vasculitis. Ann N Y Acad Sci 2007;1110: 261–270.

129 Utz JP, Limper AH, Kalra S, Specks U, Scott JP, Vuk-Pavlovic Z, Schroeder DR: Etanercept for the treatment of stage II and III progressive pulmonary sarcoidosis. Chest 2003;124:177–185.

130 Baughman RP, Drent M, Kavuru M, Judson MA, Costabel U, du Bois R, Albera C, Brutsche M, Davis G, Donohue JF, Muller-Quernheim J, Schlenker-Herceg R, Flavin S, Lo KH, Oemar B, Barnathan ES: Infliximab therapy in patients with chronic sarcoidosis and pulmonary involvement. Am J Respir Crit Care Med 2006;174:795–802.

131 Mariette X, Ravaud P, Steinfeld S, Baron G, Goetz J, Hachulla E, Combe B, Puéchal X, Pennec Y, Sauvezie B, Perdriger A, Hayem G, Janin A, Sibilia J: Inefficacy of infliximab in primary Sjögren's syndrome: results of the randomized, controlled Trial of Remicade in Primary Sjögren's Syndrome (TRIPSS). Arthritis Rheum 2004;50:1270–1276.

132 Wegener's Granulomatosis Etanercept Trial (WGET) Research Group: Etanercept plus standard therapy for Wegener's granulomatosis. N Engl J Med 2005;352:351–361.

133 Cohen S, Shoup A, Weisman MH, Harris J: Etanercept treatment for autoimmune inner ear disease: results of a pilot placebo-controlled study. Otol Neurotol 2005;26:903–907.

134 Rennard SI, Fogarty C, Kelsen S, Long W, Ramsdell J, Allison J, Mahler D, Saadeh C, Siler T, Snell P, Korenblat P, Smith W, Kaye M, Mandel M, Andrews C, Prabhu R, Donohue JF, Watt R, Lo KH, Schlenker-Herceg R, Barnathan ES, Murray J, COPD Investigators: The safety and efficacy of infliximab in moderate to severe chronic obstructive pulmonary disease. Am J Respir Crit Care Med 2007;175: 926–934.

135 Korhonen T, Karppinen J, Paimela L, Malmivaara A, Lindgren KA, Bowman C, Hammond A, Kirkham B, Järvinen S, Niinimäki J, Veeger N, Haapea M, Torkki M, Tervonen O, Seitsalo S, Hurri H: The treatment of disc-herniation-induced sciatica with infliximab: one-year follow-up results of FIRST II, a randomized controlled trial. Spine 2006;31:2759–2766.

136 Cohen SP, Bogduk N, Dragovich A, Buckenmaier CC 3rd, Griffith S, Kurihara C, Raymond J, Richter PJ, Williams N, Yaksh TL: Randomized, double-blind, placebo-controlled, dose-response, and preclinical safety study of transforaminal epidural etanercept for the treatment of sciatica. Anesthesiology 2009;110:1116–1126.

137 Hoffman GS, Cid MC, Rendt-Zagar KE, Merkel PA, Weyand CM, Stone JH, Salvarani C, Xu W, Visvanathan S, Rahman MU: Infliximab-GCA Study Group: Infliximab for maintenance of glucocorticosteroid-induced remission of giant cell arteritis: a randomized trial. Ann Intern Med 2007; 146:621–630.

138 Salvarani C, Macchioni P, Manzini C, Paolazzi G, Trotta A, Manganelli P, Cimmino M, Gerli R, Catanoso MG, Boiardi L, Cantini F, Klersy C, Hunder GG: Infliximab plus prednisone or placebo plus prednisone for the initial treatment of polymyalgia rheumatica: a randomized trial. Ann Intern Med 2007;146:631–639.

139 Rossman MD, Newman LS, Baughman RP, Teirstein A, Weinberger SE, Miller W Jr, Sands BE: A double-blinded, randomized, placebo-controlled trial of infliximab in subjects with active pulmonary sarcoidosis. Sarcoidosis Vasc Diffuse Lung Dis 2006;23:201–208.

140 Baughman RP, Lower EE, Drent M: Inhibitors of tumor necrosis factor (TNF) in sarcoidosis: who, what, and how to use them. Sarcoidosis Vasc Diffuse Lung Dis 2008;25:76–89.

141 The Lenercept Multiple Sclerosis Study Group and The University of British Columbia MS/MRI Analysis Group: TNF neutralization in MS: results of a randomized, placebo-controlled multicenter study. Neurology 1999;53:457–465.

142 van Oosten BW, Barkhof F, Truyen L, Boringa JB, Bertelsmann FW, von Blomberg BM, Woody JN, Hartung HP, Polman CH: Increased MRI activity and immune activation in two multiple sclerosis patients treated with the monoclonal anti-tumor necrosis factor antibody cA2. Neurology 1996;47: 1531–1534.

143 Mann DL, McMurray JJ, Packer M, Swedberg K, Borer JS, Colucci WS, Djian J, Drexler H, Feldman A, Kober L, Krum H, Liu P, Nieminen M, Tavazzi L, van Veldhuisen DJ, Waldenstrom A, Warren M, Westheim A, Zannad F, Fleming T: Targeted anticytokine therapy in patients with chronic heart failure: results of the Randomized Etanercept Worldwide Evaluation (RENEWAL). Circulation 2004;109: 1594–1602.

144 Chung ES, Packer M, Lo KH, Fasanmade AA, Willerson JT: Randomized, double-blind, placebo-controlled, pilot trial of infliximab, a chimeric monoclonal antibody to tumor necrosis factor-alpha, in patients with moderate-to-severe heart failure: results of the anti-TNF Therapy Against Congestive Heart Failure (ATTACH) trial. Circulation 2003;107:3133–3140.

145 Westhovens R, Yocum D, Han J, Berman A, Strusberg I, Geusens P, Rahman MU: The safety of infliximab, combined with background treatments, among patients with rheumatoid arthritis and various comorbidities: a large, randomized, placebo-controlled trial. Arthritis Rheum 2006;54: 1075–1086.

146 Listing J, Strangfeld A, Kary S, Rau R, von Hinueber U, Stoyanova-Scholz M, Gromnica-Ihle E, Antoni C, Herzer P, Kekow J, Schneider M, Zink A: Infections in patients with rheumatoid arthritis treated with biologic agents. Arthritis Rheum 2005; 52:3403–3412.

147 Dixon W, Watson K, Lunt M, Hyrich KL, Silman AJ, Symmons DP; British Society for Rheumatology Biologics Register: Rates of serious infections, including site-specific and bacterial intracellular infection, in rheumatoid arthritis patients receiving anti-TNF therapy: results from the British Society for Rheumatology Biologics Register. Arthritis Rheum 2006;54:2368–2376.

148 Nathan DM, Angus PW, Gibson PR: Hepatitis B and C virus infections and anti-tumor necrosis factor-alpha therapy: guidelines for clinical approach. J Gastroenterol Hepatol 2006;21:1366–1371.

149 Cepeda EJ, Williams FM, Ishimori ML, Weisman MH, Reveille JD: The use of anti-tumour necrosis factor therapy in HIV-positive individuals with rheumatic disease. Ann Rheum Dis 2008;67:710–712.
150 Wallis RS, Kyambadde P, Johnson JL, Horter L, Kittle R, Pohle M, Ducar C, Millard M, Mayanja-Kizza H, Whalen C, Okwera A: A study of the safety, immunology, virology, and microbiology of adjunctive etanercept in HIV-1-associated tuberculosis. AIDS 2004;18:257–264.
151 Ramos-Casals M, Brito-Zerón P, Muñoz S, Soria N, Galiana D, Bertolaccini L, Cuadrado MJ, Khamashta MA: Autoimmune diseases induced by TNF-targeted therapies: analysis of 233 cases. Medicine (Baltimore) 2007;86:242–251.
152 Williams E, Gadola S, Edwards CJ: Anti-TNF-induced lupus. Rheumatology (Oxford) 2009;48: 716–720.
153 Eriksson C, Engstrand S, Sundqvist KG, Rantapaa-Dahlqvist S: Autoantibody formation in patients with rheumatoid arthritis treated with anti-TNF alpha. Ann Rheum Dis 2005;64:403–407.
154 Elezoglou A, Kafasi N, Kaklamanis PH, Theodossiadis PG, Kapsimali V, Choremi E, Vaiopoulos G, Sfikakis PP: Infliximab treatment-induced formation of autoantibodies is common in Bechet's disease. Clin Exp Rheumatol 2007; 25(suppl 45):S65–S69.
155 Caramaschi P, Bambara LM, Pieropan S, Tinazzi I, Volpe A, Biasi D. Anti-TNF alpha blockers, autoantibodies and autoimmune disease. Join Bone Spine 2009;76:333–342.
156 Symmons DPM, Silman AJ: Anti-tumor necrosis factor therapy and the risk of lymphoma in rheumatoid arthritis: no clear answer. Arthritis Rheum 2004;50:1703–1706.
157 Keystone EC: Safety of biologic therapies-an update. J Rheumatol Suppl 2005;74:8–12.
158 Setoguchi S, Solomon DH, Weinblatt ME, Katz JN, Avorn J, Glynn RJ, Cook EF, Carney G, Schneeweiss S: Tumor necrosis factor alpha antagonist use and cancer in patients with rheumatoid arthritis. Arthritis Rheum 2006;54:2757–2764.
159 Nannini C, Cantini F, Niccoli L, Cassarà E, Salvarani C, Olivieri I, Lally EV: Single-center series and systematic review of randomized controlled trials of malignancies in patients with rheumatoid arthritis, psoriatic arthritis, and ankylosing spondylitis receiving anti-tumor necrosis factor alpha therapy: is there a need for more comprehensive screening procedures? Arthritis Rheum 2009;61:801–812.
160 Burmester GR, Mease PJ, Dijkmans BA, Gordon K, Lovell D, Panaccione R, Perez J, Pangan AL: Adalimumab safety and mortality rates from global clinical trials of six immune-mediated inflammatory diseases. Ann Rheum Dis 2009, Epub ahead of print.
161 Wallis RS, Broder MS, Wong JY, Hanson ME, Beenhouwer DO: Granulomatus infectious diseases associated with TNF antagonists. Clin Infect Dis 2004;38:1261–1265; correction Clin Infect Dis 2004;39:1254–1255.
162 Tubach F, Salmon D, Ravaud P, Allanore Y, Goupille P, Bréban M, Pallot-Prades B, Pouplin S, Sacchi A, Chichemanian RM, Bretagne S, Emilie D, Lemann M, Lorthololary O, Mariette X, Research Axed on Tolerance of Biotherapies Group: Risk of tuberculosis is higher with anti-tumor necrosis factor monoclonal antibody therapy than with soluble tumor necrosis factor receptor therapy: The three-year prospective French research axed on tolerance of biotherapies registry. Arthritis Rheum 2009;60: 1884–1894.
163 Wallis RS: Tumor necrosis factor antagonists: structure, function and tuberculosis risks. Lancet Infect dis 2008;8:601–611.
164 Furst DE, Wallis RS, Broder M, Beenhouwer DO: Tumor necrosis factor antagonists: different kinetics and/or mechanisms of action may explain differences in the risk for developing granulomatous infection. Semin Arthritis Rheum 2006;36:159–167.
165 Saliu OY, Sofer C, Stein DS, Schwander SK, Wallis RS: Tumor necrosis factor blockers: differential effects on mycobacterial immunity. J Infect Dis 2006;194:486–492.
166 Iliopoulos A, Psathakis K, Aslanidis S, Skagias L, Sfikakis PP: Tuberculosis and granuloma formation in patients receiving anti-TNF therapy. Int J Tuberc Lung Dis 2006;10:588–590.
167 Gomez-Reino JJ, Carmona L, Angel Descalzo M: Risk of tuberculosis in patients treated with tumor necrosis factor antagonists due to incomplete prevention of reactivation of latent infection. Arthritis Rheum 2007;57:756–761.
168 Sfikakis PP, Iliopoulos A, Elezoglou A, Kittas C, Stratigos A: Psoriasis induced by anti-TNF therapy: a paradoxical adverse reaction. Arthritis Rheum 2005;52:2513–2518.
169 Stratigos AJ, Sfikakis PP: Psoriasis occurring during anti-TNF therapy: causal effect or unrelated ? Future Rheumatol 2006;1:281–284.
170 Ko JM, Gottlieb AB, Kerbleski JF: Induction and exacerbation of psoriasis with TNF-blockade therapy: a review and analysis of 127 cases. J Dermatolog Treat 2009;20:100–108.

171 Moustou AE, Matekovits A, Dessinioti C, Antoniou C, Sfikakis PP, Stratigos AJ: Side effects of biologic anti-TNF therapy: a clinical review. J Am Acad Dermatol 2009, Epub ahead of print.
172 Harrison MJ, Dixon WG, Watson KD, King Y, Groves R, Hyrich KL, Symmons DP: Rates of new-onset psoriasis in patients with rheumatoid arthritis receiving anti-tumor necrosis factor alpha therapy: results from the British Society for Rheumatology Biologics Register. Ann Rheum Dis 2009;68:209–215.
173 Michaelsson G, Kajermo U, Michaelsson A, Hagforsen E: Infliximab can precipitate as well as worsen palmoplantar pustulosis: possible linkage to the expression of tumour necrosis factor alpha in the normal palmar eccrine sweat duct? Br J Dermatol 2005;153:1243–1244.
174 Saint Marcoux B, De bandt M, CRI (Club Rhumatismes et Inflammation): Vasculitides induced by TNFalpha antagonists. Joint Bone Spine 2006;76: 710–713.
175 Massara A, Cavazzini L, La Corte R, Trotta F: Sarcoidosis appearing during anti-tumor necrosis factor alpha therapy: a new 'class effect' paradoxical phenomenon. Two case reports and literature review. Semin Arthritis Rheum. 2009, Epub ahead of print.
176 Hyrich KL, Watson KD, Silman AJ, Symmons DP, British Society for Rheumatology Biologics Register: Predictors of response to anti-TNF-alpha therapy among patients with rheumatoid arthritis: results from the British Society for Rheumatology Biologics Register. Rheumatology (Oxford) 2006;45:1558–1565.
177 Esters N, Vermeire S, Joossens S, Noman M, Louis E, Belaiche J, De Vos M, Van Gossum A, Pescatore P, Fiasse R, Pelckmans P, Reynaert H, Poulain D, Bossuyt X, Rutgeerts P, Belgian Group of Infliximab Expanded Access Program in Crohn's Disease: Serological markers for prediction of response to anti-tumor necrosis factor treatment in Crohn's disease. Am J Gastroenterol 2002;97:1458–1462.
178 Bruns A, Nicaise-Roland P, Hayem G, Palazzo E, Dieudé P, Grootenboer-Mignot S, Chollet-Martin S, Meyer O: Prospective cohort study of effects of infliximab on rheumatoid factor, anti-cyclic citrullinated peptide antibodies and antinuclear antibodies in patients with long-standing rheumatoid arthritis. Joint Bone Spine. 2009;76:248–253.
179 Hueber W, Tomooka BH, Batliwalla F, Li W, Monach PA, Tibshirani RJ, Van Vollenhoven RF, Lampa J, Saito K, Tanaka Y, Genovese MC, Klareskog L, Gregersen PK, Robinson WH: Blood autoantibody and cytokine profiles predict response to anti-tumor necrosis factor therapy in rheumatoid arthritis. Arthritis Res Ther. 2009;11:R76.
180 Wong M, Ziring D, Korin Y, Desai S, Kim S, Lin J, Gjertson D, Braun J, Reed E, Singh RR: TNFα blockade in human diseases: mechanisms and future directions. Clin Immunol 2008;126:121–136.
181 Kollias G, Kontoyiannis D. Role of TNF/TNFR in autoimmunity: specific TNF receptor blockade may be advantageous to anti-TNF treatment. Cytokine Growth Factor Rev 2002;13:315–321.
182 Alexopoulou L, Kranidioti K, Xanthoulea S, Denis M, Kotanidou A, Douni E, Blackshear PJ, Kontoyiannis DL, Kollias G: Transmembrane TNF protects mutant mice against intracellular bacterial infections, chronic inflammation and autoimmunity. Eur J Immunol 2006;36:2768–2780.

Petros P. Sfikakis, MD, PhD
First Department of Propedeutic and Internal Medicine, Laikon Hospital
Athens University Medical School, 17, Ag Thoma Street
GR–11527 Athens (Greece)
Tel. +30 210 7258555, Fax +30 210 7485965, E-Mail psfikakis@med.uoa.gr

Author Index

Subject Index